T0310153

# SMART TECHNOLOGIES FOR SAFETY ENGINEERING

# SMART TECHNOLOGIES FOR SAFETY ENGINEERING

Edited by

**Professor Jan Holnicki-Szulc**
Smart Technology Centre, Institute of Fundamental Technological Research, Warsaw, Poland

John Wiley & Sons, Ltd

*Other Wiley Editorial Offices*

John Wiley & Sons Inc., 111 River Street, Hoboken, NJ 07030, USA

Jossey-Bass, 989 Market Street, San Francisco, CA 94103-1741, USA

Wiley-VCH Verlag GmbH, Boschstr. 12, D-69469 Weinheim, Germany

John Wiley & Sons Australia Ltd, 42 McDougall Street, Milton, Queensland 4064, Australia

John Wiley & Sons (Asia) Pte Ltd, 2 Clementi Loop #02-01, Jin Xing Distripark, Singapore 129809

John Wiley & Sons Canada Ltd, 6045 Freemont Blvd, Mississauga, ONT, L5R 4J3

Wiley also publishes its books in a variety of electronic formats. Some content that appears in print may
not be available in electronic books.

*Library of Congress Cataloging-in-Publication Data*

Smart technologies for safety engineering / edited by Jan Holnicki-Szulc.
    p.   cm.
  Includes bibliographical references and index.
    ISBN 978-0-470-05846-6 (cloth : alk. paper)   1. Smart materials.   2. Smart structures.
3. Automatic data collection systems.   I. Holnicki-Szulc, Jan.
TA418.9.S62S87 2008
620.8′6–dc22

                                                                2008002726

*British Library Cataloguing in Publication Data*

A catalogue record for this book is available from the British Library

ISBN  978-0-470-05846-6 (H/B)

Typeset in 10/12pt Times by TechBooks, New Delhi, India.
Printed and bound in Great Britain by Antony Rowe Ltd, Chippenham, Wiltshire

# Contents

# Preface

The contents of this book cover the following research fields:

- new concepts of *smart technologies* and their applications;
- original methods and software tools for modeling, design, simulation and control of *adaptive structures*;
- application of the *smart-tech concept* to the following hot research topics and emerging engineering issues:
  - *health monitoring* of structures and engineering systems;
  - monitoring and *prediction of environmental conditions*;
  - automatic *structural adaptation* to unpredictable, randomly changing dynamic conditions;
  - *optimal design* of adaptive structures and engineering systems.

All of the above-mentioned topics are the key issues of *safety engineering*, encompassing automatic *damage identification*, unpredictable *impact identification* (e.g. due to automotive collision, earthquake or mine explosion) and real-time *mitigation of catastrophic impact* results.

Readers of this book are assumed to have the fundamental mathematical background of an engineer. Generally, the book addresses the system identification and control problems in various fields of engineering, e.g. aeronautical, aerospace, automotive, civil, mechanical and electrical. This book presents many different case studies to provide engineers with a comprehensive source of information on damage identification, impact load absorption and damping of vibrations. For example, automotive engineers designing a car bumper or civil engineers striving to construct safe bridges (monitored and impact-resistant) may find it useful.

# About the Authors

This book presents the research results obtained over the last decade by the Smart Technology Centre (STC), http://smart.ippt.gov.pl/, operating at the Institute of Fundamental Technological Research (IPPT) of the Polish Academy of Sciences (PAN), Warsaw, Poland. It collects achievements of several PhD theses, already completed or still being elaborated.

The STC is a division of IPPT-PAN, headed by Prof. *Jan Holnicki-Szulc*. He graduated from the Warsaw University of Technology in Civil Engineering (1969) and simultaneously from the University of Warsaw in Mathematics (1972). His PhD thesis (1973) and the habilitation thesis (1983) were defended in IPPT. He spent about 5 years (in the 1980s and 1990s) visiting various research labs in the USA, France, Spain, Portugal and Mexico. In the middle of the 1990s, he started to build the STC research group at IPPT-PAN, taking advantage of the fact that Polish labs became eligible for financial support through the European research initiatives, i.e. the 4th, 5th and 6th Framework Programmes.

The first PhD thesis in the STC was defended by *Przemysław Kołakowski* (1998) and devoted to the application of the *virtual distortion method* (VDM) to optimal structural remodeling, treated as a static problem. Then, one of the next theses by *Tomasz G. Zieliński* (2003) contained a generalization of the VDM for structural dynamics and its application to damage identification via the solution of an inverse problem. *Anita Orłowska* (2007) in her thesis developed an application of the dynamic VDM to the identification of delamination in composite beams. Further development of these numerical tools, allowing for fast and effective structural remodeling and solving coupled dynamic problems (including redistribution of material, stiffness and physical nonlinearity), was done by *Marcin Wikło* (thesis just completed). Dr *Łukasz Jankowski* (PhD defended in BAM, Berlin) and Dr *Bartłomiej Błachowski* (PhD defended in IPPT-PAN, Warsaw) joined the STC in 2005 and are both involved in dynamic load identification.

The further development of VDM applications to the structural health monitoring (SHM) concepts are under development in collaboration with the current PhD students *Andrzej Świercz* and *Marek Kokot* (theses almost completed). *Małgorzata Mróz* (thesis in progress) is working on the application of the VDM to optimal remodeling of damping properties in dynamically excited structures. Another group of PhD students, *Grzegorz Mikułowski* and *Piotr Pawłowski* (theses almost completed), *Cezary Graczykowski* and *Krzysztof Sekuła* (theses in progress), *Arkadiusz Mróz* and *Marian Ostrowski* (theses in progress), have already obtained interesting research results in the field of *adaptive impact absorption* (AIA). Finally, Dr *Jerzy Motylewski* is a key person in the STC in vibroacoustic measurement techniques and hardware development.

The team of seventeen co-authors is presented below; their contributions to particular chapters are listed in the *Organization of the Book*.

|    | *(co-author)* | *(university and department of graduation)* |
|----|---------------|---------------------------------------------|
| 1. | Jan Holnicki-Szulc | Warsaw University of Technology, Civil Engineering, University of Warsaw, Mathematics |
| 2. | Tomasz G. Zieliński | Warsaw University of Technology, Civil Engineering |
| 3. | Marcin Wikło | Radom University of Technology, Mechanical Engineering |
| 4. | Cezary Graczykowski | Warsaw University of Technology, Civil Engineering |
| 5. | Piotr K. Pawłowski | Warsaw University of Technology, Civil Engineering |
| 6. | Jerzy Motylewski | Warsaw University of Technology, Electronics |
| 7. | Łukasz Jankowski | Wrocław University of Technology, Computer Science University of Wrocław, Mathematics |
| 8. | Marek Kokot | Warsaw University of Technology, Mechatronics |
| 9. | Anita Orłowska | University of Warsaw, Physics |

| 10. | Małgorzata Mróz | Kraków University of Technology, Civil Engineering |
|-----|----------------|---------------------------------------------------|
| 11. | Andrzej Świercz | Warsaw University of Technology, Civil Engineering |
| 12. | Arkadiusz Mróz | Kraków University of Technology, Civil Engineering |
| 13. | Przemysław Kołakowski | Warsaw University of Technology, Civil Engineering |
| 14. | Grzegorz Mikułowski | Warsaw University of Technology, Mechanical Engineering |
| 15. | Krzysztof Sekuła | Warsaw University of Technology, Mechanical Engineering |
| 16. | Marian Ostrowski | Warsaw University of Technology, Mechanical Engineering |
| 17. | Bartłomiej Błachowski | Gliwice University of Technology, Civil Engineering |

The major research projects of the STC include:

- various national projects in the field of SHM and AIA
- UPWIND – Integrated Wind Turbine Design, FP6-2004-Energy-3, 2006–2011
- SAFE PIPES – Safety Assessment and Lifetime Management of Industrial Piping Systems, FP6-STRP-013898, 2005–2008
- ADLAND – Adaptive Landing Gears for Improved Impact Absorption, FP6-2002-Aero-1, 2003–2006
- SAMCO – Structural Assessment Monitoring and Control, FP5-G1RT-CT-2001-05040, 2002–2006
- SMART SYSTEMS – New Materials, Adaptive Systems and their Nonlinearities: Modeling, Control and Numerical Simulation, FP5-HPRN-CT-2002-00284, 2002–2006
- PIEZODIAGNOSTICS – Smart Structural Diagnostics using Piezo-Generated Elastic Waves, FP5-GRD1-2001-40589, 2002–2005
- COPERNICUS 263 – Feasibility Study on Active Track Support, FP4, 1995–1998
- COPERNICUS 150 – Design of Adaptive Offshore Structures under Extreme Wave Loading, FP4, 1995–1999
- NEXUSEAST – European Network of Excellence in Multifunctional Microsystems, FP4, 1994–1998

The major knowledge dissemination initiatives by the STC include:

- Smart Technology Expert Courses at IPPT-PAN (since 2003)
- European workshop on Structural Health Monitoring (co-organized with AGH, Kraków and IMP-PAN, Gdańsk), Kraków, 2008
- ECCOMAS Thematic Conference SMART'07 (co-organized with IMP-PAN, Gdańsk and IST, Lisbon), Gdańsk, 2007
- ECCOMAS Thematic Conference SMART'03, Jadwisin, 2003
- AMAS Course SMART'01, Warsaw, 2001
- NATO Advanced Research Workshop SMART'98, Pułtusk, 1998

# Organization of the Book

The book has been divided into eight chapters, addressing the following problems:

- Chapter 1: **Introduction to Smart Technologies** (*J. Holnicki-Szulc, J. Motylewski and P. Kołakowski*) makes the reader briefly acquainted with some history and up-to-date trends in the fast-developing research field of Smart Technologies.
- Chapter 2: **The Virtual Distortion Method – A Versatile Reanalysis Tool** (*P. Kołakowski, M. Wikło and J. Holnicki-Szulc*) presents the basics of the method used by the authors for fast structural reanalysis. Both the static and dynamic analyses are included. The VDM-based sensitivity analysis, utilized in many subsequent chapters, is briefly announced. The versatility of the method, based on analogies between structural and nonstructural systems, is also discussed.
- Chapter 3: **VDM-Based Health Monitoring of Engineering Systems** (*P. Kołakowski, A. Świercz, A. Orłowska, M. Kokot and J. Holnicki-Szulc*) contains various applications of the VDM to the *structural health monitoring* (SHM) by solving inverse problems. First, the identification of stiffness and mass degradation in skeletal structures (both in the time and frequency domains) is discussed. Next, the very important engineering problem of the identification of delamination in composite beams (using the concept of a contact layer) is presented. Finally, two problems devoted to health monitoring of engineering systems (no longer structures) focus on leakage identification in water networks and defect identification in electrical circuits. The analogies between truss structures and other graph-modeled systems have been effectively used.
- Chapter 4: **Dynamic Load Monitoring** (*Ł. Jankowski, K. Sekuła, B.D. Błachowski, M. Wikło and J. Holnicki-Szulc*) contains research results for the problem formulated in two ways. First, on-line impact load identification is discussed as a crucial issue for *adaptive impact absorption* (AIA) systems, reacting in real time to the detected external loading. A fast response of the order of a few milliseconds is the main challenge for these systems. The second problem is related to the off-line reconstruction of an impact scenario on the basis of stored measurements of the structural response. This 'black-box' type of application is currently more frequently required in forensic engineering (e.g. in the reconstruction of transport collisions). The underlying methodology used in the second formulation is based on the VDM approach.

- Chapter 5: **Adaptive Impact Absorption** (*P. K. Pawłowski, G. Mikułowski, C. Graczykowski, M. Ostrowski, Ł. Jankowski and J. Holnicki-Szulc*) contains a description of the main AIA concept and its applications. First, the multifolding (*multifolding materials*, MFM) idea is demonstrated as a particular solution to the challenging problem of optimal topology design of AIA systems. Secondly, the problem of optimal adaptation (via the control of plastic-like stresses) of AIA systems to the identified impact is analysed. Inflatable structures with controlled release of pressure are discussed in a separate sub-chapter as a good example of AIA systems. Finally, application of the AIA concept to the design of adaptive landing gears and the related hardware issues are presented.
- Chapter 6: **VDM-Based Remodeling of Adaptive Structures Exposed to Impact Loads** (*M. Wikło, Ł. Jankowski, M. Mróz and J. Holnicki-Szulc*) presents the development of new, VDM-based numerical tools devoted to optimal design (or redesign) of AIA structures. This challenging objective requires special and original algorithms that can effectively solve the complex problem. The coupled analysis requires the plastic stress levels, element cross-sectional areas and mass distribution to be modified simultaneously. The AIA redesign tasks utilize the analytically calculated, VDM-based sensitivities, which are further employed in gradient-based optimization techniques. Remodeling of damping properties is discussed in a separate subchapter.
- Chapter 7: **Adaptive Damping of Vibration by the Pre-stress Accumulation Release Strategy** (*A. Mróz, A. Orłowska and J. Holnicki-Szulc*) presents the concept of the mentioned PAR strategy applied to adaptive damping of vibration. The effectiveness of the technique is demonstrated using smart devices able to control specially designed structural connections by switching them off and instantly back on. This semi-active approach is able to damp very effectively first modes of vibrations, transferring part of the energy to higher modes, with higher natural damping. Numerical results are presented and verified experimentally using a specially constructed demonstrator.
- Chapter 8: **Modeling and Analysis of Smart Technologies in Vibroacoustics** (*T. G. Zieliński*) discusses theoretical fundamentals of newly developed numerical tools necessary for accurate vibroacoustical modeling of structures or composites made up of poroelastic, elastic and (active) piezoelectric materials, coupled to an acoustic medium. A widespread design of such smart noise attenuators (absorbers and insulators) is still an open topic and should involve an accurate multiphysics approach. Modeling and analysis of smart multilayered panels as well as of porous layers with mass inclusions improving the acoustic absorption are presented in the second part of the chapter.

# 1

# Introduction to Smart Technologies

Jan Holnicki-Szulc, Jerzy Motylewski and Przemysław Kołakowski

## 1.1 Smart Technologies – 30 Years of History

The term *smart technologies* (Smart-Tech) is understood as a generalization of the concept of *smart structures*. Smart technologies encompass mechanical systems equipped with sensors, actuators and pre-programmed controllers, which allow a structure to adapt to unpredictable external loading conditions. The concept of smart technologies requires the knowledge about the mechanical system itself, embedded sensors and controllable devices (usually based on *smart materials*), and driving electronics with integrated software, which adds the *intelligence* to the system.

Technological developments in the field of smart materials (modifying their properties, e.g. due to variation of the electric or magnetic field) and computational techniques have reached a point in which their synergy has a significant impact on the applicability of the interdisciplinary concepts of smart technologies to real structures (*smart structures*). The material science has led to the theoretical and experimental development of multifunctional materials (e.g. piezoelectric ceramics, shape memory alloys, magnetorheological fluids (MRFs), magnetostrictive materials). Furthermore, the development of fast and miniaturized microprocessors has enabled the design of embedded systems with distributed control capabilities. The final integrated product is composed of the following items:

(1) a distributed sensing system (e.g. based on piezoelectric transducers) able to monitor the structural response;
(2) actuators (e.g. utilizing MRFs) able to modify structural properties and
(3) control units able to realize a pre-designed strategy.

Such systems can be very effective in many applications, including structural health monitoring, mechanical impact absorption, damping of vibration and noise reduction.

Intelligent structures became the object of scientific research (mainly as hypothetical solutions) as a result of new demands that came from the space engineering in the 1970s and 1980s. For example, the problem of shape preservation of a parabolic antenna, launched in space, is

*Smart Technologies for Safety Engineering*   Edited by J. Holnicki-Szulc
© 2008 John Wiley & Sons, Ltd

not a trivial task. The thermal shock that occurs while passing through the Earth's shadow can cause vibrations, which are very difficult to be damped. These vibrations are induced due to the high flexibility of space structures, resulting in their low natural damping. As a remedy, intelligent systems, able to mitigate these vibrations, had to be invented. This shape (and vibration) control problem can be formulated by making use of highly responsive piezo-sensors and piezo-based actuators. The above-mentioned shape control problem requires work to be done against the resisting structure, using actuators. In the case of large civil structures, this actuation leads to substantial energy consumption, which cannot be supplied instantly. Consequently, the applications for this type of smart structure are significantly limited. Nevertheless, an important class exists of large real structures that can be effectively controlled with smart devices, consuming little power. Such structures can be called *adaptive structures* (instead of *active structures*). They are equipped with the dissipative kind of actuators (or dissipators) only. The field of their application is, for instance, the *adaptive impact absorption* (AIA), in which a structure equipped with controllable dissipators has to absorb optimally the energy coming from external extreme loads.

In the 1990s, the smart structure concept was under development through various lab-scale demonstrations of fully active, but very flexible systems, dynamically controlled with piezo-patches. Adaptive (semi-active) systems were also applied effectively to very large real civil structures (e.g. tall buildings in seismic areas or suspension bridges with MRF-controlled dampers).

Currently, the main stream of worldwide research and development activities in the field of smart structures is focused on *structural health monitoring* (SHM) and *load identification* (mostly applied to civil, mechanical and aerospace structures). There are several periodically organized international conferences [1–5] and scientific journals devoted to the subject. These academia-industry meetings, gathering various engineering communities, demonstrate the rapid development of SHM hardware solutions, accompanied by a relatively slow progress of new software tools. Researchers do have access to huge databases collected via numerous installations (mostly the large bridges monitored in Japan and South Korea), but the *soft computing tools* (e.g. based on *neural networks* or *genetic algorithms*) seem to have encountered limitations in application to large real structures. The *structural control* is the second field of application for smart technologies (e.g. damping of vibration [6] in antiseismic structures or in suspension bridges by MRF dampers). Dedicated international conferences are periodically held on structural control [7, 8], including some SHM issues as well. There are also cyclic events strongly related to the SHM and structural control topics [9–13]. In general, the interdisciplinary solutions can provide some extraordinary properties, thus mechatronic ideas, for example, are more and more frequently applied in structural mechanics. Nevertheless, there are still new areas open to potential applications as well as possibilities of improvement in already existing solutions. An important example is the design of vehicles with improved *crashworthiness*. For example, new euro codes, imposing high survivability requirements in crash scenarios, are under preparation. An increased interest can therefore be expected in the development of effective methods (including hardware devices and software tools) for the design of safe, crash-resistant vehicles.

New, emerging areas for the application of smart technologies are autonomous systems able to:

- monitor environmental conditions;
- identify extreme loads in real time;
- adapt to overloading by proper tuning of controllable dissipators;

- monitor structural health after the reception of impact;
- facilitate a decision on further use of the structure (self-repair if possible, post-impact inspection, reinforcement or rebuilding).

Particularly challenging problems to be solved are the following:
- distributed sensor systems for monitoring large structures;
- wireless sensors and data transmission;
- fast programmable controllers for embedded, distributed systems;
- long-life power sources and energy harvesting;
- fastly responding actuators with high stroke and blocking force;
- driving electronics allowing for instantaneous generation of magnetic fields;
- software tools processing effective algorithms (e.g. solving dynamic, inverse problems), aiming at identification of the current state of structural health.

## 1.2 Smart-Tech Hardware Issues

### 1.2.1 Structual Health Monitoring

The term *structural health monitoring* (SHM) refers to an important and fast-developing area of smart technologies. Its objective is the identification of damage (or load) by investigating structural responses due to environmental (ambient) or known excitations. In SHM systems, modern sensing technologies are employed to provide accurate measurements of structural responses. The results of the measurements can be subsequently analyzed by numerical algorithms in order to identify damage in the structure. Generally, modern SHM hardware systems consist of sensors, a data acquisition and processing block, a communication item and a main computer center. Physical quantities typically measured in vibration testing include accelerations or strains. In addition, measurements of temperature, humidity and wind velocity may be required to quantify the environmental conditions. Depending on the physical quantities to be monitored, the type of sensor is selected accordingly, i.e. accelerometers, tensometers, etc. A challenging problem is to determine the minimum number of required sensors and their optimal placement.

Probably the most common measuring devices used in SHM are the piezoelectric transducers [14] because of their outstanding electromechanical properties, relatively low price, and both actuating and sensing capabilities. They have been used for years in classical *nondestructive testing and evaluation* (NDT/E) methods like ultrasonic testing or acoustic emission. The direct piezoelectric effect, utilized in sensors, is present when a mechanical deformation of the piezoelectric material produces a proportional change in the electric polarization of that material (electric charge appears on opposite faces of the material). The converse piezoelectric effect, utilized in actuators, means that an acting external electric field induces proportional mechanical stress in the piezoelectric material (the material is deformed when an electric voltage is applied). A rather restricted number of piezoelectric materials have been found to be suitable for transduction elements in piezoelectric sensors. Basically, natural crystals (e.g. quartz, tourmaline) and synthetic single crystals (e.g. gallium ortho-phosphate, crystals of the CGG group), piezoelectric ceramics (e.g. lead–zirconate–titanate, denoted as PZT) and thin films (e.g. polyvinylidene fluoride, denoted as PVDF) can be used. Piezoelectric materials used in sensors combine excellent mechanical properties with a high piezoelectric sensitivity at a low production cost. They have a number of advantages, which makes them particularly suitable for dynamic measurements. Piezoelectric sensors have extremely high stiffness (their deflections are usually in the μm range), high natural frequency (hundreds of kHz), wide measuring range and a wide operating temperature range. The crystal-based sensors have very high stability whereas the ceramic-based ones can be produced in commercial quantities. Quasi-static measurements are possible with sensors having single crystals as transduction elements. Piezoelectric sensors can directly measure force, strain, acceleration and pressure. Sometimes the piezoelectric transducers are used for both actuation and sensing of a structural response simultaneously. The electrical impedance of the transducer is directly related to the mechanical impedance of the structure on which the transducer is mounted. Changes of structural properties such as stiffness, mass and damping can be monitored this way.

The fiber optic sensors [15] have a great potential to become widely used instruments in the field of SHM. Typical electrical strain gages are small and very accurate strain-measuring devices. However, they generally have a small dynamic range and are affected by environmental conditions such as moisture and temperature. Modern optical fibers are composed of high-silica glass doped with some oxides to achieve a required refraction index. An optical fiber consists of a core encapsulated in a cladding with a smaller refraction index. This enables total internal refraction at some incidence angle of entering light. Optical fibers can be generally divided into two types: multimode and single mode. Of special importance in strain sensor technology are single-mode optical fibers. Single-mode fiber limits its guidance capability for a chosen wavelength to one mode thanks to small diameter of the core and small difference between the core and cladding refraction indices. In practice, however, instead of just one mode, two orthogonally polarized, strongly coupled modes are carried by most commercially produced fibers. This effect, called birefringence, is often preferable as it helps the fiber to maintain the polarization of a guided wave and transmit it long distances. Light launching into a single-mode fiber of small core is difficult and favors a light source with highly directional output, e.g. a light emitting diode or injection laser diode. The major division of the fiber optic sensors distinguishes the interferometric sensors for outside (i.e. surface) application and *fiber Bragg grating* (FBG) sensors for inside application (i.e. embedded in the structure). The most widespread interferometric sensors interrogate a measurand-induced change of phase in the light propagating along a single-mode optical fiber. Several different layouts of interferometric sensors may be used, depending on the arrangement of optical paths. The most common configurations, based on phase change analysis, are the Michelson and Fabry–Perot interferometers. Interferometric optical fiber sensors provide high-sensitivity measurements. They require a special signal recovery technique (demodulation) to perform absolute measurements. Development of low-cost fabrication methods that do not compromise the strength and fatigue life of the optical fibers should facilitate the wide use of the sensors. The FBG sensors are highly sensitive devices as well. Their manufacturing process is automated and ensures no strength loss of the optical fiber. The intracore Bragg grating fiber optic sensor relies on the narrowband reflection from a fiber segment of periodic variations (gratings) in the core index of refraction of a single-mode fiber. If FBG sensors are used with the ratiometric demodulation system, they can build a robust, absolute-measurement, low-cost sensing system, which can be integrated on an optoelectronic chip, easily interconnected with the structure.

The *laser Doppler vibrometer* (LDV) can also be applied to reliable measuring of vibration characteristics (e.g. displacements, velocities) of large structures, even under a low-level excitation such as ambient vibration. The LDV is an indispensable tool whenever direct contact with the investigated object is impossible or undesirable.

Today's conventional monitoring systems have their measuring points wire-connected to the centralized data acquisition unit through coaxial cables. The sensors provide analog time signals, which need to be sampled and digitized for use in modern discrete-signal-processing

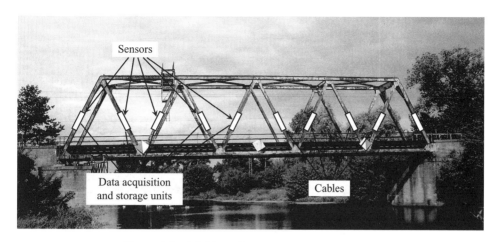

**Figure 1.1** A bridge with a cable-based SHM system

systems. As the analog signal sometimes travels a long distance from the sensor to the data aquisition unit, it may become noisy and degrade with the increasing cable length. A conventional SHM system, consisting of sensors, preamplifiers, a preliminary signal processing unit, an analog–digital (A/D) converter, centralized data acquisition and storage unit, is depicted in Figure 1.1. An alternative for the conventional SHM system can be obtained by merging some up-to-date technologies, which enable improved data acquisition and transmission. *Microelectromechanical systems* (MEMSs) for SHM are miniature electromechanical sensor and actuator systems fabricated in numerous series, using the processes of *very large system integration* (VLSI) technologies. Combining the advanced MEMS sensors with microcontrollers, an inexpensive, compact and intelligent sensing system can be made without the necessity to rely on the expensive data logger and the A/D transform interface. Modern, intelligent MEMSs equipped with microcontrollers should evolve towards devices able to:

- provide wireless comunication between the sensor and the data acquisition unit by installing special chips (already accomplished);
- handle large amonuts of data and filter out only the relevant information by utilizing programmable microcontrollers with dedicated software (to be accomplished).

Such remote intelligent monitoring system is expected to reduce the monitoring cost significantly. An example of a modern SHM system, with enhanced functionality thanks to the intelligent data processing plus wireless communication between sensors and the centralized data storage unit, is shown in Figure 1.2. However, many challenges associated with developing wireless SHM systems include restricted power consumption, data fidelity during wireless communication, real-time data acquisition from multiple sensing points under a limited communication bandwidth and the difficulties in time synchronization.

Although many real-time monitoring and control systems are designed to be computer based, the final decision on continued or suspended operation of a structure/system should be taken by qualified staff. The fast-growing databases, containing different kinds of data, have been the incentive to develop the *decision support systems* (DSSs), which facilitate on-line monitoring of the structure/system by using artificial intelligence methods, e.g. *case-based reasoning* (CBR).

**Figure 1.2**   A bridge with a wireless SHM system

## 1.2.2 Adaptive Impact Absorption

The term *adaptive structures* refers to *controllable structures*, equipped with pliant actuators, which modify (in real time) the mechanical properties of the structure exposed to extreme loads in such a way that no externally generated control forces are introduced. As a consequence, there is no need for an energy source to control the performance of the actuators (except for the low-energy source required to power the pliant dissipators). This class of *smart structures* is much more likely to be applied in civil and mechanical engineering, where the control process is restricted to a 'smart' strategy of dissipation of the strain energy accumulated due to the external load. Actuators take the form of energy dissipators with controlled characteristics (e.g. with the *yield stresses* as control parameters), which trigger plastic-like behavior (yielding) in the overloaded zones. The predetermined optimal distribution of plastic-like properties in structural elements (realized by *structural fuses* e.g. in the form of controlled pneumatic or hydraulic valves) can significantly increase the total energy dissipation level and improve the overall structural performance.

Consider an example of the adaptive chassis of a passenger car, shown in Figure 1.3(a). Here, the plastic-like behavior in the truss-like suspension system can be controlled by quick opening and closing of the hydraulic ultimate valves. For example, assume that the plastic limits in the adaptive elements cannot exceed the value $\sigma^u$. If the limit pressures, which trigger the opening of the valves $A_1$ and $A_2$ (see Figure 1.3(b)), make the condition $\sigma^\star < \sigma^u$ satisfied, then the overall characteristic of the adaptive element will take the form shown in Figure 1.3(c). A similar effect can be realized by using an MRF instead of controllable valves.

Numerical tests [16] have demonstrated high efficiency of the above concept for the impact energy absorption of a railway car crashing against a rigid wall. The obtained results show that for an impact velocity of 25 m/s, safe adaptation of the chassis is possible using just passive elements, i.e. where the plastic limits $\sigma^\star$ are assumed to be constant and not controlled. However, by adopting a control strategy for the process, which allows the plastic limits $\sigma^\star$ to be tuned in each adaptive element, the maximum safe impact velocity is increased to 38 m/s. Applicability of the adaptive concept to the design of structures for extreme loads, especially for impacts (cf. Chapter 5) and for vibration control (cf. Chapter 7), is demonstrated in this book. Some applications to the design of civil and mechanical structures in particular are also presented.

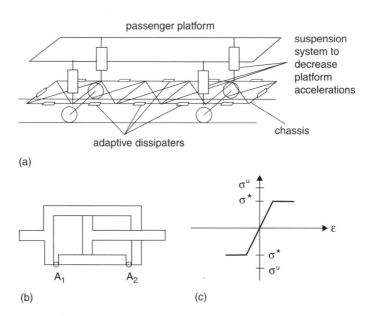

**Figure 1.3**   (a) Two-layer adaptive suspension system, (b) model of an adaptive hydraulic absorber, (c) characteristic of the adaptive element

The hardware required for AIA systems includes accelerometers, force, pressure and strain sensors. Some parts of the instrumentation are similar to SHM systems, but the necessary items are a microcontroller, feedback units and activating elements, i.e. actuators (see Figure 1.4). The microcontroller is a computer on a chip and is used to control electronic devices. A typical microcontroller contains all the memory and interfaces needed for a simple application, whereas a general-purpose microcontroller requires additional chips to provide these functions. Thus the typical microcontroller is a kind of microprocessor characterized by self-sufficiency and cost-effectiveness, in contrast to the general-purpose microcontroller (the kind used in a personal computer (PC)). The use of a microcontroller significantly reduces the number of chips and the amount of wiring in an AIA system.

For the last decade a vivid debate has run about the advantages and disadvantages of programmable logic controllers (PLCs) compared to PC-based ones. While the technological differences between the PLCs and PCs disappear, with PLCs using commercial off-the-shelf

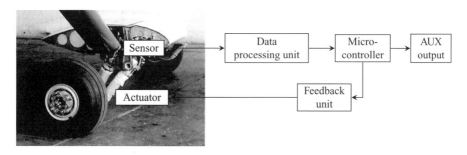

**Figure 1.4**   Block diagram of a basic AIA system

hardware and PCs incorporating real-time operating systems, a new class of controllers, the programmable automation controller (PAC), is emerging. Although many industrial applications are solved with traditional tools, there is a strong demand for simple low-cost PLCs. This has been an incentive for the growing production of low-cost micro-PLCs with digital input/output (I/O) using ladder logic. Most PLC users create PAC software by adding new functionality such as Ethernet communication, motion control and advanced algorithms into the existing architecture. PACs can directly take high-accuracy measurements, which are then passed directly into their control systems for immediate processing. In the AIA applications, it is extremely important to collect high-speed measurements for loads and vibrations. PACs represent the latest achievement in programmable controllers. The future for PACs is the incorporation of embedded technology. A challenging issue is the ability to design software in order to define hardware.

The field programmable gate arrays (FPGAs) [17] are electronic components commonly used by manufacturers to create custom chips, allowing the 'intelligence' to be placed in new devices. These devices consist of:

- configurable logic blocks able to perform a variety of functions;
- programmable interconnections acting as switches to connect the function blocks;
- I/O blocks that pass data in and out of the chip.

FPGAs are comparable to having a computer that literally rewires its internal circuitry to run a specific application. For example, one of the PAC platforms, named the Compact FieldPoint (http://www.ni.com/pac/cfp.htm), consists of hot-swappable analog and digital I/O modules and controllers with Ethernet and serial interfaces. The Compact FieldPoint network communication interfaces automatically publish the measurements through an Ethernet network. Another PAC platform named CompactRIO (http://www.ni.com/pac/crio.htm) is an FPGA-based reconfigurable control and acquisition system, designed for applications that require a high degree of customization and high-speed control. The architecture combines a real-time embedded processor for complex algorithms and custom calculations with a reconfigurable I/O (RIO) FPGA core.

## 1.3 Smart-Tech Software Issues

Smart technologies include not only the hardware devices but also the accompanying software tools, realizing smart strategies of structural monitoring and control. The numerical tools for the SHM and the design of adaptive structures, presented in this book, are often based on the *virtual distortion method* (VDM) (see Chapter 2). The VDM has been developed especially for a fast reanalysis, sensitivity analysis and control of dynamic structural responses. The presented software packages (see Chapters 3, 4 and 6) are numerically very effective and original.

In the design of structures for crashworthiness, present-day numerical simulation tools used in routine engineering practice focus on precise calculation of the crashing response of the base structures, but offer little guidance in the design process of optimal, energy-absorbing systems. On the contrary, as a result of the approach presented in this book (see Chapter 6), the optimal material distribution and nonlinear material characteristics could be designed (with the help of suitable software packages) for a predicted set of crash scenarios, including side impacts, which are difficult to be absorbed.

The numerical tools developed to design dynamically responding (also in the case of physical nonlinearities) systems can also be applied for damage identification and structural health monitoring problems. Consequently, Chapter 3 of this book is dedicated to inverse dynamic problems, where the VDM-based sensitivity analysis allows the damage location and intensity to be determined by observation of perturbations in structural responses. A part of Chapter 4 presents a solution of an SHM-related problem of load (location and intensity in time) reconstruction via inverse analysis.

# References

1. *International Conference on Damage Assessment of Structures DAMAS*, published as Key Engineering Materials, held in Pescara (Italy), Sheffield (UK), Dublin (Ireland), Cardiff (UK), Southampton (UK), Gdańsk (Poland), Torino (Italy), 7th event in 2007, every 2 years.
2. *American Workshop on Structural Health Monitoring*, held in Stanford University, California (USA), 6th event in 2007, every 2 years.
3. *European Workshop on Structural Health Monitoring*, held in Cachan (France), Munich (Germany), Granada (Spain), 3rd event in 2006, every 2 years.
4. *Structural Health Monitoring and Intelligent Infrastructure*, held in Tokyo (Japan), Shenzhen (China), Vancouver (Canada), 3rd event in 2007, every 2 years.
5. *Asia-Pacific Workshop on Structural Health Monitoring*, held in Yokohama (Japan), 1st event in 2006, every 2 years.
6. A. Preumont, *Vibration Control of Active Structures, An Introduction*, 2nd edition, Kluwer, Dordrecht, 2002.
7. *World Conference on Structural Control and Monitoring*, held in Los Angeles (USA), Kyoto (Japan), Como (Italy), San Diego (USA), 4th event in 2006, every 4 years.
8. *European Conference on Structural Control*, held in Barcelona (Spain), Champs-sur-Marne (France), Vienna (Austria), 3rd event in 2004, every 4 years.
9. *International Conference on Noise and Vibration Engineering*, held in Katholieke Universiteit Leuven (Belgium), 16th event in 2006, every 2 years.
10. *ECCOMAS Thematic Conference on Smart Structures and Materials*, held in Jadwisin (Poland), Lisbon (Portugal), Gdańsk (Poland), 3rd event in 2007, every 2 years.
11. *Experimental Vibration Analysis for Civil Engineering Structures EVACES*, held in Bordeaux (France), Porto (Portugal), 2nd event in 2007, every 2 years.
12. *IMAC Conference on Structural Dynamics*, held in Florida (USA), 25th event in 2007, annual.
13. *SPIE, Smart Structures and Materials/NDE*, held in San Diego (USA), 11th event in 2007, annual.
14. G. Gautschi, *Piezoelectic Sensorics*, Springer, Berlin, 2002.
15. E. Udd, *Fibre Optic Smart Structures*, John Wiley & Sons, Inc., New York, 1995.
16. J. Holnicki-Szulc and L. Knap, Adaptive crashworthiness concept, *International Journal of Impact Engineering*, **30**(6), 2004, 639–663.
17. U. Meyer-Baese, *Digital Signal Processing with Field Programmable Gate Arrays*, 2nd edition, Springer, Berlin, 2004.

# 2

# The Virtual Distortion Method – A Versatile Reanalysis Tool

Przemysław Kołakowski, Marcin Wikło and Jan Holnicki-Szulc

## 2.1 Introduction

The *virtual distortion method* (VDM) has been extensively developed since the mid 1980s in the Institute of Fundamental Technological Research (see the authors' affiliation). The term *distortions* was adopted from the book entitled *Theory of Elasticity* written by Nowacki [1], who used them to model material dislocations and to describe thermoelastic interactions. Then, Holnicki-Szulc and Gierlinski [2] initiated the VDM as such, introducing the notion of *virtual distortions* and proposing the idea of *influence matrix*, which is the essence of the method.

The concept of VDM is similar to the previously existing approach, mainly *initial strains*, which was first theoretically considered by Kroener [3]. Subsequently, Argyris [4] and Maier [5] used initial strains in order to model the phenomenon of plasticity in structures. The introduction of an initial strain to the structure causes disturbance in the total equilibrium conditions. As there is no relation between the imposed strain and the global response of the structure, the redistribution of stresses in the initial strains approach takes place in iterations. This is essentially the difference between initial strains and VDM, where local–global relations between elements of the structure are gathered in the influence matrix and further utilized in computations. The influence matrix stores information about the whole mechanical knowledge of the structure (topology, materials, boundary conditions). Thanks to this, the redistribution of stresses due to the introduction of a virtual distortion (equivalent to initial strain) is performed simply in one step, without iterations.

The VDM belongs to fast reanalysis methods, which basically means that a primary response of the structure (obtained via a finite element mehtod (FEM) analysis) is further modified by introducing fields of virtual distortions in a fast and efficient way. It was proved by Akgun *et al.* [6] that there is an equivalence between the VDM and the general Sherman–Morrison–Woodbury formulas derived in 1949/50, telling how to compute efficiently an inverse of a

*Smart Technologies for Safety Engineering*   Edited by J. Holnicki-Szulc
© 2008 John Wiley & Sons, Ltd

matrix subject to a variation. A brief overview of other fast reanalysis methods can be found in the following section.

The VDM was first applied by Holnicki-Szulc to induce pre-stress in elastic structures. His works include analysis, design (e.g. remodeling) and control applications [7, 8]. The same formal framework has recently been used in other applications related to smart structure technologies, e.g. adaptive structures or inverse problems of identification. The VDM can work effectively in the plastic regime, provided that the nonlinearity is approximated by piecewise linear sections.

One-dimensional models (trusses, beams) are the most effective in VDM, as the number of distortions to be imposed in a finite element is small (just one for trusses and three for beams). Plate or shell elements require more distortion states and consequently the composition of the influence matrix becomes more complex and time-consuming. Thanks to the analogies between trusses and nonstructural systems, i.e. water networks or electrical circuits, the VDM has recently been extended to model these systems. Truss structures are quite popular in civil engineering, so the method presented in this chapter is readily applicable for optimal design and health monitoring of real structures. For all these reasons, the strength of VDM is further demonstrated for the truss model.

The purpose of the chapter is to demonstrate the capabilities of VDM by providing an overview of major developments of the method done so far in structural statics and dynamics [9]. VDM-based sensitivity analysis as a useful tool for optimization is briefly described. Versatility of the method appears to be its greatest advantage. It is interesting that the framework of VDM is general enough.to solve problems from other technical fields too, e.g. hydraulic or electrical engineering, utilizing the idea of a graph.

## 2.2 Overview of Reanalysis Methods

A few articles [10, 11] reviewing the static methods of structural reanalysis have appeared in the literature in the last 20 years. The most recent one has been published by Akgun *et al.* [6], who describe and compare three methods of structural reanalysis – the combined approximation (CA) method, theorems of structural variation (TSV) and the virtual distortion method (VDM). It is shown that all the methods stemming from structural analysis are equivalent to the Sherman–Morrison [12] and Woodbury [13] (SMW) formulas originating from purely mathematical considerations of linear modifications of matrices. Akgun *et al.* admit that the capability of handling physically nonlinear problems by the VDM was the incentive for them to extend the SMW formulas to the nonlinear range as well. Unlike in the VDM, the nonlinear reanalysis by the SMW formulas requires an iteration procedure (e.g. Newton-like methods).

Fox and Miura [14] and Noor and Lowder [15] presented the idea of the reduced basis approach (also called the Ritz vector approach in model reduction or eigenproblems) in structural reanalysis. The point is that the displacement vector of the modified structure is approximated with a linear combination of only a few (significantly less than the number of the degrees of freedom) linearly independent vectors (similar to the influence vectors in the VDM) of a previously analyzed structure. Kirsch and Liu [16] continued to develop the reduced basis idea in the combined approximation (CA) method. The basis vectors in the CA method are calculated from a recurrent formula using an inverse of a decomposed stiffness matrix. The number of basis vectors in reanalysis is arbitrarily selected, but rarely exceeds ten, even for large problems. Satisfactory accuracy of response of the system reanalyzed by the CA method is usually assured with only a few basis vectors. If the basis vectors come close to being linearly dependent, then the solution becomes nearly exact. The approach was primarily developed for linear static analysis. An extension of CA to geometrically nonlinear problems [17] is worth noting.

The theorems of structural variation (TSV) [18] are in fact very similar to the VDM and were initiated at the same time. Instead of applying unit strains for building the influence matrix, unit loadings are used. Like the VDM, the method provides exact results. The first theorem expresses element forces and nodal displacements in a modified structure in terms of forces for the original structure and forces due to unit loadings. The second theorem concerns analogous expressions for displacements. The TSV method has been extended to two-dimensional [19] and three-dimensional [20] finite elements. Elastoplastic analysis can also be performed by TSV [21]. No development of the TSV method in dynamics has been done, as far as the authors know.

Deng and Ghosn [22] developed the pseudoforce method (PM) to perform reanalysis. The concept of pseudoforces, analogous to virtual distortions in VDM and pseudoloads in TSV, is used to model structural modifications. Based on the SMW formulas, which require the inverse of an initial stiffness matrix, an algorithm is proposed for solving both the linear and nonlinear reanalysis problems. It is noted that at some point of nonlinear incremental analysis, factorization of the stiffness matrix may be necessary. Otherwise the PM solution will prove to be costlier than a standard solver. Linear reanalysis of optimal placement of bracing for a two-dimensional frame and an elastoplastic analysis of a bridge deck are presented.

Bae and Grandhi [23] use the successive matrix inversion (SMI) method for reanalysis of structural systems. For initialization, the inverse of an initial stiffness matrix $K$ is required. Subsequently, the applied structural modification $\Delta K$ is decomposed into submodifications $\Delta K_j$ ($j = 1$ DOF (degree of freedom)), each one of which has only the $j$th nonzero column for the DOF $\times$ DOF system. This allows advantage to be taken of the Neumann (binomial) series expansion at the element level in order to obtain a recursive formula for finding the inverse of the modified stiffness matrix $K + \Delta K$ instead of inverting it directly. The SMI method is applied to a truss, frame and plate in linear statics. Approximate (not exact) solutions are obtained.

An approach proposing improvement of accuracy to the Neumann series expansion was proposed by Hurtado [24]. To this end, Shanks transformation (ST) is used to handle large modifications effectively. A significant improvement compared to the Pade approximation, described in Chen *et al.* [25], is demonstrated. Comparison with CA shows that the presented method is equally accurate, but exhibits faster convergence with the increase of expansion terms in the Neumann series. Linear examples of trusses are presented.

The term reanalysis in the nonlinear range may be understood in two ways. The first way is the standard modification to a structural parameter like in linear problems. The second way is different – it is rather an improvement (reduction of operations) of the Newton–Raphson procedure, which performs iterations to follow a nonlinear path. Examples of the different understanding of reanalysis are applications of the ST method and the Leu and Tsou [26] method.

Most of the existing reanalysis methods in dynamics concentrate on resolving the modal problem, in which only modifications to eigenvalues and eigenmodes are considered. This problem is solved quasi-statically in the frequency domain (no dependence on time has been investigated). A review of some eigenvalue reanalysis methods can be found in Chen *et al.* [25].

Recent methods dealing with reanalysis of the eigenproblem are generally named in the literature as structural dynamic modification (SDM). For solving the SDM problem, Ravi *et al.* [27] propose the single-step perturbation method as an alternative to the previously developed multiple-step perturbation. The single-step approach seems to outrank the multistep one, both in terms of accuracy for large modifications and computational effort. Yap and Zimmermann [28] prove that their iterative SDM method provides better estimates of both natural frequencies and mode shapes than the sensitivity-based methods. It can also provide a reasonable trade-off between accuracy and computational effort. The McDonnell–Douglass test space structure was used to demonstrate the validity of their approach. Chen [29] proposes an efficient iterative SDM for large modifications of modal parameters, based only on a limited knowledge of the original mode shapes (neither the original stiffness nor mass matrix is required). His noniterative high-order approximation approach also gives good estimations of the modified modal parameters. Reduced eigenvalue reanalysis presented by Grissom et al. [30] is used to predict the behavior of a structure with multiple absorbers, on the basis of the response of the structure without absorbers. The method is confronted with impedance-based approaches. The obtained results agree with the ones produced by the finite element (FE) code NASTRAN and measured in experiment. Recently, Kirsch et al. extended the CA method to nonlinear dynamic problems [31]. Similarly to SDM, the CA approach in structural dynamics is also limited to recalculation of an eigenproblem. The procedure involves shifts of the basis vectors as well as Gram–Schmidt orthogonalizations. The effectiveness very much depends on a proper choice of the basis vectors. The approach has been validated against the FE code ADINA. Huang et al. [32] propose a reanalysis method based on the Rayleigh–Ritz analysis, which handles extension of the basis vectors. This enables performance of an eigenproblem reanalysis in the case of topological changes, i.e. addition of members and joints to the structure. Accuracy depends greatly upon the number of eigenmodes analyzed for the original structure. All the above-mentioned SDM methods neglect the damping matrix in the analysis. With the perturbation approach, proposed by Cronin [33] and Tang and Wang [34], it is possible to analyze modifications to the damping characteristics of a structure as well. The assumption is that the original structure exhibits classical (proportional) damping, which means that it has the same modes as the corresponding undamped structure. Thus, the perturbation reanalysis can be performed in the configuration space by using the known real modes.

The only reanalysis method, known to the authors, producing a dynamic response in the time domain is the one based on the dynamic modification method (DMM) proposed by Muscolino [35]. Cacciola et al. [36] continue to develop this method proving its numerical efficiency and accuracy. For performing dynamic analysis, the equations of motion for a classically damped structure are uncoupled by the modal coordinate transformation, which also reduces the modal space (similarly to the Ritz vector approach). As a result, diagonal instead of full matrices enter the equations of motion. The second step is reformulation of the reduced problem in the state space. A tridiagonal transition matrix has to be defined. This allows a relatively simple solution procedure to be employed for the state variables involving operations (including inversions) on tridiagonal matrices. Finding the solution back in the original modal space is straightforward. For performing a dynamic reanalysis with this method, an analogous procedure is used, in which the increment of modification has to be specified explicitly. All other matrices appearing in the reanalysis are related to the original structure. It is claimed that nonproportional damping can be handled by the same method by treating it as a system modification. Accurate results of a response in the time domain with only five modes in the reduced basis are presented for a truss structure. Both the deterministic and stochastic loads are considered. The DMM fails only for drastic modifications of the original system, for which the order of the reduced modal space is not significantly lower than the original one and the numerical gain vanishes.

## 2.3 Virtual Distortion Method – The Main Idea

In the whole chapter, the lower case subindices refer to elements in the local coordinate system. The upper case subindices refer to nodes in the global coordinate system. Einstein's summation convention is used. Underlined indices are exempt from summation.

A simple two-bar truss has been chosen for demonstration of the main idea. Figure 2.1 schematically depicts an *original structure* consisting of two parallel truss elements, suppressed at the upper common node and exhibiting identical deformation at the lower common node. Assume that the left-hand element of the original structure has been subject to a modification (e.g. due to a change of its cross-sectional area).

Let us call the corresponding *initial strain* of the left-hand element (in isolation, i.e. out of structure) a *virtual distortion* $\varepsilon_1^0$. This initially deformed member has to comply with the continuity constraints of the structure. Thus, placing the element back into the structure provokes a self-equilibrated state of residual stresses $\sigma_i^R$ and a compatible state of strains $\varepsilon_i^R$ (see the *prestressed structure* in Figure 2.1)

Then, let us apply an external force-type load $P$ to the analyzed structure. It generates the deformation denoted by $\varepsilon_i^L$ in the *loaded*, linearly elastic structure. Superposing these two states of the prestressed and the loaded structure gives, as a result, a *distorted (modelled) structure* (with a combination of linearly elastic responses to initial strains and external load). It is now postulated that (as marked in Figure 2.1) the distorted structure should be identical in terms of final strains $\varepsilon_i$ and internal forces $A_{\underline{i}}\sigma_{\underline{i}}$ with a *modified structure* (with a modified cross-sectional area in the left-hand element from $A_1$ to $\hat{A}_1$).

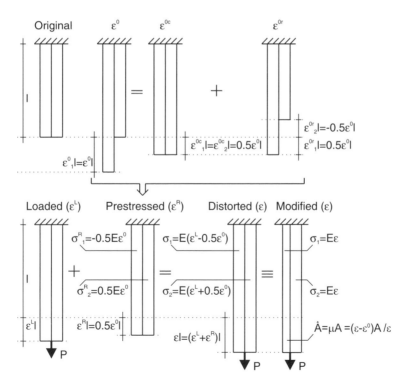

**Figure 2.1**   Virtual distortion method scheme

Virtual distortions can be used to simulate not only modifications of material distribution but also material nonlinearities, i.e. plastic effects. An arbitrary state of distortions can be uniquely decomposed $\varepsilon_i^0 = \varepsilon_i^{0c} + \varepsilon_i^{0r}$ (cf. Reference [8]; see Figure 2.1). The component $\varepsilon^{0c}$ is responsible for the compatible, stress-free deformation of the structure (e.g. caused by homogenous heating of both elements of the truss) while the component $\varepsilon^{0r}$ causes the self-equilibrated, strain-free stress state in the structure (e.g. caused by heating of the left element with simultaneous cooling of the right one). The components $\varepsilon_i^{0r}$ are presented in Figure 2.1 prior to satisfying the continuity constraints.

## 2.4 VDM in Structural Statics

### 2.4.1 Influence Matrix in Statics

The main feature distinguishing the VDM from the initial strains approach is the influence matrix $D_{ij}^{\varepsilon}$. It describes strains in the truss member $i$ caused by the unit virtual distortion $\varepsilon_i^0 = 1$ (unit initial strain) applied to the member $j$. The unit virtual distortion is practically imposed as a pair of *self-equilibrated compensative forces* of reverse signs (equivalent to a unit strain as in Figure 2.2) applied to the nodes of the strained element. The influence matrix $D_{ij}^{\varepsilon}$ collects $m$ influence vectors, where $m$ denotes the number of truss elements. In order to build an influence vector, a solution of a standard linear elastic problem (equilibrium condition) by the finite element method has to be found:

$$K_{MN}u_N = f_M \tag{1}$$

with $K_{MN}$ being the stiffness matrix, $u_N$ the displacement vector and $f_M$ the force vector in global coordinates. Usually, the obtained displacements serve to calculate a corresponding

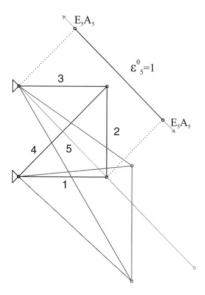

**Figure 2.2**   Influence of the unit distortion applied in a chosen location

response in strains:

$$\varepsilon_i = G_{iN} u_N \tag{2}$$

with $G_{iN}$ being the geometric matrix, which transforms global degrees of freedom to local strains (continuity condition). The response in strains is a standard for building an influence vector. However, storage of any other required response is also useful, i.e. displacements, stresses or forces.

The external force vector $f$ in Equation (1) corresponds to two compensative forces (axial forces in the case of truss structures) applied to a structural member, equivalent to application of a unit strain to the unconstrained member (see the diagonal element in Figure 2.2 after applying the pair of forces). The response of the structure to the imposition of the unit virtual distortion $\varepsilon_5^0 = 1$ is depicted by the deformed configuration in Figure 2.2.

Thus to build the influence matrix $D_{ij}^\varepsilon$, $m$ solutions of a linear elastic problem have to be found. The set (1) has to be solved with $m$ different right-hand sides corresponding to $m$ pairs of compensative forces applied successively in each structural member. In this way, the influence matrix stores information about the entire structure properties including topology, material characteristics and boundary conditions, in calculation of the structural response.

Note that the static influence matrix for statically determinate structures becomes an identity matrix (zero redundancy means no interrelations between members) and the VDM loses its major tool. For truss structures, the strain influence matrix $D_{ij}^\varepsilon$ is quadratic, nonsymmetric and singular. Making use of Betti's mutual work principle it can easily be proved that the matrix $A_i l_i D_{ij}^\varepsilon$ becomes symmetric, where $A_i$ and $l_i$ denote the initial cross-sectional area and element length, respectively. The rank of the $m \times m$ symmetric matrix is rank $[A_i l_i D_{ij}^\varepsilon] = m - k$, where $m$ and $k$ denote the number of all elements and the structural redundancy, respectively. It means that there are $m$–$k$ linearly independent components $\varepsilon_i^{0c}$ causing stressless compatible strains $\varepsilon_i^R$ and $k$ linearly independent components $\varepsilon_r^{0r}$, causing strainless self-equilibrated stresses $\sigma_i^R$. All nonvanishing eigenvalues of the matrix $A_i l_i D_{ij}^\varepsilon$ are positive; thus it is nonnegative definite.

Analogously, it can be proved that the influence matrix storing responses in stresses $A_i l_i D_{ij}^\sigma = A_i l_i E_i (D_{ij}^\varepsilon - \delta_{ij})$ is symmetric, nonpositive definite of the rank $[A_i l_i D_{ij}^\sigma] = k$. $E_i$ denotes Young's modulus and $\delta_{ij}$ is the Kronecker's delta. This matrix (originally called $Z$) was introduced by Maier (cf. Reference [5]) and applied to the initial strains approach, allowing for the elastoplastic analysis of stress redistribution through a quadratic programming procedure. In the VDM approach, however, the distortions $\beta_j^0$ modelling plastic permanent deformations can be simply determined by solving a set of linear equations, while satisfying the condition that yield stress should be reached in all overloaded members (cf. Reference [8]).

### 2.4.2 Stiffness Remodeling in Statics

Let us confine our considerations to truss structures in the elastic range first. Consider introducing a field of initial strains $\varepsilon^0$ (called *virtual distortions*) into a truss structure, which induces residual displacements and strains in the structure, expressed as follows (cf. Reference [7] and [8]):

$$u_N^R = B_{Nj}^\varepsilon \varepsilon_j^0, \tag{3}$$

$$\varepsilon_i^R = D_{ij}^\varepsilon \varepsilon_j^0. \tag{4}$$

The matrix $B_{Nj}^{\varepsilon}$ collects structural responses in displacements to the unit virtual distortion $\varepsilon_j^0 = 1$ equivalent to the unit strain. The matrix $D_{ij}^{\varepsilon}$ collects the corresponding structural responses in strains. Both matrices are marked with the superscript $\varepsilon$, which indicates the fact that they were built due to imposition of strain-like virtual distortions. The two influence matrices are related by the geometric matrix $G_{iN}$ (cf. Equation (2)) as follows:

$$D_{ij}^{\varepsilon} = G_{iN} B_{Nj}^{\varepsilon} \tag{5}$$

Residual stresses are expressed by

$$\sigma_i^{R} = E_{\underline{i}} (D_{\underline{i}j}^{\varepsilon} - \delta_{\underline{i}j}) \varepsilon_j^0 \tag{6}$$

where $E_i$ denotes the Young's modulus and $\delta_{ij}$ the Kronecker's delta.

Assume that application of external load to the structure provokes elastic *linear response* $\varepsilon_i^{L}$, $\sigma_i^{L}$, which will be superposed with the *residual response* $\varepsilon_i^{R}, \sigma_i^{R}$. Thus, in view of Equations (4) and (6),

$$u_N = u_N^{L} + u_N^{R} = u_N^{L} + B_{Nj}^{\varepsilon} \varepsilon_j^0 \tag{7}$$

$$\varepsilon_i = \varepsilon_i^{L} + \varepsilon_i^{R} = \varepsilon_i^{L} + D_{ij}^{\varepsilon} \varepsilon_j^0 \tag{8}$$

$$\sigma_i = \sigma_i^{L} + \sigma_i^{R} = E_{\underline{i}}\varepsilon_{\underline{i}}^{L} + E_{\underline{i}} \left( D_{\underline{i}j}^{\varepsilon} - \delta_{\underline{i}j} \right) \varepsilon_j^0 = E_{\underline{i}} \left( \varepsilon_{\underline{i}} - \varepsilon_{\underline{i}}^0 \right) \tag{9}$$

The relation between element forces $p_i$ and stresses $\sigma_i$ is known via the cross-sectional areas $A_i$:

$$p_i = A_{\underline{i}}\sigma_{\underline{i}} \tag{10}$$

Internal $p_i$ and external $f_N$ forces are linked via the transposed geometric matrix $G_{Ni}$ (cf. Equation (2)):

$$f_N = G_{Ni} l_i p_i \tag{11}$$

Let us now take into account structural stiffness modifications exemplified by changes of Young's modulus. This means considering a modified value $\hat{E}_i$. In view of Equations (9) and (10) element forces can be expressed in the modified structure and original structure with an introduced virtual distortion field (i.e. distorted structure), as follows:

$$\hat{p}_i = \hat{E}_{\underline{i}} A_{\underline{i}} \hat{\varepsilon}_{\underline{i}} \tag{12}$$

$$p_i = E_{\underline{i}} A_{\underline{i}} \left( \varepsilon_{\underline{i}} - \varepsilon_{\underline{i}}^0 \right) \tag{13}$$

The main postulate of the VDM in static remodeling requires that local strains (including plastic strains) and forces in the modified and distorted structure are equal:

$$\hat{\varepsilon}_i = \varepsilon_i \tag{14}$$

$$\hat{p}_i = p_i \tag{15}$$

This postulate leads to the following relation:

$$\hat{E}_i A_i \varepsilon_i = E_i A_i \left(\varepsilon_i - \varepsilon_i^0\right) \qquad (16)$$

Equation (16) provides the coefficient of the stiffness change for each truss element $i$ as the ratio of the modified Young's modulus to the original one:

$$\mu_i^{\mathrm{E}} \stackrel{\mathrm{def}}{=} \frac{\hat{E}_i}{E_i} = \frac{\varepsilon_i - \varepsilon_i^0}{\varepsilon_i} = \frac{\hat{A}_i}{A_i} \stackrel{\mathrm{def}}{=} \mu_i^{\mathrm{A}} \qquad (17)$$

Note that the coefficient $\mu_i^{\mathrm{E}}$ may be equivalently expressed as the ratio of the original to modified cross-sectional area $\mu_i^{\mathrm{A}}$ of a truss element. If $\mu_i^{\mathrm{E}} = 1$ we deal with an intact structure. Variation of the coefficient in the range $0 \le \mu_i^{\mathrm{E}} \le 1$ means a reduction of stiffness and in the range $\mu_i^{\mathrm{E}} \ge 1$ an increase of stiffness. Substituting Equation (8) into (17) gives a set of equations for $\varepsilon_i^0$, which must be solved for an arbitrary number of modified elements (usually small compared to all elements in the structure), described by coefficients $\mu_i^{\mathrm{E}}$ different from 1:

$$\left[\delta_{ij} - (1 - \mu_i^{\mathrm{E}}) D_{ij}^\varepsilon \right] \varepsilon_j^0 = (1 - \mu_i^{\mathrm{E}}) \varepsilon_i^{\mathrm{L}} . \qquad (18)$$

### 2.4.3 Plasticity in Statics

The virtual distortion field introduced into the structure may be twofold. We shall distinguish between purely virtual distortions $\varepsilon_i^0$ (having no physical meaning) used for modeling structural parameter modifications (e.g. changes of cross-sectional area) and plastic-like distortions $\beta_i^0$ used for simulating physical nonlinearities in the structure. The plastic-like distortions are identified with plastic strains:

$$\beta_i^0 \equiv \varepsilon_i^{\mathrm{pl}} \qquad (19)$$

and have no virtual character. Thus the plastic behavior of members is effectively included in the strain and stress formulas in the following way (cf. Equations (8) and (9)):

$$\varepsilon_i = \varepsilon_i^{\mathrm{L}} + D_{ik}^\varepsilon \beta_k^0 \qquad (20)$$

$$\sigma_i = E_i \varepsilon_i^{\mathrm{L}} + E_i \left(D_{ik}^\varepsilon - \delta_{ik}\right) \beta_k^0 = E_i \left(\varepsilon_i - \beta_i^0\right) \qquad (21)$$

The VDM can be used to model a nonlinear constitutive relation provided that it is piecewise linear (see Figure 2.3). Consequently, assume the behavior of material, after reaching the yield limit $\sigma_i^\star$, as a linear section with inclination $\gamma_i E_i$ to the horizontal axis, which is less than the original Young's modulus $E_i$:

$$\sigma_i - \sigma_i^\star = \gamma_i E_i \left(\varepsilon_i - \varepsilon_i^\star\right) \qquad (22)$$

The coefficient $\gamma_i$ determines isotropic hardening of the material. If $\gamma_i = 0$, perfectly plastic behavior occurs.

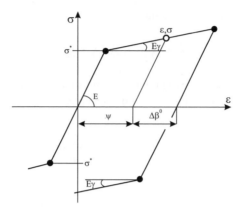

**Figure 2.3**   Nonlinear constitutive law (piecewise linear)

Substituting Equations (20) to (22), a local set of equations is assembled to be solved for plastic distortions $\beta_i^0$:

$$\left[\delta_{ik} - (1 - \gamma_{\underline{i}})D_{ik}^\varepsilon\right]\beta_k^0 = (1 - \gamma_{\underline{i}})(\varepsilon_{\underline{i}}^{\mathrm{L}} - \varepsilon_{\underline{i}}^\star). \tag{23}$$

### 2.4.4 Example 1 in Statics

The five-element, $1\,m$ by $1\,m$ truss structure, shown in Figure 2.4, has been chosen for demonstration of the VDM capabilities in statics. All elements have the same Young's modulus $E = 210\,\mathrm{GPa}$ and cross-sectional area $A = 1.0 \times 10^{-5}\,\mathrm{m}^2$. The structure is subjected to a static vertical force of $F = 2.5\,\mathrm{kN}$ in node 2. Buckling is not taken into account. The strain influence matrix $\mathbf{D}^\varepsilon$ of the structure and an equivalent of the stress influence matrix $\mathbf{D}^\varepsilon - \mathbf{I}$

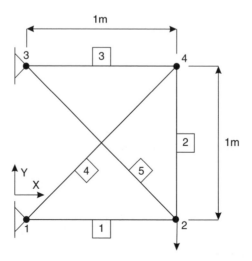

**Figure 2.4**   Five-element truss structure for testing VDM algorithms

(see Section 2.4.1) take the following values:

$$
\mathbf{D}^\varepsilon =
\begin{bmatrix}
0.8845 & -0.1155 & -0.1155 & 0.2310 & 0.2310 \\
-0.1155 & 0.8845 & -0.1155 & 0.2310 & 0.2310 \\
-0.1155 & -0.1155 & 0.8845 & 0.2310 & 0.2310 \\
0.1634 & 0.1634 & 0.1634 & 0.6733 & -0.3267 \\
0.1634 & 0.1634 & 0.1634 & -0.3267 & 0.6733
\end{bmatrix}
\tag{24}
$$

$$
\mathbf{D}^\varepsilon - \mathbf{I} =
\begin{bmatrix}
-0.1155 & -0.1155 & -0.1155 & 0.2310 & 0.2310 \\
-0.1155 & -0.1155 & -0.1155 & 0.2310 & 0.2310 \\
-0.1155 & -0.1155 & -0.1155 & 0.2310 & 0.2310 \\
0.1634 & 0.1634 & 0.1634 & -0.3267 & -0.3267 \\
0.1634 & 0.1634 & 0.1634 & -0.3267 & -0.3267
\end{bmatrix}
\tag{25}
$$

Note that the degree of redundancy of the structure is 1, which is also the rank of the stress influence matrix $\mathbf{D}^\varepsilon - \mathbf{I}$ (there is only one state of pre-stress available for the truss – all columns of the matrix $\mathbf{D}^\varepsilon - \mathbf{I}$ are linearly dependent). The rank of the strain influence matrix $\mathbf{D}^\varepsilon$ is equal to 4 (there may be four states of strains accompanying the single pre-stress state).

Let us first demonstrate how the topology of the structure can be quickly remodeled, simulating elimination of elements 2, 3 and 4 by virtual distortions. To this end, the condition $\mu_i = 0$ is imposed in the mentioned members. This leads to a set of equations ($3 \times 3$) to be solved for $\varepsilon^0$ in one step (cf. Equation (18)). The results of the analysis are presented in Table 2.1. The remaining members 1 and 5 form a statically determinate structure, which can be further optimized to become isostatic (i.e. of zero redundancy and uniformly strained). To achieve this goal in this example, the stress in element 5 should be reduced (increasing the cross-section by $\sqrt{2}$) to match the stress in element 1.

Next, an elastoplastic analysis of the structure will be performed, assuming the yield limit $\sigma_i^* = 294$ MPa, and perfectly plastic ($\gamma_i = 0$) post-critical behavior. The nominal load $F$ is gradually increased by the factor $\alpha > 1$. Only one member can enter the plastic zone without

**Table 2.1** Results of elimination of three members from the original truss structure

| | $\varepsilon^L$ | $\varepsilon^R$ | $\varepsilon$ | $\varepsilon^0$ |
|---|---|---|---|---|
| 1 | −0.664E−3 | −0.526E−3 | −0.119E−2 | 0.000E+0 |
| 2 | 0.526E−3 | 0.397E−3 | 0.923E−3 | 0.923E−3 |
| 3 | 0.526E−3 | 0.397E−3 | 0.923E−3 | 0.923E−3 |
| 4 | −0.744E−3 | −0.610E−3 | −0.135E−2 | −0.135E−2 |
| 5 | 0.938E−3 | 0.744E−3 | 0.168E−2 | 0.000E+0 |

| | $\sigma^L$ | $\sigma^R$ | $\sigma$ | $\mu$ |
|---|---|---|---|---|
| 1 | −0.139E+9 | −0.110E+9 | −0.250E+9 | 1.000 |
| 2 | 0.110E+9 | −0.110E+9 | −0.642E−7 | 0.000 |
| 3 | 0.110E+9 | −0.110E+9 | 0.104E−7 | 0.000 |
| 4 | −0.156E+9 | 0.156E+9 | −0.234E−7 | 0.000 |
| 5 | 0.197E+9 | 0.156E+9 | 0.353E+9 | 1.000 |

**Table 2.2**  Results of elastoplastic analysis of the truss at the
stage preceding collapse ($\alpha = 1.66$)

|   | $\varepsilon^{L}$ | $\varepsilon^{R}$ | $\varepsilon$ | $\varepsilon^{0}$ |
|---|---|---|---|---|
| 1 | −0.110E−2 | 0.112E−3 | −0.990E−3 | 0.000E+0 |
| 2 | 0.873E−3 | 0.112E−3 | 0.985E−3 | 0.000E+0 |
| 3 | 0.873E−3 | 0.112E−3 | 0.985E−3 | 0.000E+0 |
| 4 | −0.124E−2 | −0.158E−3 | −0.139E−2 | 0.000E+0 |
| 5 | 0.156E−2 | 0.325E−3 | 0.188E−2 | 0.483E−3 |

|   | $\sigma^{L}$ | $\sigma^{R}$ | $\sigma$ | $\mu$ |
|---|---|---|---|---|
| 1 | −0.231E+9 | 0.234E+8 | −0.208E+9 | 1.000 |
| 2 | 0.183E+9 | 0.234E+8 | 0.207E+9 | 1.000 |
| 3 | 0.183E+9 | 0.234E+8 | 0.207E+9 | 1.000 |
| 4 | −0.259E+9 | −0.332E+8 | −0.293E+9 | 1.000 |
| 5 | 0.327E+9 | −0.332E+8 | 0.294E+9 | 1.000 |

violating the integrity of the structure. It is the diagonal element 5. Any other plastic hinge
(element) will provoke a kinematic mechanism. The results of the final stage (just before
collapse at $\alpha = 1.66$ when the other diagonal 4 is very close to the yield limit $\sigma_i^{*}$) are presented
in Table 2.2. The use of the VDM for optimal static design of more complex structures, including
beams and in-plane loaded plates, is amply exemplified in Reference [8].

### 2.4.5  Example 2 in Statics

The problem of optimal remodeling of truss structures in statics, using the VDM, was previously presented in Reference [37]. From that article, a medium-size truss example was chosen
to provide an insight into topological optimization capabilities of the VDM. The *ground structure* approach, utilizing a regular $5 \times 5$ grid of nodes and considering 300 possible connections
between them, was adopted. In order to reduce the computational effort associated with the lit-
eral ground structure, many members (i.e. overlapping and between supports) were disregarded
to start up with only 136 connections. The horizontal-to-vertical aspect ratio of the grid is 8 : 5
($24\,\text{m} \times 15\,\text{m}$). Uniform cross-section $A = 2.55\,\text{cm}^2$ is assumed for all initial members.

The problem is posed in a classical manner as finding the minimum volume of the ground
structure subjected to one static force $P$, provided that limit stresses are not exceeded in any
member. An equivalent formulation of the problem is to find maximum stiffness at a constant
volume. It is clear that most of the initial members should be eliminated in the process of
structural remodeling and the remaining ones should be resized. For one load case, presented
here, the resultant optimal truss is an isostatic (i.e. statically determinate, uniformly stressed)
structure. Using a VDM-based gradient optimization, two solutions, complying with the requirement, were found. The reason for having two solutions is that the threshold for elimination
of members in the remodeling process – the coefficient of stiffness change (cf. Equation (17)) –
was arbitrarily adjusted in the range $0.05 \leq \mu_i \leq 0.10$.

The first topology, shown in Figure 2.5, consists of only six members and the corresponding
final volume is $80.61\,\text{dm}^3$. The second topology, depicted in Figure 2.6, consists of 12 members
and the corresponding final volume is $80.26\,\text{dm}^3$. Formally, the performance of the algorithm
was equally good in both cases. However, from the application point of view, the first topology
is more attractive.

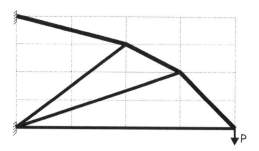

**Figure 2.5**   First optimal topology for the 136-element ground structure

## 2.5 VDM in Structural Dynamics

### 2.5.1 Influence Matrices in Dynamics

For dynamic problems, the influence matrix has to be given one more dimension – time. The imposition of unit virtual distortion takes place in the first instant of an analyzed period of time. This corresponds to the first time step in numerical algorithms, where certain time discretization is assumed. Similarly to statics, the unit virtual distortion $\varepsilon_i^0 = 1$ in the first time step is applied to an element as a pair of self-equilibrated compensative forces, causing a unit strain of the element when taken out of the structure (see Figure 2.7). Such action has the character of an impulse excitation, which is consecutively imposed in all elements of the structure to compose the whole influence matrix. In practice, the response of the structure to the *impulse virtual distortion* in the element is calculated using the Newmark integration algorithm over a chosen period of time. Two influence matrices are of interest here, one of them storing the structural response in displacements (further denoted by $\mathbf{B}^\varepsilon$) and the other – in strains (further denoted by $\mathbf{D}^\varepsilon$).

The above-described influence matrices in dynamics are full analogies to the influence matrix in statics (expressed in displacements or strains), able to model stiffness changes and material nonlinearities. An important parameter in dynamics is inertia and the ability to model the mass changes. To this end, another kind of influence matrice (further denoted by $\mathbf{B}^f$, $\mathbf{D}^f$) must be introduced. This time, the virtual distortions are successively imposed in degrees of freedom of the structure (see Figure 2.8). The principal difference is that the nodal virtual distortion has the form of an unequilibrated unit impulse force, contrary to the unit virtual distortion applied as a pair of self-equilibrated forces. This kind of *impulse force distortion* at a node aims at

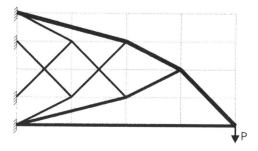

**Figure 2.6**   Second optimal topology for the 136-element ground structure

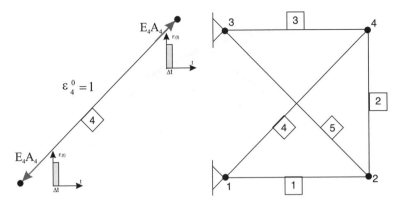

**Figure 2.7**   Impulse virtual distortion in an element, showing the process of matrix $\mathbf{B}^\varepsilon$, $\mathbf{D}^\varepsilon$ composition

capturing the influence of inertia on structural response. Again, the response of the structure to the impulse force distortion at nodes is obtained by the Newmark algorithm.

It will be demonstrated in subsequent sections that by establishing the influence matrices $\mathbf{B}^\varepsilon$, $\mathbf{D}^\varepsilon$, $\mathbf{B}^f$, $\mathbf{D}^f$, the remodeling of stiffness and mass in the structure becomes feasible. A nonlinear constitutive relation can also be accounted for. However, linear geometric relations (small strains) are assumed. Examples of VDM applications in dynamics are demonstrated in Chapters 3 and 6.

### 2.5.2  Stiffness Remodeling in Dynamics

In signal processing performed in many fields of engineering, the output response of a system is expressed as an integral of the product of the input excitation and transfer function (i.e. the system's response to an impulse function like Dirac's delta) over some period of time.

For a simple harmonic oscillator of mass $m$ and natural frequency $\omega$, the convolution of the two functions determines the displacement $u(t)$ due to a series of impulses $f(\tau)\mathrm{d}\tau$ over the

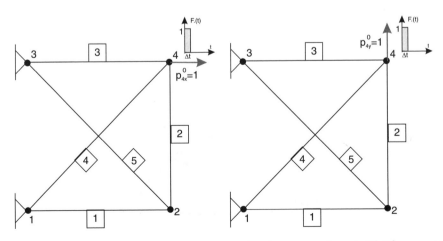

**Figure 2.8**   Impulse virtual distortion at a node, showing the process of matrix $\mathbf{B}^f$, $\mathbf{D}^f$ composition

time period $\langle 0, t \rangle$, as

$$u(t) = \frac{1}{m\omega} \int_0^t f(\tau) \sin \omega(t - \tau) d\tau \qquad (26)$$

Equation (26) is called the Duhamel's integral. Its range of validity is limited by the assumption of the system's linearity, i.e. exhibiting small strains by a structural system.

Similarly to the Duhamel's integral, the VDM residual response in displacement, modeling some modifications in the structure, can be written as a discrete convolution of the influence matrix (in displacements) $B^\varepsilon_{Mj}$ and virtual distortions $\varepsilon^0_j$. Superposing the residual with a linear (no modifications for an elastic structure) response gives (cf. Equation (7))

$$u_M(t) = u^L_M(t) + \sum_{\tau=0}^{t} B^\varepsilon_{Mj}(t - \tau)\varepsilon^0_j(\tau) \qquad (27)$$

The summation (not integral) over time in Equation (27) indicates that the considered period of time $\langle 0, t \rangle$ was discretized in order to use the finite element method (FEM). For performing time integration, the authors have chosen the Newmark algorithm.

Using Equations (2) and (5), the total strain, composed of the linear and residual parts, can be conveniently written as

$$\varepsilon_i(t) = \varepsilon^L_i(t) + \sum_{\tau=0}^{t} D^\varepsilon_{ij}(t - \tau)\varepsilon^0_j(\tau) \qquad (28)$$

The corresponding stresses take the following form:

$$\sigma_i(t) = E_{\underline{i}} \left( \varepsilon_{\underline{i}}(t) - \varepsilon^0_{\underline{i}}(t) \right)$$
$$= \sigma^L_i(t) + E_{\underline{i}} \left( \sum_{\tau=0}^{t-1} D^\varepsilon_{\underline{i}j}(t - \tau)\varepsilon^0_j(\tau) + \left( D^\varepsilon_{\underline{i}j}(0) - \delta_{\underline{i}j} \right) \varepsilon^0_j(t) \right) \qquad (29)$$

Retrieving the valid static postulate (cf. Equation (16)) of equivalence between the distorted and modified structure in terms of strains and internal forces, the following modification coefficient can be derived:

$$\mu^E_i \overset{\text{def}}{=} \frac{\hat{E}_{\underline{i}}}{E_{\underline{i}}} = \frac{\varepsilon_{\underline{i}}(t) - \varepsilon^0_{\underline{i}}(t)}{\varepsilon_{\underline{i}}(t)} = \frac{\hat{A}_{\underline{i}}}{A_{\underline{i}}} \overset{\text{def}}{=} \mu^A_i \qquad (30)$$

Note that by using the formula (30), structural stiffness can be modified here either as a change of Young's modulus or of the cross-sectional area, analogously to statics. Another observation is that the coefficient $\mu_i$, constant in time, is expressed in dynamics in terms of the time-dependent components $\varepsilon_i(t)$ and $\varepsilon^0_i(t)$. After reshaping Equation (30), the system of equations to be solved for virtual distortions $\varepsilon^0_i$ is obtained:

$$\left[ \delta_{ij} - (1 - \mu^E_{\underline{i}})D^\varepsilon_{\underline{i}j}(0) \right] \varepsilon^0_j(t) = (1 - \mu^E_{\underline{i}})\varepsilon^{\neq t}_{\underline{i}}(t) \qquad (31)$$

where $\varepsilon_i^{\neq t}(t)$ denotes strains aggregated in all time steps preceding the current time instant $t$:

$$\varepsilon_i^{\neq t}(t) = \varepsilon_i^L(t) + \sum_{\tau=0}^{t-1} D_{ij}^{\varepsilon}(t - \tau)\varepsilon_j^0(\tau) \tag{32}$$

Note that the governing matrix on the left-hand side of Equation (31) is time-independent; hence it remains constant throughout all time steps. Only the right-hand side vector varies. The set is local, i.e. limited to the elements (in local coordinates), for which stiffness is remodeled.

### 2.5.3 Plasticity in Dynamics

Determination of the VDM strain and stress formulas for the dynamic plastic range can be quickly done by replacing the virtual distortion $\varepsilon_j^0(t)$ in Equations (28) and (29) with the plastic distortion $\beta_k^0(t)$, to produce

$$\varepsilon_i(t) = \varepsilon_i^L(t) + \sum_{\tau=0}^{t} D_{ik}^{\varepsilon}(t - \tau)\beta_k^0(\tau) \tag{33}$$

$$\sigma_i(t) = E_{\underline{i}}\left(\varepsilon_{\underline{i}}(t) - \beta_{\underline{i}}^0(t)\right)$$
$$= \sigma_i^L(t) + E_{\underline{i}}\left(\sum_{\tau=0}^{t-1} D_{\underline{i}k}^{\varepsilon}(t - \tau)\beta_k^0(\tau) + \left(D_{\underline{i}k}^{\varepsilon}(0) - \delta_{\underline{i}k}\right)\beta_k^0(t)\right) \tag{34}$$

As in statics, a piecewise linear constitutive law (see Figure 2.3) is also adopted in dynamics. This time the relation is written in the incremental form, enabling an increment of plastic distortion $\Delta\beta_k^0(t)$ to be determined in every time step:

$$\left(1 - \gamma_{\underline{i}}\right)\left(\sigma_{\underline{i}}(t) - \text{sign}(\sigma_{\underline{i}}^{TR})\sigma_{\underline{i}}^*\right) = \gamma_{\underline{i}}E_{\underline{i}}\left(\text{sign}(\sigma_{\underline{i}}^{TR})\Psi_{\underline{i}} + \Delta\beta_{\underline{i}}^0(t)\right) \tag{35}$$

The vector $\sigma_i^{TR}$ in Equation (35) denotes trial stresses, necessary to determine elements entering the plastic zone, according to the formula

$$\sigma_i^{TR} = E_{\underline{i}}(\varepsilon_{\underline{i}}(t) - \beta_{\underline{i}}^0(t - 1)) \tag{36}$$

The vector $\Psi_i$ in Equation (35) denotes an equivalent (total) plastic strain at isotropic hardening, expressed as

$$\Psi_i = \sum_t \left|\Delta\beta_i^0(t)\right| \tag{37}$$

For performing stepwise plastic analysis, the increment of strains needs to be explicitly specified in the strain formula (33):

$$\varepsilon_i(t) = \varepsilon_i(t - 1) + \Delta\varepsilon_i(t) \tag{38}$$

The strain increment in the current time step is given as

$$\Delta \varepsilon_i(t) = \Delta \varepsilon_i^{\mathrm{L}}(t) + \sum_{\tau=0}^{t-1} D_{ik}^{\varepsilon}(t - \tau)\Delta \beta_k^0(\tau) + D_{ik}^{\varepsilon}(0)\Delta \beta_k^0(t) \tag{39}$$

The stresses (34), expressed in incremental form, yield

$$\sigma_i(t) = \sigma_i(t - 1) + E_{\underline{i}}\Delta \varepsilon_{\underline{i}}(t) - E_{\underline{i}}\Delta \beta_{\underline{i}}^0 \tag{40}$$

Substituting Equation (39) in (40) and reshaping, using Equation (36) gives a local set of equations (limited to plastic elements), which has to be solved for $\Delta \beta_k^0$

$$\left[E_i \delta_{ik} - \left(1 - \gamma_{\underline{i}}\right) E_{\underline{i}} D_{\underline{i}k}^{\varepsilon}(0)\right]\Delta \beta_k^0(t)$$
$$= \left(1 - \gamma_{\underline{i}}\right)\sigma_{\underline{i}}^{\mathrm{TR}} - \mathrm{sign}(\sigma_{\underline{i}}^{\mathrm{TR}})\left(\left(1 - \gamma_{\underline{i}}\right)\sigma_{\underline{i}}^{\star} + \gamma_{\underline{i}} E_{\underline{i}} \Psi_{\underline{i}}\right) \tag{41}$$

Equation (41) is known elsewhere as the return mapping algorithm for rate-independent plasticity.

### 2.5.4 Mass Remodeling in Dynamics

The inertia effects (mass remodeling) influencing structural behavior are inherent in dynamic analysis. If we want to account for mass modifications, it is necessary to introduce another quantity – an impulse force distortion $f_M^0$ (pseudoload). Unequilibrated distortions, each one in the form of a unit impulse force, are successively applied to global degrees of freedom, producing a corresponding influence matrix $\mathbf{B}^{\mathrm{f}}$ (in displacements) or $\mathbf{D}^{\mathrm{f}}$ (in strains) (see Section 2.5.1). It is an important distinction from the matrices $\mathbf{B}^{\varepsilon}$, $\mathbf{D}^{\varepsilon}$, in which a self-equilibrated pair of forces (equivalent to a unit strain) was applied. This time, however, it is necessary to collect the out-of-balance influences in order to reflect the changes in inertia.

Equations of motion for the modified structure subject to a change of mass and the structure modelled by impulse force distortions are given by

$$\hat{M}_{MN}\ddot{u}_N(t) + K_{MN}u_N(t) = f_M(t) \tag{42}$$
$$M_{MN}\ddot{u}_N(t) + K_{MN}u_N(t) = f_M(t) + f_M^0(t) \tag{43}$$

Subtracting Equation (43) from (42) gives

$$\hat{M}_{MN}\ddot{u}_N(t) = M_{MN}\ddot{u}_N(t) - f_M^0(t) \tag{44}$$

Equation (44) constitutes the dynamic postulate of the VDM (cf. the static postulate of Equation (16)), showing that the inertia forces and accelerations in the modified and distorted structure are equal. Rearrangment of Equation (44) leads to

$$\Delta M_{MN}\ddot{u}_N(t) + f_M^0(t) = 0 \tag{45}$$

where

$$\Delta M_{MN} = \hat{M}_{MN} - M_{MN} = \sum_i \left( \mu_i^\rho - 1 \right) M_{MN}^i \tag{46}$$

defines a modification of the global mass matrix. The summation in Equation (46) denotes aggregation over all finite elements $i$ and $M_{MN}^i$ is the element consistent mass matrix after the transformation to the global coordinate system (i.e. the element matrix expanded to the global dimensions). The coefficient $\mu_i^\rho$ defines the ratio of the modified density to the original one (or the modified cross-sectional area to the original one):

$$\mu_i^\rho \stackrel{\text{def}}{=} \frac{\hat{\rho}_i}{\rho_i} = \frac{\hat{A}_i}{A_i} \stackrel{\text{def}}{=} \mu_i^A \tag{47}$$

Determination of the influence matrix $B_{MN}^f$ and force distortions $f_N^0$ enables nodal displacements for the mass remodeling problem to be expressed in the following way:

$$u_M(t) = u_M^L(t) + \sum_{\tau=0}^{t} B_{MN}^f(t - \tau) f_N^0(\tau) \tag{48}$$

The corresponding nodal acceleration (second derivative of Equation (48) with respect to time) takes the form

$$\ddot{u}_M(t) = \ddot{u}_M^L(t) + \sum_{\tau=0}^{t} \ddot{B}_{MN}^f(t - \tau) f_N^0(\tau) \tag{49}$$

Substituting Equation (49) in (45) and rearranging, the following set of equations is obtained:

$$\left[ \delta_{MK} + \Delta M_{MN} \ddot{B}_{NK}^f(0) \right] f_K^0(t) = -\Delta M_{MN} \ddot{u}_N^{\neq t}(t) \tag{50}$$

where

$$\ddot{u}_M^{\neq t}(t) = \ddot{u}_M^L(t) + \sum_{\tau=0}^{t-1} \ddot{B}_{MN}^f(t - \tau) f_N^0(\tau) \tag{51}$$

collects the contribution from the preceding time steps. Note that again (cf. Equation (31)) the governing matrix in Equation (50) is constant in all time steps and only the right-hand side varies. The set is local, i.e. limited to the degrees of freedom (in global coordinates) corresponding to the remodeled mass in neighboring elements.

In combined reanalyses, when both stiffness and mass changes are considered and the nonlinear constitutive law is adopted, it is necessary to define all relations on the element level. Therefore there is a need to define the matrix $D_{iN}^f$ (in strains), which is related to $B_{MN}^f$ (in displacements) as follows (cf. Equation (5)):

$$D_{iN}^f = G_{iM} B_{MN}^f \tag{52}$$

The process of mass remodeling at structural nodes (e.g. modeling an impacting mass) proceeds analogously to the one presented above for elements.

## 2.6 VDM-Based Sensitivity Analysis

The analytical formulation of the VDM allows the sensitivities to be calculated of mechanical quantities, e.g. stresses with respect to virtual distortions $\varepsilon^0$, modeling stiffness-related changes of selected parameters $\mu$, directly. It is assumed that an objective function $F$, depending upon a mechanical quantity (symbolically denoted by $\sigma$) is defined as follows:

$$F = F(\sigma(\varepsilon^0(\mu))) \tag{53}$$

If the linear constitutive relation is dealt with, the sensitivies of the objective function (53) with respect to parameters $\mu$ can be calculated using the chain rule of differentiation in the following form:

$$\frac{\partial F}{\partial \mu} = \frac{\partial F}{\partial \sigma} \frac{\partial \sigma}{\partial \varepsilon^0} \frac{\partial \varepsilon^0}{\partial \mu} \tag{54}$$

The component $\partial F/\partial \sigma$ can be obtained thanks to the definition of the objective function (53) itself, the component $\partial \sigma/\partial \varepsilon^0$ – thanks to posing the problem in the framework of the VDM (cf. Equation (9)) and the component $\partial \varepsilon^0/\partial \mu$ by utilizing appropriate conditions (e.g. Equation (17)) describing the problem.

If a nonlinear constitutive law (piecewise linear) is concerned (cf. Figure 2.3), there is an influence of the whole plastic zone on the structural behavior and the sensitivities must reflect the fact, taking the following form:

$$\frac{\partial F}{\partial \mu} = \frac{\partial F}{\partial \sigma} \left( \frac{\partial \sigma}{\partial \varepsilon^0} + \sum_{\text{plastic}} \frac{\partial \sigma}{\partial \beta^0} \frac{\partial \beta^0}{\partial \varepsilon^0} \right) \frac{\partial \varepsilon^0}{\partial \mu} \tag{55}$$

Note that the second component in brackets in (55) refers to the whole plastic zone, which is symbolically denoted by $\sum$. For determination of the component $\partial \beta^0/\partial \varepsilon^0$, both the virtual $\varepsilon^0$ (cf. Equation (8)) and plastic $\beta^0$ (cf. Equation (20)) distortions are introduced into the piecewise linear relation (22), which is subsequently differentiated with respect to $\varepsilon^0$. For the distortions $\mathbf{f}^0$, modelling inertia changes, the sensitivity formulas (54) and (55) become valid by replacing $\varepsilon^0$ with $\mathbf{f}^0$.

In dynamics, sensitivity analysis is performed using the Newmark integration algorithm. Even with reasonable time discretization, calculation of gradients appears to be the major numerical cost in optimization. Nevertheless, the VDM-based accurate sensitivities allow various engineering analyses to be performed effectively (see Chapters 3 and 6).

## 2.7 Versatility of VDM in System Modeling

The VDM framework is general enough to encompass problems concerning systems (see Table 2.3) other than structural as well. This is due to the analogies between structural mechanics, hydraulics and electrical engineering valid for truss-like systems, which were discovered by Cross [38]. Regardless of the type of such a system, modeled by a *graph*, the VDM is able to capture all system features in the influence matrix.

**Table 2.3**   Analogies between truss-like engineering systems including VDM modeling

|  | **Structural mechanics** | **Hydraulic engineering** | **Electrical engineering** |
|---|---|---|---|
| System | Truss structure of some redundancy | Closed-loop water network | Closed-loop electrical network |
| Internal constraints | Degree of redundancy | Number of closed loops | Number of closed loops |
| Model | Graph | Oriented graph | Oriented graph |
| Relations | Equilibrium of forces | Balance of flow | Equilibrium of currents (First Kirchhoff's) |
|  | Constitutive (Hooke's) | Constitutive (nonlinear) | Constitutive (Ohm's) |
|  | Continuity of strains | Continuity of flow | Continuity of voltage (Second Kirchhoff's) |
| Quantities | Force | Water flow | Current |
|  | Displacement | Water head | Potential |
|  | Strain | Pressure head | Voltage |
|  | Elastic modulus | Hydraulic compliance | Conductance |
| Type of virtual distortions in the VDM | Strain Force | Pressure head | Current |

Using the general system theory [39] and the oriented graph approach, it turns out that similar relations govern constitutive, continuity and equilibrium conditions for truss structures, water networks and electrical circuits. Taking advantage of this fact, the VDM idea, originating from structural mechanics, was adapted to model both types of nonstructural systems (see Sections 3.4 and 3.5 for details).

## 2.8   Recapitulation

### 2.8.1   General Remarks

The virtual distortion method has proved to be a versatile tool of structural and system reanalysis for 20 years of its development. In the authors' opinion, the principal advantages of the VDM, distinguishing it from other reanalysis methods, are:

- exact, analytical formulation, capturing all system features in the influence matrix;
- handling nonlinear constitutive law;
- dynamic reanalysis in the time domain enabling modification of both stiffness and inertia parameters.

The VDM is an analytical approach, producing exact (not approximate) results, unlike the majority of the structural reanalysis methods presently used. The exactness of the method is due to the influence matrix capturing all relations between the local disturbance and a global response of a structure (or system). Any response of a structure subject to modifications is simply a linear combination of components of the influence matrix and virtual distortions (design variables), even for physically nonlinear problems. The exact formulation of VDM is especially important in calculating precisely the sensitivities, which are subsequently utilized in gradient-based optimization. This feature also enables effective handling of large modifications of parameters (see Section 2.4.4). If the number of design variables to be modified is small compared to the degrees of freedom, the VDM solution is fast, as the set of equations to be solved is always local (referring only to the modified locations).

The inverse of the stiffness matrix is required in the VDM only at the stage of building the influence matrix. For some applications, e.g. progressive collapse analysis, only selected influence vectors need to be built, for others e.g. identification problems, it is convenient to have a full influence matrix at the start. In statics, once assembled for a structure with some redundancy, the influence matrix remains constant throughout the whole reanalysis. Nevertheless, building the full influence matrix, especially for dynamics, may involve some initial computational cost for large structures. However, creating the full influence matrix has the advantage of forming a computational basis for the VDM. With this matrix and VDM-based analytical sensitivities, various types of analyses, sometimes addressing really complex problems, can be performed.

As shown in Reference [6], reanalysis methods are variations of the SMW formulas. For the VDM in the linear regime, the equivalence to SMW is restricted with the condition that only the required influence vectors (not the whole influence matrix), corresponding to the modified locations, are constructed. For physical nonlinearities, the VDM algorithm progresses without iterations due to consideration of a piecewise linear (in particular bilinear) constitutive law. It seems that for the extension of SMW formulas to the nonlinear regime, proposed in Reference [6], iterations would not be necessary either if an analogous piecewise linear relation were assumed instead of the general nonlinear one.

Apart from the DMM approach [35], the VDM appears to be exceptional in performing dynamic reanalysis in the time domain. Thanks to consideration of two fields of virtual distortions – one modeling stiffness modifications as in statics and the other modeling mass modifications – various changes of structural parameters can be effectively tracked in dynamics. The time-domain dynamic VDM turns out to be a powerful tool when analyzing combined problems of design and adaptation for structures subjected to impact loading (see Section 6.3).

Thus far the VDM is basically limited to skeletal structures. An extension of the method for continuum, similarly to what has been done within the TSV approach (cf. References [19] and [20]), will be the subject of future research.

### 2.8.2 Applications of the VDM to Structures

The list of main application areas of the VDM in structural mechanics includes:

- in statics:
  - stiffness remodeling (direct problem): topology optimization,
  - piecewise linear constitutive law (direct problem): noniterative plasticity;
- in dynamics:
  - stiffness remodeling (inverse problem): identification of stiffness degradation in structural health monitoring,
  - mass remodeling (inverse problem): identification of dynamic load history,
  - combined stiffness, mass remodeling and piecewise linear plasticity (direct problem): optimal design of adaptive structures for dynamic loads of known characteristics.

In static analysis, the VDM can be successfully used for remodeling of structures. Analytically derived sensitivities for trusses [40] and frames [41] allow for gradient-based topological optimization [37]. Another useful accomplishment in statics is the ability to analyze a nonlinear (piecewise linear) constitutive relation, practically employed in the progressive collapse analysis [8]. A preliminary, quasi-static study of optimal design of adaptive structures was presented in Reference [42]. An overview of the VDM in statics can be found in Reference [43].

In dynamic analysis, the VDM, contrary to most reanalysis methods, is able to model the response of a modified structure in the time domain. Using stiffness degradation as a damage modeling parameter, an inverse dynamic analysis, examining the response due to an impulse excitation, was proposed for damage identification [44]. In this way, the VDM has been applied to a new field of structural health monitoring (cf. Chapters 3 and 4). A comparison of performance of the gradient-based VDM approach with a softcomputing method was made in Reference [45]. The VDM damage identification philosophy can also be transferred to the frequency domain by applying a harmonic excitation and looking only at amplitudes [46]. A similar problem of load identification (location and magnitude) using the VDM was successfully handled in Reference [47] (cf. Chapter 4). Using the VDM, first attempts to evaluate the crashworthiness of structures [48] and to design adaptive structures [49], effectively dissipating the energy of impact load (cf. Chapter 5), were undertaken. However, the problems have to be analyzed with the assumption of nonlinear geometry (large strains) and still remain a research challenge.

The virtual distortion method has also been the subject of study of other researchers. Makode and Ramirez [50] describe an extension of the VDM to frame structures. In the subsequent paper by Makode and Corotis [51] (using the name 'pseudo-distortion method'), simultaneous modification of the moment of inertia and cross-sectional area is included. Also, the elastoplastic analysis with multiple hinges in the structure, located either at one or two ends of the frame element, is presented. In the companion paper [51], the VDM is used to account for secondary geometric effects caused by large axial forces influencing the bending moments in frame structures.

Recently, an application of the VDM to probabilistic analysis has been developed. Di Paola *et al.* [52, 53] use the VDM to model uncertainty of parameters in truss structures.

### 2.8.3 Applications of the VDM to Nonstructural Systems

In Reference [54], closed-loop water networks, assuming a steady-state flow, are modeled using the VDM. For water heads measured in all nodes of the network, an algorithm for detecting leakage in the midpoint of a branch was proposed. VDM sensitivity enabled a quadratic programming tool to be employed to solve the problem. As a result, the algorithm gives the location and intensity of leakage. Multiple leakage detection is feasible. Continuation of the research will include precise location of leakage along the branch, optimal location of the measuring nodes as well as consideration of transient effects (unsteady flow).

In Reference [55], closed-loop electrical networks are modeled using the VDM. A static-like approach with constant current intensities is proposed. As an extension, a quasi-static approach for harmonic current sources has been developed, where only amplitudes and phase shifts are analyzed in the complex numbers domain. Finally, a dynamic-like time-dependent approach, able to reflect transient behaviour of electrical networks as a response to an impulse current, is described in Reference [56]. An 'electrical finite element' has been elaborated, enabling a FEM-like analysis to be performed. Defects in electrical networks are effectively modeled in all the mentioned approaches as a loss of conductance in branches. The changes of conductance (defect coefficient) may be tracked in the continuous range from zero (break in the network) to one (intact conductance).

Sections 3.4 and 3.5 give a more detailed insight into the nonstructural applications of the VDM.

# References

1. W. Nowacki, *Teoria sprezystosci* (in Polish), PWN, Warsaw, 1970.
2. J. Holnicki-Szulc and J. Gierliński, Structural modifications simulated by virtual distortions, *International Journal of Numerical Methods in Engineering*, **28**, 1989, 645–666.
3. E. Kroener, Kontinuumstheorie der Versetzungen und Eigenspannungen, in editors, *Ergebnisse der Angewandten Mathematik* (in German) (eds L. Collatz and F. Loesch,), Vol. 5, Springer, Berlin, 1958.
4. J.H. Argyris, Elasto-plastic matrix displacement analysis of three-dimensional continua, *Journal of Royal Aeronautical Society*, **69**, 1965, 633–636.
5. G. Maier, A matrix structural theory of piecewise-linear plasticity with interacting yield planes, *Meccanica*, **7**, 1970, 51–66.
6. M.A. Akgun, J.H. Garcelon and R.T. Haftka. Fast exact linear and non-linear structural reanalysis and the Sherman–Morrison–Woodbury formulas, *International Journal Numerical Methods in Engineering*, **5**, 2001, 1587–1606.
7. J. Holnicki-Szulc, Virtual distortion method, in *Lecture Notes in Engineering*, Springer, Heidelberg–Berlin, 1991.
8. J. Holnicki-Szulc and J.T. Gierliński, *Structural Analysis, Design and Control by the Virtual Distortion Method*, John Wiley & Sons, Ltd, Chichester, 1995.
9. P. Kołakowski, M. Wikło and J. Holnicki-Szulc, The virtual distortion method – a versatile reanalysis tool for structures and systems, *Structural and Multidisciplinary Optimization*, (to appear).
10. A.M.A. Kassim and B.H.V. Topping, Static reanalysis: a review, *Journal of Structural Engineering ASCE*, **113**(6), 1987, 1029–1045.
11. J.F.M. Barthelemy and R.T. Haftka, Approximation concepts for optimum design – a review, *Structural Optimization*, **5**, 1993, 129–144.
12. J. Sherman and W.J. Morrison, Adjustment of an inverse matrix corresponding to changes in the elements of a given column or a given row of the original matrix, *Annals of Mathematical Statistics*, **20**, 1949, 621.
13. M. Woodbury, Inverting modified matrices, Memorandum report 42, Statistical Research Group, Princeton University, NewJersey, 1950.
14. R.L. Fox and H. Miura, An approximate analysis technique for design calculations. Technical note, *American Institute of Aeronautical and Astronautical Journal*, **9**(1), 1971, 177–179.
15. A.K. Noor and H.E. Lowder, Approximate techniques of structural reanalysis, *Computers and Structures*, **4**, 1974, 801–812.
16. U. Kirsch and S. Liu, Exact structural reanalysis by a first-order reduced basis approach, *Structural Optimization*, **10**, 1995, 153–158.
17. U. Kirsch, A unified reanalysis approach for structural analysis, design and optimization, *Structural and Multidisciplinary Optimization*, **25**, 2003, 67–85.
18. K.I. Majid and D.W.C. Elliott, Forces and deflections in changing structures, *Structural Engineering*, **51**, 1973, 93–101.
19. B.H.V. Topping and A.M.A. Kassim, The use and efficiency of the theorems of structural variation for finite element analysis, *International Journal of Numerical Methods in Engin.*, **24**, 1987, 1900–1920.
20. M.P. Saka, The theorems of structural variation for solid cubic finite elements, *Computers and Structures*, **68**, 1998, 89–100.
21. K.I. Majid and T. Celik. The elastic–plastic analysis of frames by the theorems of structural variation, *International Journal of Numerical Methods in Engineering*, **21**, 1985, 671–681.
22. L. Deng and M. Ghosn, Pseudoforce method for nonlinear analysis and reanalysis of structural systems, *Journal of Structural Engineering ASCE*, **127**(5), 2001, 570–578.
23. H.R. Bae and R.V. Grandhi, Successive matrix inversion method for reanalysis of engineering structural systems, *American Institute of Aeronautical and Astronautical Journal*, **42**(8), 2004, 1529–1535.

24. J.E. Hurtado, Reanalysis of linear and nonlinear structures using iterated Shanks transformation, *Computing Methods in Applied Mechanical Engineering*, **191**, 2002, 4215–4229.
25. S.H. Chen, X.W. Yang and H.D. Lian, Comparison of several eigenvalue reanalysis methods for modified structures, *Structural Multidisciplinary Optimization*, **20**, 2000, 253–259.
26. L.J. Leu and C.H. Tsou, Applications of a reduction method for reanalysis to nonlinear dynamic analysis of framed structures, *Computing Mechanics*, **26**, 2000, 497–505.
27. S.S.A. Ravi, T.K. Kundra and B.C. Nakra, Reanalysis of damped structures using the single step perturbation method, *Journal of Sound and Vibration*, **211**(3), 1998, 355–363.
28. K.C. Yap and D.C. Zimmermann, A comparative study of structural dynamic modification and sensitivity method approximation, *Mechanical Systems and Signal Processing*, **16**(4), 2002, 585–597.
29. H.P. Chen, Efficient methods for determining modal parameters of dynamic structures with large modifications, *Journal of Sound and Vibration*, **298**, 2006, 462–470.
30. M.D. Grissom, A.D. Belegundu and G.H. Koopmann, A reduced eigenvalue method for broadband analysis of a structure with vibration absorbers possessing rotatory inertia, *Journal of Sound and Vibration*, **281**, 2005, 869–886.
31. U. Kirsch, M. Bogomolni and I. Sheinman, Nonlinear dynamic reanalysis of structures by combined approximations, *Computing Methods in Applied Mechanical Engineering*, **195**, 2006, 4420–4432.
32. C. Huang, S.H. Chen and Z. Liu, Structural modal reanalysis for topological modifications of finite element systems, *Engineering Structures*, **22**, 2000, 304–310.
33. D.L. Cronin, Eigenvalue and eigenvector determination for nonclassically damped dynamic systems, *Computing Structures*, **36**, 1990, 133–138.
34. J. Tang and W.L. Wang, Structural reanalysis in configuration space by perturbation approach: a real mode method, *Computing Structures*, **58**(4), 1996, 739–746.
35. G. Muscolino, Dynamically modified linear structures: deterministic and stochastic response, *Journal of Engineering Mechanics, ASCE*, **122**(11), 1996, 1044–1051.
36. P. Cacciola, N. Impollonia and G. Muscolino, A dynamic reanalysis technique for general structural modifications under deterministic and stochastic input, *Computing Structures*, **83**, 2005, 1076–1085.
37. P. Kołakowski and J. Holnicki-Szulc, Optimal remodelling of truss structures – simulation by virtual distortions, *Computing Assistence of Mechanical Engineering Science*, **4**(2), 1997, 257–281.
38. H. Cross, Analysis of flow in networks of conduits or conductors, Engineering Experiment Station bulletin, 1936, 286.
39. N.C. Lind, Analysis of structures by system theory, *Journal of Structural Divison, ASCE*, **88**, ST2 (Apr.), 1962.
40. P. Kołakowski and J. Holnicki-Szulc, Sensitivity analysis of truss structures – virtual distortion method approach. *International Journal of Numerical Methods in Engineering* **43**(6), 1998, 1085–1108.
41. J.T. Putresza and P. Kołakowski, Sensitivity analysis of frame structures – virtual distortion method approach. *International Journal of Numerical Methods in Engineering*, **50**(6), 2001, 1307–1329.
42. J. Holnicki-Szulc, A. Maćkiewicz and P. Kołakowski, Design of adaptive structures for improved load capacity, *American Institute of Aeronautical and Astronautical Journal*, **36**(3), 1998, 471–476.
43. J. Holnicki-Szulc and T. Bielecki, Virtual distortion method – review. Applications to fast redesign and sensitivity analysis, *International Journal of Nonlinear Science and Numerical Simulation*, **1**(2), 2000, 71–98.
44. P. Kołakowski, T.G. Zieliński and J. Holnicki-Szulc, Damage identification by the dynamic virtual distortion method, *Journal of Intelligent Material System Structures*, **15**(6), 2004, 479–493.
45. P. Kołakowski, L.E. Mujica and J. Vehi, Two approaches to structural damage identification: model updating vs. soft computing, *Journal of Intelligent Material System Structures*, **17**(1), 2006, 63–79.
46. A. Świercz, P. Kołakowski and J. Holnicki-Szulc. Identification of damage using low frequency harmonics in trusses and beams, In *Proceedings of the International Conference on Noise and Vibration Engineering*, ISMA2006, CD-ROM edition, 2006.

47. Ł. Jankowski, Off-line identification of dynamic loads, Structural and Multidisciplinary Optimization (to appear).
48. J. Holnicki-Szulc and L. Knap, Adaptive crashworthiness concept, *International Journal of Impact Engineering*, **30**(6), 2004, 639–663.
49. J. Holnicki-Szulc and P. Pawłowski, High-performance impact absorbing materials – the concept, design tools and applications, *Smart Materials and Structures*, **12**, 2003, 461–467.
50. P.V. Makode and M.R. Ramirez, Reanalysis of rigid frame structures by the virtual distortion method, *Structural Multidisciplinary Optimization* **11**(1–2), 1996, 71–79.
51. P.V. Makode and R.B. Corotis, Non-linear analysis of frame structures by pseudodistortions, *Journal of Structural Engineering*, ASCE, **125**(11), 1999, 1309–1327.
52. M. Di Paola, A. Pirrotta and M. Zingales, Stochastic dynamics of linear elastic trusses in presence of structural uncertainties (virtual distortion approach), *Probability Engineering Mechanics*, **19**(1–2), 2004, 41–51.
53. M. Di Paola, Probabilistic analysis of truss structures with uncertain parameters (virtual distortion method approach). *Probability Engineering Mechancis*, **19**(4), 2004, 321–329.
54. J. Holnicki-Szulc, P. Kołakowski and N. Nasher, Leakage detection in water networks, *Journal of Intelligent Material System Structures*, **16**(3), 2005, 207–219.
55. M. Kokot and J. Holnicki-Szulc, Health monitoring of electric circuits, in *Proceedings of the 6th International Conference on Damage Assessment of Structures – DAMAS, 2005*, Gdansk, Poland, 4–6 July 2005, pp. 669–676.
56. M. Kokot and J. Holnicki-Szulc, Shm concept applied to dynamical defect identification in electrical circuits, in *Proceedings of the 3rd European Workshop on Structural Health Monitoring*, 5–7 July 2006, Granada, Spain, pp. 1111–1118.

# 3

# VDM-Based Health Monitoring of Engineering Systems

Przemysław Kołakowski, Andrzej Świercz, Anita Orłowska, Marek Kokot and Jan Holnicki-Szulc

## 3.1 Introduction to Structural Health Monitoring

*Structural health monitoring* (SHM) is a fast-developing interdisciplinary field of research having its roots in *vibroacoustics* and *nondestructive testing and evaluation* (NDT/E). The fast development of the area is due to the fact that SHM is heavily stimulated by the engineering problems of maintenance and safe operation of technical infrastructure. The use of SHM is now becoming a standard in high-cost modern infrastructure. Thus far, the majority of SHM applications have been demonstrated in civil, aerospace and mechanical engineering.

The SHM process involves three major stages [1]:

(1) collection of measurements from an array of *sensors* (e.g. piezoelectric [2], optical fiber [3]), using periodically sampled dynamic responses of a system due to known (e.g. induced by a shaker) or *ambient* (e.g. induced by wind) *excitation* [4] over a defined period of time;
(2) the proper *signal processing* [5,6] and extraction of damage-sensitive features from these measurements;
(3) the subsequent analysis of these features to assess possible degradation of the system's health.

Rytter [7] proposed that the damage assessment of the system should be considered as a stepwise process, consisting of the following three consecutive steps:

(1) Detection – is there damage in the system or not?
(2) Identification – what kind of damage has occurred in the system, where is it located, how severe is it?
(3) Prognosis – how much useful life remains for the system?

The first step is relatively easy to accomplish by tracking changes in the *structural signature* (response in terms of a selected quantity) over time. The second step, on which most of the research is focused, is a real challenge as it often involves sophisticated methods, able to produce the correct results of identification. The third step should not be difficult provided that the second step has been completed successfully.

The importance of diagnostics was probably first appreciated by the community of mechanical engineers dealing with *rotating machinery*, where damage involves high risk to the staff and high cost of repair, e.g. in turbines. Natke and Cempel [8] present vibroacoustics as an effective tool for machine diagnostics. The specific feature of rotating machinery is that it is self-exciting and relatively compact (not of huge dimensions), which is often not the case in SHM. The machine diagnostics is usually a low-frequency problem.

Many SHM methods originate from the NDT/E methods [9], i.e. ultrasonic testing, radiographic testing, acoustic emission, penetrant testing, magnetic particle inspection, eddy currents and optical holography, which are successfully applied in the industry for local detection of flaws in structural components. Most of the NDT/E methods require external excitation, e.g. ultrasonic testing, but some of them do not, e.g. acoustic emission. The NDT/E methods usually operate in high frequencies.

The objective of SHM is to create a monitoring system (possibly for the whole structure, which is sometimes of complicated topology and considerable dimensions), able to track changes in structural condition continually and raise appropriate alerts if a defect is detected. As a consequence of the evolution from the two major streams, i.e. machine diagnostics and NDT/E, the SHM methods for identifying structural damage can be roughly split into low-frequency methods (nonultrasonic) and high-frequency methods (ultrasonic), respectively.

Low-frequency methods [10], relying on vibration-based global approaches, need numerical models in order to monitor health changes of a structure. Damage identification consists in subsequent *model updating* [11], which is a process of finding the best-fit solution to the *inverse problem* of identification, often performed by classical gradient-based optimization methods using damage-sensitive parameters. The greatest challenge is *system identification* [12], i.e. matching responses of the numerical model to experiments.

High-frequency methods [13], analyzing a structure locally, usually do not require structural models. They widely rely on the local phenomenon of *elastic wave propagation* [14] in solid media and make use of *machine learning* and *pattern recognition* [15] to identify damage, using *artificial intelligence (soft computing)* methods, e.g. neural networks, case-based reasoning. The greatest challenge is to build a representative database in order to be able to retrieve patterns that are the most similar to the analyzed case reliably.

In both low- and high-frequency methods, it is an important issue to discriminate actual damage from the influence of environment on a structure [16]. As application examples, low-frequency methods are used in civil engineering for examining stiffness degradation of a bridge [17], while high-frequency methods are used in aerospace engineering for crack identification in aircraft components [18].

The point of the chapter is to demonstrate the capabilities of the *virtual distortion method* in SHM, applied not only to structures but also to nonstructural systems, i.e. water networks and electrical circuits. Einstein's summation convention has again been adopted as in Chapter 2. Underlined indices are exempt from summation. The lower case subindices refer to elements (branches) and the upper case ones refer to nodes.

## 3.2 Damage Identification in Skeletal Structures

### 3.2.1 Introduction

Vibration-based (low-frequency) SHM has gained a lot of researchers' attention in recent years. The analysis can be carried out directly in the time domain, consindering sampled time signals. Most frequently, however, it is transferred to the frequency domain by performing the *fast fourier transform* (FFT) for the time signals and considering only frequency spectra. The main stream of the low-frequency SHM is concentrated on utilizing ambient excitation and applying the *operational modal analysis* (OMA) [19] to localize and quantify the damage. The environmental type of excitation has its weak points, but the newest trend is to enrich the OMA with eXogenous input (OMAX) of known characteristics to facilitate damage identification [20]. Degradation of stiffness is usually examined in SHM and damage-sensitive parameters (e.g. beam curvature [21], kinetic energy [22]) are defined to track stiffness changes in the structures. Besides the modal parameters, the antiresonance frequencies [23] may successfully contribute to damage identification as well. A number of effective and robust methods for system identification and subsequent vibration-based SHM have been elaborated, e.g. the eigensystem realization algorithm (ERA) [24], stochastic subspace identification (SSI) [25], complex mode indicator function (CMIF) [26] and PolyMAX [27]. Some of them, e.g. the least squares complex exponential (LSCE) [28], have already become essential tools in industry applications.

Novel methodology of SHM based on the *virtual distortion method* (VDM) in terms of an identification algorithm (software) and on piezoelectric sensors (hardware) in terms of a structural response observation was proposed in Reference [29]. The formulation in the time domain was general enough to handle an arbitrary shape of the excitation signal, but the major obstacle was a large computational effort. Thus in order to reduce the numerical cost, the problem has been reformulated in the frequency domain [30] due to the assumption of applying harmonic excitation. This type of excitation is inherent in some vibroacoustic problems, e.g. rotating machinery, or can be applied to real structures by shakers. Apart from the *in situ* applications, it is also commonly used in laboratory experiments.

The main point of this chapter is to present a VDM-based methodology of damage identification (location and intensity) for skeletal structures (trusses and beams) by considering not only stiffness but also mass modifications. It is assumed that the system does not vary in time during measurements. A simplified dynamic problem with no damping is considered. Sensitivity information is effectively utilized in an optimization algorithm, leading to damage identification as a result of solving an inverse problem. Piezoelectric sensors are used to capture a structural response to a low-frequency (up to 1 kHz) excitation. Experimental verification of the proposed approach has been carried out for a space truss demonstrator.

### 3.2.2 Time Domain (VDM-T) versus Frequency Domain (VDM-F)

The first attempt to handle the problem of damage identification using the VDM was made during the project PiezoDiagnostics [31]. Low-frequency impulse excitation (e.g. a windowed sine pulse) was applied and time responses of a truss structure were captured by piezoelectric transducers. The electrical responses (voltage) of the transducers were proportional to mechanical strains.

The first formulation assumed only stiffness degradation (cf. Equation (30) in Chapter 2) as a damage modeling parameter. Consequently, only the static postulate of equivalence of internal forces and strains (cf. Equation (16) in Chapter 2) was utilized. The extension of the VDM remodeling approach, including mass modifications in the time domain, was developed later (see Chapters 2 and 6).

Inverse analysis performed on the time responses enabled the location and intensity of stiffness degradation in truss elements to be identified successfully. To this end, the local set of equations (31) in Chapter 2 had to be solved in every time step processed by the Newmark integration algorithm. However, it turned out that the cost of gradient calculation is significant, so another approach in the frequency domain was subsequently developed by assuming harmonic excitation only.

A steady-state problem is characterized by continuous-in-time harmonic excitation of the form

$$f(t) = f \, \sin(\omega t) \tag{1}$$

where $f$ and $\omega$ denote amplitude and frequency of the exciting force, respectively. This assumption significantly simplifies considerations of the equation of motion as all mechanical quantities and their derivatives vary in the same harmonic manner, e.g. displacements are expressed as

$$u(t) = u \, \sin(\omega t) \tag{2}$$

Therefore, it is quite sufficient to analyze only time-independent amplitude values of the quantities. Consequently, the dynamic problem becomes quasi-static, e.g. relations (28) and (48) in Chapter 2 have the same form, but with no time dependence.

Now, both stiffness and mass modifications in the structure should be considered; e.g. in the case of displacements both residual responses (27) and (48) in Chapter 2 have to be included. Assuming the two kinds of modifications and taking into account relation (1), the equations of motion for the modified (with mass and stiffness changes) and distorted structures can be expressed as (cf. Equations (42) and (43) in Chapter 2):

$$-\omega^2 \, \hat{M}_{KL} \, \hat{u}_L + \hat{K}_{KL} \, \hat{u}_L = f_K \tag{3}$$

$$-\omega^2 \, M_{KL} \, u_L + K_{KL} \, (u_L - u_L^0) = f_K + f_K^0 \tag{4}$$

where the distortion $u_L^0$, expressed in global (system) coordinates and introduced solely for the simplicity of notation in Equation (4), corresponds to the stiffness modeling distortion $\varepsilon_j^0$, expressed in local (element) coordinates. Subtracting Equation (4) from (3) leads to the relation for $f_K^0$, which is similar to (45) in Chapter 2:

$$-\omega^2 \, \Delta M_{KL} \, u_L + f_K^0 = 0 \tag{5}$$

Note that to obtain Equation (5), the static postulate of VDM (equivalence of element strains and forces; cf. Equations (14) to (16) in Chapter 2), expressed in global coordinates, was used:

$$u_L = \hat{u}_L \tag{6}$$

$$K_{KL} \, (u_L - u_L^0) = \hat{K}_{KL} \, \hat{u}_L \tag{7}$$

When both stiffness and mass changes are analyzed, strains in elements depend on both the distortions $\varepsilon_j^0$ and $f_L^0$:

$$\varepsilon_i = \varepsilon_i^L + D_{ij}^{\varepsilon}\,\varepsilon_j^0 + D_{iL}^{f}\,f_L^0 \tag{8}$$

Analogously, displacements corresponding to global degrees of freedom are expressed by the following relation:

$$u_M = u_M^L + B_{Mj}^{\varepsilon}\,\varepsilon_j^0 + B_{ML}^{f}\,f_L^0 \tag{9}$$

Substituting Equation (8) in (17) in Chapter 2 and (9) in (5), a set of $j + L$ equations is assembled:

$$\begin{bmatrix} \delta_{ij} - (1 - \mu_{\underline{i}})\,D_{ij}^{\varepsilon} & -(1 - \mu_{\underline{i}})D_{iL}^{f} \\ -\omega^2 \Delta M_{KM}\,B_{Mj}^{\varepsilon} & \delta_{KL} - \omega^2 \Delta M_{KM}\,B_{ML}^{f} \end{bmatrix} \begin{bmatrix} \varepsilon_j^0 \\ f_L^0 \end{bmatrix} = \begin{bmatrix} (1 - \mu_{\underline{i}})\,\varepsilon_i^L \\ \omega^2 \Delta M_{KM}\,u_M^L \end{bmatrix}. \tag{10}$$

The set is local, i.e. variables $\varepsilon_j^0$ and $f_L^0$ are confined to the modified locations only (elements or corresponding degrees of freedom). It is solved using a singular value decomposition (SVD) solver.

### 3.2.3 Modifications in Beams

For a two-dimensional beam element there are three components of virtual distortions that have to be applied. The three distortion components correspond to three states of deformation (see Figure 3.1) in the orthogonal base, obtained through the solution of the eigenproblem of the two-dimensional beam element stiffness matrix. Apart from the axial type of distortion $\varepsilon$ (as for trusses), the beam distortions also include pure bending $\kappa$ and bending plus shear $\chi$ terms. Further in the section, if a Latin lower case index describes a quantity for a beam element, it should be remembered that it refers to a triplet of the above-described distortions (not a single axial strain). Thus, for the two-dimensional beam model, apart from the modification coefficient corresponding to axial strain (cf. Equation (30) in Chapter 2), analogous coefficients for bending (the ratio of the modified $\hat{J}$ to initial $J$ moment of inertia) may also be defined:

$$\mu_i^{\varepsilon} = \mu_i^{A} = \frac{\hat{A}_{\underline{i}}}{A_{\underline{i}}} = \frac{\varepsilon_{\underline{i}}(t) - \varepsilon_{\underline{i}}^0(t)}{\varepsilon_{\underline{i}}(t)} \tag{11}$$

$$\mu_i^{\kappa} = \mu_i^{J} = \frac{\hat{J}_{\underline{i}}}{J_{\underline{i}}} = \frac{\kappa_{\underline{i}}(t) - \kappa_{\underline{i}}^0(t)}{\kappa_{\underline{i}}(t)} \tag{12}$$

$$\mu_i^{\chi} = \mu_i^{J} = \frac{\hat{J}_{\underline{i}}}{J_{\underline{i}}} = \frac{\chi_{\underline{i}}(t) - \chi_{\underline{i}}^0(t)}{\chi_{\underline{i}}(t)} \tag{13}$$

As the axial response for the two-dimensional beam element is independent of bending/shear, it is possible to distinguish practically two coefficients for the analysis, i.e. $\mu_i^{\varepsilon}$ denoting axial stiffness change and $\mu_i^{\kappa}$ denoting bending stiffness change. As a consequence, the consistent mass matrix $6 \times 6$ is divided into two parts – the first one containing only the cross-sectional area $A$ of a beam and the second one containing only its moment of inertia $J$. The global mass

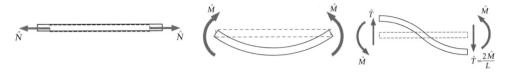

**Figure 3.1**  Virtual distortion states for a beam element

matrix is assembled as

$$M_{KL} = \sum_i \left( \overset{A}{\hat{M}}{}^i_{KL} + \overset{J}{\hat{M}}{}^i_{KL} \right) \tag{14}$$

where $i$ denotes a part of the mass matrix in global coordinates, corresponding to the beam element $i$. The increment of mass (cf. Equation (14)) is then expressed as

$$\Delta M_{KL} = \hat{M}_{KL} - M_{KL} = \sum_i \left( (\mu_i^A - 1) \overset{A}{\hat{M}}{}^i_{KL} + (\mu_i^J - 1) \overset{J}{\hat{M}}{}^i_{KL} \right) \tag{15}$$

### 3.2.4 Problem Formulation and Optimization Issues

In most approaches to damage identification, the measured quantity is acceleration, because it is relatively easy to obtain. However, the raw acceleration signal in time is never used directly – it requires FFT processing to transfer the analysis into the frequency domain. In the proposed approach a different quantity is measured. It is namely strain in time, measured by piezo-transducers, which is then directly used in the VDM-T approach (cf. Reference [29]). For harmonic excitation (VDM-F) only frequency-dependent amplitudes of strains are examined. Thus it is possible to speak about the VDM frequency-domain approach. It should be noted that there is no need for an FFT processing of the time signal as performed in standard frequency-domain methods.

As often happens in parameter estimation procedures, the identification task is posed as a nonlinear least squares minimization problem. The objective function expressed in strains collects time responses for the VDM-T approach:

$$F^t(\mu) = \sum_t \left( \frac{\varepsilon_k - \varepsilon_k^M}{\varepsilon_k^M} \right)^2 \tag{16}$$

and amplitude responses from selected $n_\omega$ frequencies of operation for the VDM-F approach:

$$F^\omega(\mu) = \sum_{n_\omega} \left( \frac{\varepsilon_k - \varepsilon_k^M}{\varepsilon_k^M} \right)^2 \tag{17}$$

Both functions (16) and (17) collect responses from $k$ sensors, placed in those elements where nonzero strains of high signal-to-noise ratio are measured. The strain $\varepsilon_k$ in an arbitrarily selected location $k$ is influenced by virtual distortions $\varepsilon_i^0$, which may be generated in any element $i$ of the structure (cf. Equation (8) in Chapter 2). One should also note that the modification coefficient $\mu_i$, quantifying potential damage and used as a variable in optimization, depends upon the virtual distortions $\varepsilon_i^0$ non linearly (cf. Equation (17) in Chapter 2). The VDM-T approach is dynamic, so only a few sensors are able to provide enough information due to relatively dense time sampling of the signal (many time steps in the numerical analysis). The VDM-F approach is quasi-static, so the number of sensors has to be equal to the number of potentially modified locations for the sake of uniqueness of solution. The preceding statement is valid only when one frequency of excitation is used; for more frequencies, the number of sensors can be proportionally reduced (at least formally). Hence, if the whole structure is to be inspected for a given frequency, the strain in every member must be measured. This is the major drawback of the VDM-F approach.

Natural constraints are imposed on the modification coefficient $\mu_i$, which is nonnegative by definition (cf. Equation (17) in Chapter 2):

$$\mu_i \geq 0 \tag{18}$$

If degradation of a member is considered, another constraint has to be imposed on $\mu_i$:

$$\mu_i \leq 1 \tag{19}$$

Using Equations (28) and (8) in Chapter 2, the gradient of the VDM-T objective function (16) with respect to the optimization variable $\mu_i$ is expressed as

$$\nabla F_i^{\mathrm{t}} = \frac{\partial F^{\mathrm{t}}}{\partial \mu_i} = \frac{\partial F^{\mathrm{t}}}{\partial \varepsilon_k} \frac{\partial \varepsilon_k}{\partial \varepsilon_j^0} \frac{\partial \varepsilon_j^0}{\partial \mu_i} = \sum_t \frac{2}{(\varepsilon_k^{\mathrm{M}}(t))^2} \left( \varepsilon_k(t) - \varepsilon_k^{\mathrm{M}}(t) \right) \sum_{\tau=0}^{t} D_{kj}^{\varepsilon}(t - \tau) \frac{\partial \varepsilon_j^0(\tau)}{\partial \mu_i} \tag{20}$$

and the gradient of the VDM-F objective function (17) yields

$$\nabla F_i^{\omega} = \frac{\partial F^{\omega}}{\partial \mu_i} = \frac{\partial F^{\omega}}{\partial \varepsilon_k} \left( \frac{\partial \varepsilon_k}{\partial \varepsilon_j^0} \frac{\partial \varepsilon_j^0}{\partial \mu_i} + \frac{\partial \varepsilon_k}{\partial f_L^0} \frac{\partial f_L^0}{\partial \mu_i} \right) = \sum_{n_\omega} \frac{2}{(\varepsilon_k^{\mathrm{M}})^2} \left( \varepsilon_k - \varepsilon_k^{\mathrm{M}} \right) \left( D_{kj}^{\varepsilon} \frac{\partial \varepsilon_j^0}{\partial \mu_i} + D_{kL}^{\mathrm{f}} \frac{\partial f_L^0}{\partial \mu_i} \right) \tag{21}$$

Note that the gradient (20) involves calculation of convolution and summation over all time steps, which significantly increases the computational effort. This is the major drawback of the VDM-T approach. The partial derivatives $\partial \varepsilon_j^0 / \partial \mu_i$ and $\partial f_L^0 / \partial \mu_i$ can be easily calculated by differentiating relation (31) in Chapter 2 with respect to $\mu_i$, for the VDM-T approach:

$$\left[ \delta_{ij} - (1 - \mu_i) D_{ij}^{\varepsilon}(0) \right] \frac{\partial \varepsilon_j^0(t)}{\partial \mu_i} = -\varepsilon_i(t) \tag{22}$$

and by differentiating relation (10), performed $n_\omega$ times, independently for each considered frequency $\omega$, for the VDM-F approach:

$$\begin{bmatrix} \delta_{ij} - (1 - \mu_i) D_{ij}^{\varepsilon} & -(1 - \mu_i) D_{iL}^{\mathrm{f}} \\ -\omega^2 \Delta M_{KM} B_{Mj}^{\varepsilon} & \delta_{KL} - \omega^2 \Delta M_{KM} B_{ML}^{\mathrm{f}} \end{bmatrix} \begin{bmatrix} \dfrac{\partial \varepsilon_j^0}{\partial \mu_i} \\ \dfrac{\partial f_L^0}{\partial \mu_i} \end{bmatrix} = \begin{bmatrix} -\varepsilon_i \\ \omega^2 M_{KM} u_M \end{bmatrix} \tag{23}$$

Note that the left-hand side matrices in Equations (31) in Chapter 2, (22) and (10), (23) are respectively alike, which simplifies the computations. Only the right-hand sides vary. The optimization variable $\mu$ is updated according to the steepest descent method:

$$\mu^{(n+1)} = \mu^{(n)} - \alpha F^{(n)} \frac{\nabla \mathbf{F}^{(n)}}{[\nabla \mathbf{F}^{(n)}]^{\mathrm{T}} \nabla \mathbf{F}^{(n)}} \tag{24}$$

The superscript $(n)$ denotes values in current iteration and $(n + 1)$ in subsequent iteration. The constant $\alpha$ varies in the range 0.1–0.3.

### 3.2.5 Numerical Algorithm

The numerical algorithm solving the inverse problem of parameter identification performs several steps in two stages: (A) initial calculations and (B) iterative calculations. Stage A, slightly varying for VDM-T and VDM-F approaches, basically involves the following numerical operations:

A. Initial calculations:

(1) Calculate the response $\varepsilon_k^{\mathrm{L}}$ of the intact structure subjected successively to $n_\omega$ harmonic excitations of different frequencies, using a numerical model.
(2) Determine the measured response $\varepsilon_k^{\mathrm{M}}$ of the structure with introduced modifications using $k$ sensors in experiment (alternatively, simulate the measured response numerically).
(3) Compute the influence matrices $D_{ij}^\varepsilon$, $B_{Lj}^\varepsilon$, $D_{iL}^{\mathrm{f}}$, $D_{ML}^{\mathrm{f}}$ for $n_\omega$ frequencies of excitation.
(4) Set the initial value of the optimization variable to unity $\mu_i = 1$, which implies $\Delta M_{KL} = 0$, $\varepsilon_i^0 = 0$, $f_L^0 = 0$, $\varepsilon_k = \varepsilon_k^{\mathrm{L}}$, $u_L = u_L^{\mathrm{L}}$.

Stage B is also similar in structure but has been intentionally separated to show the use of specific formulas. Thus for VDM-T and VDM-F, stage B respectively requires:

B (VDM-T). Iterative calculations employing the Newmark procedure:

(1) Store the current value of the objective function (16) as the former value $F_{\mathrm{for}}^t$.
(2) Solve the set (22) for partial derivatives in every time step $t$.
(3) Calculate the gradient $\nabla F_i^t$ using Equation (20).
(4) Determine the next value of the variable $\mu_i^{(n+1)}$ using Equation (24).
(5) Solve for distortions $\varepsilon_i^0$ using Equation (31) in Chapter 2 in every time step $t$.
(6) Update $\varepsilon_k$ using Equation (28) in Chapter 2.
(7) Calculate the current value of the objective function $F_{\mathrm{cur}}^\omega$.
(8) Check the termination criterion. If $F_{\mathrm{cur}}^\omega / F_{\mathrm{for}}^\omega \leq 10^{-3}$ then STOP or else go to B(VDM-T)-1.

B (VDM-F). Iterative calculations:

(1) Store the current value of the objective function (17) as the former value $F_{\mathrm{for}}^\omega$.
(2) Solve the set (23) for partial derivatives $n_\omega$ times.
(3) Calculate the gradient $\nabla F_i^\omega$ using Equation (21).
(4) Determine the next value of the variable $\mu_i^{(n+1)}$ using Equation (24).
(5) Update $\Delta M_{KL}$ using Equation (15).
(6) Solve for distortions $\varepsilon_i^0$, $f_L^0$ using Equation (10) $n_\omega$ times.

(7) Update $\varepsilon_k$, $u_L$ using Equations (7) and (8) in Chapter 2 $n_\omega$ times.
(8) Calculate the current value of the objective function $F^\omega_{\mathrm{cur}}$.
(9) Check the termination criterion. If $F^\omega_{\mathrm{cur}}/F^\omega_{\mathrm{for}} \leq 10^{-3}$ then STOP or else go to B(VDM-F)-1.

Note that the above algorithm can be first used at the stage of numerical model calibration, when both the measured $\varepsilon^M_k$ and calculated $\varepsilon^L_k$ responses refer to the intact structure. Subsequently, the same algorithm can be applied for identification of stiffness/mass modifications introduced to the structure. Then the measured response $\varepsilon^M_k$ refers to the analyzed modification scenario.

## 3.2.6 Numerical Examples

### 3.2.6.1 Two-dimensional Truss Structure

As a numerical example, consider the two-dimensional truss structure shown in Figure 3.2. The initial structure consists of 20 steel elements, for which the following parameters are assumed:

- cross-sectional area: $A = 10^{-4}\,\mathrm{m}^2$;
- Young's modulus: $E = 210\,\mathrm{GPa}$;
- density: $\rho = 7800\,\mathrm{kg/m}^3$;
- height and width of a single section: 1 m;
- amplitude of the harmonic load: $P = 200\,\mathrm{N}$.

*Steady-State Approach (VDM-F)*
In the frequency-domain approach, strain responses (amplitudes) are measured by sensors located in all elements using different excitation frequencies $\omega$.

The modifications of cross-sectional area $\mu_i = \hat{A}_i/A_i$ in some truss elements were applied (see Figure 3.2). The response of the modified structure, stored in the vector $\varepsilon^M_k$, was computed numerically and then employed for the calculation of the objective function (17). The vector of the modification parameters $\mu_i$ was iteratively recomputed according to part B (VDM-F) of the numerical algorithm. For case 1, only one harmonic frequency, $\omega = 700\,\mathrm{rad/s}$, is used. In this case, 157 iterations had to be performed to fulfill the termination condition B(9). The identified modification parameters indicate possible defects and for some elements are under- or overestimated (see Figure 3.3). For case 2, it is assumed that the measurements are

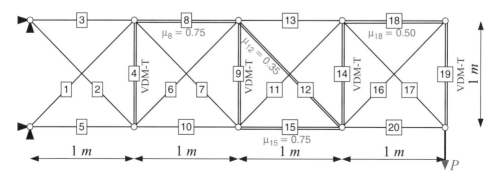

**Figure 3.2** Two-dimensional truss structure with introduced modifications

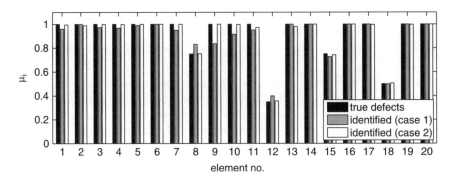

**Figure 3.3**  Identification results for one (case 1) and four (case 2) excitation frequencies

collected for four frequencies, $\omega = [100, 700, 2100, 3650]^T$, rad/s, independently. Thus the dimension of the 'vector' $\varepsilon_k^M$ is $20 \times 4$ and for each frequency $\omega$, respective influence matrices have to be calculated. This time, the sufficient number of iterations was only 51 to fulfill the termination condition. A comparison of the identified modification parameters in both cases is shown in Figure 3.3. The computational time of damage identification in both cases was similar (ca. 1 min on a 36 Hz Intel processor). In practice, measured responses are accompanied by noise, thus it is advisable to include it in numerically computed responses for the modified structure. Therefore random noise affecting every strain amplitude was applied according to the following formula:

$$\tilde{\varepsilon}_k^M = \varepsilon_{\underline{k}}^M \left(1 + r_{\underline{k}}\right) \tag{25}$$

where $r_k$ is an arbitrary value from interval $\langle -0.1, \ 0.1 \rangle$. A comparison of the simulated response (for a chosen frequency) for the original and modified structures with and without noise is presented in Figure 3.4. The described damage identification for case 2 (cf. Figure 3.3) was repeated for noisy data. The termination criterion of the inverse analysis could not be met and the process stopped after 300 iterations ($F_{cur}/F_{for} \approx 10^{-2}$). However, the obtained modification parameter distribution indicates possible defects in the structure (see Figure 3.5).

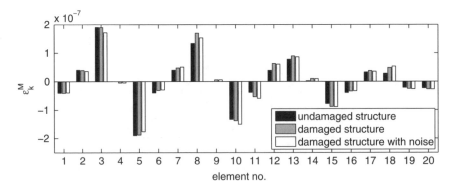

**Figure 3.4**  Influence of noise on simulated strain responses

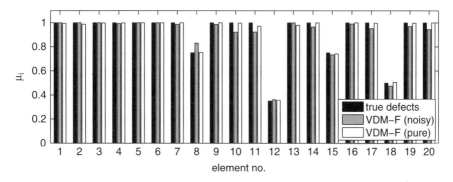

**Figure 3.5**   Identification results for pure and noisy data (case 2)

### Time-Domain Approach (VDM-T)

The previously presented VDM-F approach is a frequency-domain analysis based on the amplitude responses measured in every element of the structure. On the contrary, the time-domain approach (VDM-T) allows the stiffness modification parameters $\mu_i$ to be identified using only a few sensors collecting time-dependent responses. The same damage scenario is considered for the VDM-T approach (cf. Figure 3.2). However, the stiffness modifications ($\mu_i = \hat{E}_i/E_i$) are related to Young's modulus reductions in truss elements (not cross-sectional areas, as in VDM-F), i.e. no mass modifications are accounted for. In Figure 3.6(a), the excitation signal – a windowed sine pulse with the maximum amplitude of $P = 200\,\text{N}$ – is depicted. Strain responses are collected from only four selected vertical elements (cf. Figure 3.2). An example of strain response for the original and modified structures in element 1 is shown in Figure 3.6(b). For the other three sensing elements, the collected responses are similar. The inverse analysis using VDM-T leads to correct values, but considerable computational time is required to complete it (ca. 300 min). The results of damage identification for VDM-T and VDM-F are compared in Figure 3.6(c).

### 3.2.6.2  Two-Dimensional Beam Structure (VDM-F Only)

The next numerical example is a beam structure loaded as illustrated in Figure 3.7. The investigated modification parameters are related to cross-sectional areas as well as the moment of inertia (independently) for each finite element. Identical material and geometry parameters are assumed as in the truss example. The moment of inertia is equal to $J = 2.0833 \times 10^{-6}\,\text{m}^4$. The structure is divided into 25 finite elements. For each one, the modification parameters $\mu_i^A$ and $\mu_i^J$ are evaluated. The measured strain responses (in this case axial and bending components), collected by sensors located in every finite element, are used for evaluation of the modification parameters. The responses are numerically simulated for harmonic load with amplitudes $P = 100\,\text{N}$ (axial force) and $M = 1\,\text{Nm}$ (bending moment) and frequency $\omega = 2\pi[10, 40, 155, 250]^\text{T}\,\text{rad/s}$.

Two scenarios of damage identification (VDM-F only) were performed, namely with and without noise. Similarly to the previous example, the numerically calculated responses of the modified structure were subjected to random perturbations up to $10\,\%$ according to formula (25). The results of inverse analysis are presented in Figures 3.8 and 3.9, for both modification parameters $\mu_i^A$ and $\mu_i^J$, respectively. The analysis without noise was stopped when the decrease of the objective function achieved the level $10^{-4}$ (after 417 iterations). For the case with noise,

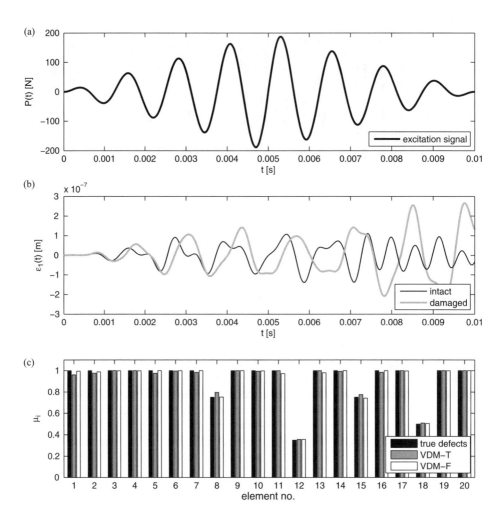

**Figure 3.6**  Two-dimensional truss structure: (a) excitation signal, (b) strain responses in element 1, (c) results of identification via an inverse analysis

the same number of iterations were performed. In both cases, the duration of inverse analysis was approximately 10 minutes on a 3 GHz Intel processor.

### 3.2.7 Experimental Verification

The experimental stand of a simply supported three-dimensional truss structure is presented in Figure 3.10. The truss consists of 70 steel elements (eight segments) of circular cross-sections

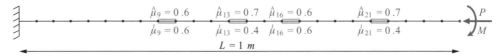

**Figure 3.7**  Beam structure with introduced modifications

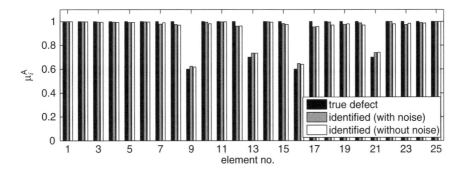

**Figure 3.8**    Identification of parameters $\mu_i^A$ for pure and noisy data

$A = 0.660\,\mathrm{cm^2}$ and two lengths $L_1 = 0.5\,\mathrm{m}$ and $L_2 = 0.71\,\mathrm{m}$. To join the structural elements 26 nodes have been used. Four nodes serve as supports – two of them cannot move in the horizontal plane (out-of-plane rotations are allowed) while the other two can freely slide along (rotations are also allowed). The overall dimensions of the truss are 4 m length, 0.5 m width, 0.35 m height.

Harmonic excitation is realized by a piezo-actuator in the middle of the structure (see the lower left corner in Figure 3.10(a)). A piezo-patch sensor (see Figure 3.10(b)), supposed to measure axial strain, is glued to each element. As some problems occurred with the strain sensors due to inappropriate adhesion, 22 accelerometers in every node of the truss structures have also been mounted. The identification problem was reformulated to handle vertical acceleration measurements; i.e. the objective function was expressed in terms of accelerations.

First, the stage of tuning the numerical model to the experiment was completed. This was achieved for the frequency of excitation $\omega = 35\,\mathrm{Hz}$, just below the first natural frequency of the structure. The material data adopted for the numerical model are Young's modulus $E = 205\,\mathrm{GPa}$ and density $\rho = 7850\,\mathrm{kg/m^3}$. Figure 3.11(a) shows vertical acceleration amplitudes in all 22 nodes measured in the experiment and modeled numerically for the intact structure. The biggest deviation is around 20 %, so in order to have good correspondence with experiment, just eight measurements were preserved. In this way, the deviation of the model from experiment is reduced to less than 3 % (see Figure 3.11(b)).

From a purely mathematical point of view, if just one excitation frequency is dealt with, a sensor in every element suspected of damage is needed. Otherwise the quasi-static inverse

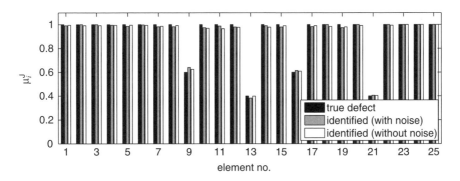

**Figure 3.9**    Identification of parameters $\mu_i^J$ for pure and noisy data

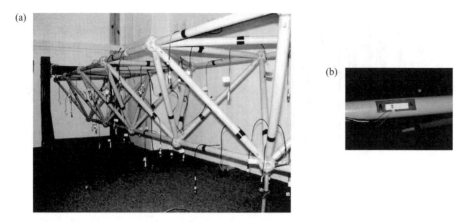

**Figure 3.10** Experimental stand for a three-dimensional truss structure: (a) general view, (b) piezo-sensor

problem becomes underdetermined. Therefore the search zone needs to be confined to as many members as there are accurate measurements collected. This means that the limit is one eight-element segment of the truss structure.

A replacement of two elements in the selected segment has been investigated as a modification scenario. Two horizontal steel members (see Figure 3.12) were replaced with other ones of significantly increased cross-section $\mu_{46} = \mu_{48} = 2.04$ ($A_{46} = A_{48} = 1.344\,\text{cm}^2$). Results of the inverse analysis for the experimental data recorded for this modification scenario are depicted in Figure 3.13. It is apparent that the location of the introduced modifications has been found correctly. The intensity in one member is very close to the assumed change (4 % difference). The deviation in the other member reaches 20 %. This result can be improved by measuring strains, which are less sensitive to noise. Also, a more sophisticated optimization algorithm should lead to a better quality of identification in a shorter time.

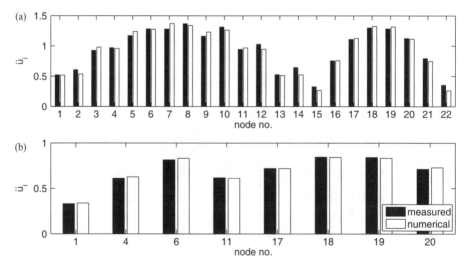

**Figure 3.11** Matching the model to experiment: (a) initial 22 measurements, (b) selected 8 measurements

**Figure 3.12** Modification scenario analyzed in the experiment (two longitudinal elements 46 and 48 are replaced in the selected segment)

## 3.2.8 Conclusions

This section presents an application of the virtual distortion method to structural health monitoring of skeletal structures (trusses and beams). Modifications of stiffness and mass are considered for simulating damage. In the VDM-T approach, impulse excitation is applied and dynamic analysis is involved. Responses (time histories) due to excitation close to natural frequencies are preferred as they have a good signal-to-noise ratio and discrimination of damage is then simpler. Harmonic excitation is assumed to transfer the identification problem to the frequency domain (VDM-F) with quasi-static analysis (amplitudes of quantities to be analyzed only). As damping is neglected, nonresonance frequencies are the subject of analysis in VDM-F.

Strain is the quantity contributing to the objective function. The reason is that strains vary much more smoothly in time compared to accelerations, giving reliable information even with poor sampling. On the other hand, voltage proportional to strains can be measured with simple piezoelectric patch sensors, although there is always the matter of scaling.

The VDM-T approach, requiring only a few sensors because the time histories were known, turned out to be quite time-consuming. The VDM-F approach is indeed much more efficient computationally, but nevertheless requires many more sensors to perform a successful inverse analysis. The number of sensors can be compensated, with more than one excitation frequency

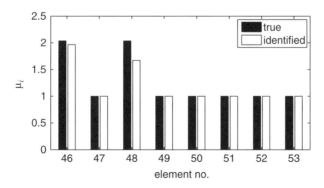

**Figure 3.13** Results of identification for experimental data

contributing to the objective function. Also, the position of the actuator can be changed for a given frequency, providing better data for the identification algorithm.

The VDM-based health monitoring approaches for skeletal structures proved to handle noisy data quite well. The results of identification are then inferior to pure simulation, but still satisfactory. Experimental sessions were carried out for acceleration measurements, proving the validity of the VDM-based damage identification in skeletal structures.

Further research will be concentrated on:

- conducting a series of laboratory tests (see Section 3.2.7) with strain measurements collected by piezo-patch sensors;
- analysis of combinations of many excitation frequencies (VDM-F) with various excitation positions to calculate optimal parameters for excitation;
- application of a more efficient optimization method (currently the steepest descent) to the identification algorithm;
- improvements to the time-domain approach (VDM-T), extending it to handle arbitrary, real-life excitations (e.g. induced by a train passing over a bridge);
- verification of the methodology in field tests.

## 3.3 Modeling and Identification of Delamination in Double-Layer Beams

### 3.3.1 Introduction

Fast development of composites has provided the incentive for dealing with one of the most severe defects in such structures – delamination. A review of methods dealing with delamination in the 1990s is provided by Zou *et al.* [32].

Many recent papers devoted to delamination are focused on the problem of proper modeling of delamination initiation and growth. Meo and Thieulot [33] compare four different ways (cohesive zone, nonlinear springs, birth and death elements, and tiebreak contact) of modeling delamination growth applied to a double cantilever beam test. Only the tiebreak contact method failed to match experimental results due to the adopted stress-based failure criterion. Iannucci [34] proposes an interface modeling technique for explicit finite element (FE) codes using not only a stress threshold for damage commencement but also a critical energy release rate for the particular delamination mode. Conventional interface modeling methods suffer from several shortcomings, i.e. interface elements have to be introduced *a priori*, spurious deformation occurs at the onset of delamination, traction oscillations accompany the process of delamination growth and finite elements have to be aligned with a potential delamination surface. These unwanted features can be avoided by modeling delamination at a mesoscopic level, proposed by De Borst and Remmers [35], who use the partition-of-unity property of the finite-element shape functions.

The next important problem is the search for dynamic responses of the delaminated structure that are sensitive to detecting the damage. Kim and Hwang [36] examine the influence of debonding in face layers of honeycomb sandwich beams on the frequency response functions. Li *et al.* [37] describe the potential of random decrement signatures along with neural networks in delamination detection. Zak [38] demonstrates that damped nonlinear vibrations are very sensitive to delamination location and length, depending upon excitation.

Another group of papers propose solutions to the inverse problem of identification of delamination, belonging to the main stream of SHM. Bois *et al.* [39] inversely identify delamination by applying a model combining the delaminated zone with a piezoelectric ply. Electromechanical impedance of the transducer working both as actuator and sensor was measured. Ishak *et al.* [40] use soft-computing methods to detect delamination via inverse analysis. The strip element method was employed to train the multilayer neural network. Schnack *et al.* [41] propose a method of identification of single delamination by transforming a general ill-posed problem, described by the Kohn–Vogelius functional, into a coupled system of well-posed Euler–Lagrange equations. A smooth iterative numerical solution was proposed. Ramanujam *et al.* [42] apply the Nelder–Mead gradient-free optimization method to find a solution to the inverse problem, with many starting points required to obtain a minimum. Generally, only the sensors close to delamination showed a deviation from the reference response.

This section takes up two important problems regarding defects in composite beams. The first one is proper modeling of delamination between the layers of laminate; the other one is effective identification of delamination zones [43].

For modeling delamination, a concept of the contact layer between laminates has been proposed. The layer consists of truss elements supposed to model vanishing of the shear forces in delamination zones and to provide appropriate contact conditions between the laminate layers. The two features of the contact layer, modifying its properties in selected zones, are easy to model by virtual distortions. Sensitivity information with an underlying model updating procedure is effectively used in an optimization algorithm solving the inverse problem of *a posteriori* identification. A concept of on-line identification, not related to the VDM, is also proposed.

## 3.3.2 Modeling of Delamination

### 3.3.2.1 Interconnection between Laminates (Contact Layer)

Delamination of a double-layer beam is understood here as an existing defect of certain extension in the structure. The process leading to the existence of the defect (crack growth) is not the subject of interest here. The authors have proposed the simplest possible model of delamination, taking into account the contact between the laminate layers.

This section puts forward a proposition of introducing a special interconnection between the layers of the laminate for modeling delamination in two aspects. The principal one is that the shear forces in the interconnection, joining two laminate layers, should vanish in delamination zones. The secondary aspect, less frequently taken into account in delamination modeling, is that proper contact between laminates should exist in delamination zones. This section will focus on an explanation of how the interconnection between the laminate layers is constructed and how the VDM can be used to model its behavior.

When modeling delamination in multilayer composites, a Timoshenko beam is appropriate to account for interactions of laminates (cf. Reference [44]). In the proposed approach, however, the Bernoulli beam is the subject of consideration for two reasons: (1) two-dimensional beams consisting of only two layers are considered and (2) a simplified Bernoulli model enables identification of delamination to be performed by solving an inverse problem at an affordable numerical cost. In the FEM model, standard two-noded beam elements are used for the laminate layers. The two-layer beam with the interconnection for modeling delamination is shown in Figure 3.14. Middle axes of the beams are depicted with continuous horizontal lines (shown in white in Figure 3.14(a)). In between, the interconnection of the laminates, consisting of the diagonal and vertical truss elements joining the middle axes of the beams, is introduced. The role of the diagonal truss elements is to simulate vanishing of the shear forces in delamination

(a)

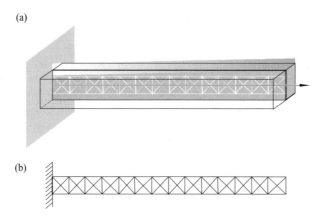

(b)

**Figure 3.14**  (a) Double cantilever beam with the contact layer supposed to model delamination, (b) contact layer to be analysed by the VDM

zones. The role of the vertical truss elements is to ensure proper contact conditions between the laminates. Stiffness of the vertical truss elements is assumed in an arbitrary way (see Figure 3.15), enabling some penetration of one laminate layer into another. In fact, this is not the case in reality, but this adjustment of stiffness of the vertical elements turned out to be very important when matching the numerical model to experiments. Friction is not taken into account, although there is no formal obstacle to include it within the framework of the VDM. The reason is again to minimize the numerical effort.

### 3.3.2.2  Modeling of Delamination by the VDM

The contact layer has a modular structure. Each section of the layer consists of three elements: two diagonals denoted by A and B and one vertical denoted by C in Figure 3.16.

If delamination is to be modeled in selected sections of the contact layer, then the following conditions in the diagonal elements A, B need to be fulfilled (cf. definition (17) in Chapter 2):

$$\mu_i^A = 0 \;\Rightarrow\; \varepsilon_i^{0A} = \varepsilon_i^A \tag{26}$$
$$\mu_i^B = 0 \;\Rightarrow\; \varepsilon_i^{0B} = \varepsilon_i^B \tag{27}$$

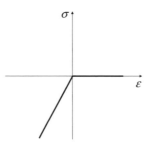

**Figure 3.15**  Stress–strain relationship for the vertical elements of the contact layer

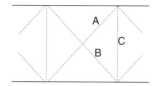

**Figure 3.16**  Zoomed section of the contact layer

The superscripts A, B at a quantity refer to corresponding diagonal elements. Depending upon the kind and direction of load applied to the structure, the two laminates will either be in contact or not. This is accounted for by examining the sign of strain in the vertical element C. If tensile strain is encountered in the vertical elements, then appropriate distortions are generated, modeling no contact between the laminate layers:

$$\mu_i^C = 0 \implies \varepsilon_i^{0C} = \varepsilon_i^C \qquad \text{if } \varepsilon_i^C > 0 \tag{28}$$

If compressive strain is encountered in the vertical elements, then no distortions are generated and the layers stay in full contact (in spite of local delamination in between):

$$\mu_i^C = 1 \implies \varepsilon_i^{0C} = 0 \quad \text{if } \varepsilon_i^C \le 0 \tag{29}$$

It should be remembered that the assumed characteristic (see Figure 3.15) allows for some penetration of one layer into another, which gives the possibility of tuning the numerical model to experiment.

Implications of the conditions (26) and (27) are that distortions are equal to total strains, but these relations cannot be directly used because total strains are *a priori* unknown. In order to determine distortions the general system of equations (18) in Chapter 2 need to be solved, which allows for an arbitrary change of the coefficient $\mu_i$:

$$\begin{bmatrix} \delta_{ij} - (1 - \mu_i^A)D_{ij}^A & -(1 - \mu_i^B)D_{ij}^B \\ -(1 - \mu_i^A)D_{ij}^A & \delta_{ij} - (1 - \mu_i^B)D_{ij}^B \end{bmatrix} \begin{bmatrix} \varepsilon_j^{0A} \\ \varepsilon_j^{0B} \end{bmatrix} = \begin{bmatrix} (1 - \mu_i^A)\varepsilon_i^{LA} \\ (1 - \mu_i^B)\varepsilon_i^{LB} \end{bmatrix} \tag{30}$$

Simultaneously, the conditions $\mu_i = 0$ (cf. Equations (26) and (27)) must be substituted into the system to model delamination in the required zone. The distortion vector and influence matrix have been divided into parts corresponding to elements A and B, which facilitates the organization of the algorithm and code. As soon as distortions in the diagonal elements are generated, it is necessary to check the sign of strain in the vertical elements, using a prediction, which is supposed to reflect its sign (not the value) correctly (see Section 3.3.2.3 for comments):

$$\varepsilon_i^C = \varepsilon_i^{LC} + D_{ij}^A \varepsilon_j^{0A} + D_{ij}^B \varepsilon_j^{0B} \tag{31}$$

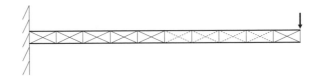

**Figure 3.17**   Double cantilever beam subjected to first static loading

If it turns out that tensile stress is present, then the system of equations (30) has to be extended, to include the calculation of distortion in vertical elements as well:

$$
\begin{bmatrix}
\delta_{ij} - (1 - \mu_i^A)D_{ij}^A & -(1 - \mu_i^B)D_{ij}^B & -(1 - \mu_i^C)D_{ij}^C \\
-(1 - \mu_i^A)D_{ij}^A & \delta_{ij} - (1 - \mu_i^B)D_{ij}^B & -(1 - \mu_i^C)D_{ij}^C \\
-(1 - \mu_i^A)D_{ij}^A & -(1 - \mu_i^B)D_{ij}^B & \delta_{ij} - (1 - \mu_i^C)D_{ij}^C
\end{bmatrix}
\begin{bmatrix}
\varepsilon_j^{0A} \\
\varepsilon_j^{0B} \\
\varepsilon_j^{0C}
\end{bmatrix}
$$
$$
=
\begin{bmatrix}
(1 - \mu_i^A)\varepsilon_i^{LA} \\
(1 - \mu_i^B)\varepsilon_i^{LB} \\
(1 - \mu_i^C)\varepsilon_i^{LC}
\end{bmatrix}
\tag{32}
$$

Thus, the algorithm of delamination modeling consists of the following steps:

(1) Initialize: influence matrix, linear response.
(2) Solve the set (30) to determine distortions for vanishing of shear forces in the contact layer.
(3) Check the sign of strain in vertical elements of the contact layer using Equation (31).
(4) If no contact between the laminate layers appears, solve the extended set (32) to determine the distortions.
(5) Update strains in the whole structure.

In dynamics, the algorithm repeats the same stages in every time step processed by the Newmark integration procedure. All quantities, except for the modification coefficient $\mu_i$, are then time-dependent. Computations are costlier compared to statics, because the amount of data is multiplied by the number of time steps.

### 3.3.2.3 Numerical Example in Statics

A double cantilever beam (1 m long and 0.02 m high) has been chosen to demonstrate modeling of delamination by the VDM in statics (see Figure 3.17). The contact layer is divided into 10 sections with three elements of type A, B, C in each one, resulting in 30 connecting elements altogether. The assumed geometrical and material data are listed in Table 3.1.

**Table 3.1**   Material data for delamination modeling in statics

| Quantity | Beam element | Connecting truss element |
| --- | --- | --- |
| Young's modulus (GPa) | 70 | 30 |
| Cross-sectional area (m$^2$) | $5 \times 10^{-5}$ | $5 \times 10^{-5}$ |
| Density (kg/m$^3$) | 3300 | 1 |

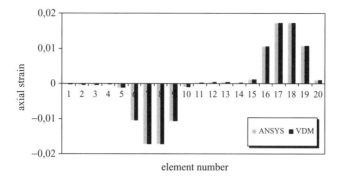

**Figure 3.18**   Axial strains in elements A and B of the contact layer for the first static loading

A delamination zone extending through four sections, 6 to 9, is analyzed (cf. Figure 3.17). Two cases of static loading are considered. Calculations performed by the FEM package ANSYS serve as verification of the VDM results. The first load case (see Figure 3.17) is a single vertical force applied to the top beam at the cantilever's free end. The intention is to examine the appropriateness of delamination modeling by the VDM for the case of a closed crack. Figure 3.18 depicts axial strains in diagonal elements A (numbered 1 to 10) and B (numbered 11 to 20) of the contact layer for the structure with the assumed delamination. It is natural that strains in the contact layer are larger in these elements A and B, which are the closest to the delamination zone. Axial strains in vertical elements C are calculated according to Equation (31). For the considered loading, the first term in formula (31) is negligibly small and the second and third terms cancel out, due to almost identical values and reverse signs of strains in elements A and B. Axial strains in vertical elements C, oscillating around numerical zero, are shown in Figure 3.19. The second case of loading, corresponding to the structure with an open crack, is presented in Figure 3.20. Axial strains in elements A and B appear only within the delamination zone and are of the same sign and value (see Figure 3.21). Contrary to the first loading case (cf. Figure 3.17), axial strains in contact elements C, depicted in Figure 3.22, are now two orders of magnitude larger than in diagonal elements A and B.

For calculations with ANSYS, the contact elements CONTA171 and the associated TARGE169 were used. The normal contact stiffness factor was FKN = 0.1 and the sliding

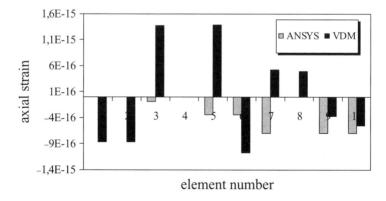

**Figure 3.19**   Axial strains in elements C of the contact layer for the first static loading

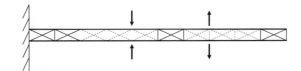

**Figure 3.20**  Double cantilever beam subjected to the second static loading

contact stiffness factor was FKT = 1.0. Agreement of the VDM results with ANSYS is very good. Figure 3.19 shows values close to numerical zero, so the discrepancies between the VDM results and ANSYS are self-explanatory in this case.

### 3.3.2.4  Numerical Example in Dynamics

Another double cantilever beam (1 m long and 0.02 m high) with slightly different material data, shown in Table 3.2 (cf. Table 3.1 in Section 3.3.2.3), is considered in the dynamic test. The delamination zone now extends through three sections, 5 to 7 (see Figure 3.23). An impulse force of magnitude $P$ and duration equal to one period of sine was applied within the delamination zone.

Figure 3.24 shows the time history of strain in the element C under the applied force, for the intact and delaminated structure. It is apparent that the strains grow considerably (see the time range between 0.002 s and 0.004 s) when the crack between the upper and lower beams opens. The presented VDM results overlap the ones generated by ANSYS (the latter are not depicted in Figure 3.24). The effect is also visualized in Figure 3.25, presenting the shape of the deformed structure while working in the open crack mode at the point of maximum strain.

### 3.3.2.5  Experimental Validation in Dynamics

An experimental validation of the numerical model of delamination was carried out. A double cantilever structure consisting of two aluminum beams of 0.8100 m length, 0.0250 m width and 0.0024 m height (each), schematically depicted in Figure 3.26(a), was investigated. The two beams were joined by 10 screws (Figure 3.27(a)), placed in uniform distances over the length and marked by vertical lines in Figure 3.26(a). This division was naturally adopted in

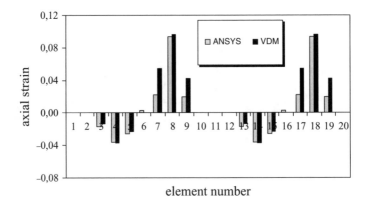

**Figure 3.21**  Axial strains in elements A and B of the contact layer for the second static loading

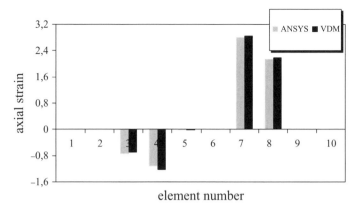

**Figure 3.22**  Axial strains in elements C of the contact layer for the second static loading

**Table 3.2**  Material data for delamination modeling in dynamics

| Quantity | Beam element | A, B elements | C element |
|---|---|---|---|
| Young's modulus (GPa) | 70 | 10 | 1 |
| Cross-section area (m$^2$) | $5 \times 10^{-5}$ | $5 \times 10^{-5}$ | $5 \times 10^{-5}$ |
| Density (kg/m$^3$) | 3300 | 1 | 1 |

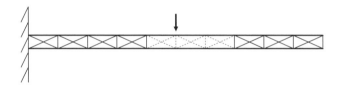

**Figure 3.23**  Double cantilever beam subjected to dynamic loading

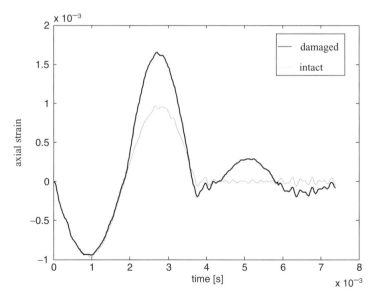

**Figure 3.24**  Axial strains in a chosen element C within the delamination zone

**Figure 3.25**    Structural deflection in the open crack mode

structuring the VDM-based contact layer (see Figure 3.26(b)), whose vertical elements are placed exactly at the location of screws, resulting in 10 sections altogether. The height of the contact layer, joining middle axes of the beams, is equal to the height of one beam, i.e. 0.0024 m. The contact layer consists of 20 truss elements of type A or B (diagonal) and 10 truss elements of type C (vertical). Each aluminum beam is divided into 80 finite beam elements, so there are 8 beam elements in each of the 10 sections of the contact layer.

The measuring system, presented in Figure 3.28(a), was used in the experiment. It consisted of:

(1) an activation line, including a signal generator, amplifier and piezoelectric actuator (applying a bending moment to the beam), shown in Figure 3.27(b);
(2) a detection line, including a piezoelectric sensor, depicted in Figure 3.27(c), conditioning amplifier and oscilloscope.

Both lines of the measuring system were coupled by a controling computer.

A windowed sine signal (see Figure 3.28(b)) induced by the actuator was applied to the structure and its response (voltage proportional to strains) was captured by the sensor for two cases: the intact structure with no delamination and the damaged structure with two screws removed in the middle of the beam (marked by dashed lines in Figure 3.26(a)), corresponding to the delamination extending through three sections of the contact layer.

A numerical model of the structure was created and tuned to experimental response using the intact structure configuration. The tuned data read: Young's modulus $E = 41$ GPa for the beam elements and stiffness $EA = 925 \times 10^3$ N for the truss elements in the contact layer.

Arbitrary delamination was introduced to the structure in experiment by removing two screws. Responses of the damaged structure were collected and compared with a numerical analysis using the previously tuned model. A comparison of the first eigenfrequencies between the experiment and numerical analysis is presented in Table 3.3.

Good agreement of results between the experiment and the VDM model is shown in Figure 3.29, presenting time histories for the intact and damaged structure.

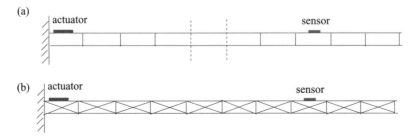

**Figure 3.26**    (a) Layout of a double-layer cantilever beam, (b) the corresponding contact layer

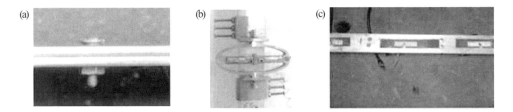

**Figure 3.27** (a) Screw connection, (b) actuator, (c) sensors

(a)

(b)

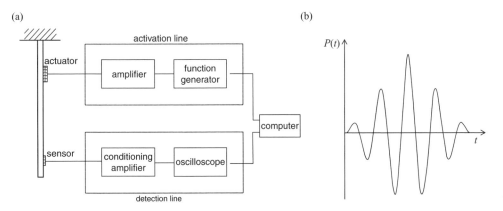

**Figure 3.28** (a) Measuring system, (b) excitation signal

**Table 3.3** Comparison of eigenfrequencies obtained in the experiment and numerical analysis

| Eigenfrequency | Experimental | Numerical |
|---|---|---|
| 1 | 5.00 | 5.05 |
| 2 | 29.70 | 29.36 |
| 3 | 70.60 | 73.27 |
| 4 | 140.00 | 129.05 |
| 5 | 199.00 | 199.30 |

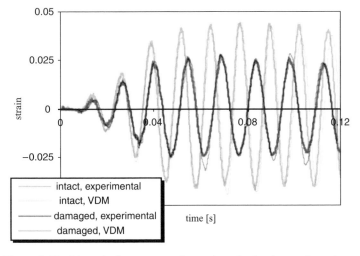

**Figure 3.29** Numerical versus experimental results for the cantilever beam

### 3.3.3  Identification of Delamination

#### 3.3.3.1  Formulation of an Off-Line Problem

Depending upon the application, the important problem of defect identification can be handled in an off-line (*post factum*) or on-line (*real-time*) procedure. This section is devoted to the problem of off-line identification of delamination by employing VDM-based gradient optimization. Within the framework of VDM, an analytical sensitivity analysis can be effectively performed. Therefore the inverse problem of identification can be solved using classical optimization tools, i.e. gradients of the objective function with respect to a design variable.

Usually, frequencies and mode shapes obtained via modal analysis enter the objective function in the inverse problem of identification. In the presented approach, strains have been chosen for the purpose. The reason is that on the one hand strains can be easily modeled in the VDM (cf. Equation (8) in Chapter 2) and on the other they can be measured by piezoelectric sensors. One more advantage is that the variation of strains in dynamics is relatively smooth (compared to accelerations for instance), which is convenient from the signal processing point of view. Thus, the identification task is posed as a standard nonlinear least squares minimization problem, with the objective function expressed as follows:

$$F(\boldsymbol{\mu}) = \left( \frac{\varepsilon_k^{\text{beam}} - \varepsilon_k^{\text{beamM}}}{\varepsilon_k^{\text{beamM}}} \right)^2 \tag{33}$$

The function (33) collects responses from selected $k$ sensors placed in beam elements, where flexural strains are measured. Note that the measurements are taken in horizontal beam elements (see Figures 3.14 and 3.16), whereas modifications to the structure are introduced only in truss elements A, B, C of the contact layer. This is an important distinction, which implies building an extended influence matrix $\mathbf{D}_{kj}^{\text{ext}}$, collecting not just the interrelations within the contact layer but also the influence of the truss members on the connected beam members.

The axial strain $\varepsilon_i$ in truss elements (of type A, B or C) depends nonlinearly upon the modification coefficient $\mu_i$ (see Equation (17) in Chapter 2), which is further used as a variable in optimization. Natural constraints are imposed on the modification coefficient $\mu_i$, which is nonnegative by definition:

$$\mu_i \geq 0 \tag{34}$$

If delamination has been identified, the constraints (34) are active. Another formal constraint has to be imposed on $\mu_i$, related to structural degradation:

$$\mu_i \leq 1 \tag{35}$$

Using Equation (8) in Chapter 2 and building the extended influence matrix, the gradient of the objective function (33) with respect to the optimization variable $\mu_i$ is expressed as

$$\nabla F_i = \frac{\partial F}{\partial \mu_i} = \frac{\partial F}{\partial \varepsilon_k^{\text{beam}}} \frac{\partial \varepsilon_k^{\text{beam}}}{\partial \varepsilon_j^0} \frac{\partial \varepsilon_j^0}{\partial \mu_i} = \frac{2}{(\varepsilon_k^{\text{beamM}})^2} \left( \varepsilon_k^{\text{beam}} - \varepsilon_k^{\text{beamM}} \right) D_{kj}^{\text{ext}} \frac{\partial \varepsilon_j^0}{\partial \mu_i} \tag{36}$$

The partial derivative $\partial \varepsilon_j^0 / \partial \mu_i$ can be easily calculated by differentiating relation (30) or (32) with respect to $\mu_i$:

$$
\begin{bmatrix}
\delta_{ij} - (1 - \mu_i^A)D_{ij}^A & -(1 - \mu_i^B)D_{ij}^B & -(1 - \mu_i^C)D_{ij}^C \\
-(1 - \mu_i^A)D_{ij}^A & \delta_{ij} - (1 - \mu_i^B)D_{ij}^B & -(1 - \mu_i^C)D_{ij}^C \\
-(1 - \mu_i^A)D_{ij}^A & -(1 - \mu_i^B)D_{ij}^B & \delta_{ij} - (1 - \mu_i^C)D_{ij}^C
\end{bmatrix}
\begin{bmatrix}
\dfrac{\partial \varepsilon_j^{0A}}{\partial \mu_i} \\
\dfrac{\partial \varepsilon_j^{0B}}{\partial \mu_i} \\
\dfrac{\partial \varepsilon_j^{0C}}{\partial \mu_i}
\end{bmatrix}
=
\begin{bmatrix}
-\varepsilon_i^A \\
-\varepsilon_i^B \\
-\varepsilon_i^C
\end{bmatrix}
\quad (37)
$$

Note that the left-hand side matrices in Equations (32) and (37) are alike, which simplifies the computations. Only the right-hand sides are different.

The optimization variable $\mu_i$ is updated according to the steepest descent method (cf. Equation (24)). The algorithm solving the problem of identification of delamination contains the following two stages:

### A. Initial Calculations

(1) Calculate responses $\varepsilon_i^L$ (contact layer, truss elements) and $\varepsilon_k^{\text{beamL}}$ (sensors, beam elements) of the intact structure subjected to external load, using a numerical model.
(2) Determine the measured response $\varepsilon_k^{\text{beamM}}$ of the structure with introduced modifications using $k$ sensors in the experiment (alternatively, simulate the measured response numerically).
(3) Compute the influence matrix for the contact layer $D_{ij}$ and the extended influence matrix $D_{kj}^{\text{ext}}$ including the truss–beam interactions.
(4) Set the initial value of optimization variable to unity $\mu_i = 1$, which implies $\varepsilon_i^0 = 0$, $\varepsilon_i = \varepsilon_i^L$ and $\varepsilon_k = \varepsilon_k^{\text{beamL}}$.

### B. Iterative Calculations

(1) Store the current value of the objective function (33) as the former value $F_{\text{for}}$.
(2) Solve the set (37) for partial derivatives $\partial \varepsilon_j^0 / \partial \mu_i$.
(3) Calculate the gradient $\nabla F_i$ using Equation (36).
(4) Determine the next value of the variable $\mu_i^{(n+1)}$ using Equation (24).
(5) Solve for distortions $\varepsilon_i^0$ using Equation (32).
(6) Update $\varepsilon_k^{\text{beam}}$, using Equation (8) in Chapter 2.
(7) Calculate the current value of the objective function $F_{\text{cur}}$.
(8) Check the termination criterion – if $F_{\text{cur}}/F_{\text{for}} \leq 10^{-3}$ then STOP or else go to B(1).

In dynamics, the stages are handled in each time step by the Newmark integration procedure. The corresponding gradient has a similar form to Equation (20).

### 3.3.3.2 Numerical Off-Line Identification in Statics and Dynamics

For checking the effectiveness of the VDM-based identification algorithm, examples of double cantilever beams, presented in Sections 3.3.2.3 and 3.3.2.4, are used.

In statics, a cantilever beam, shown in Figure 3.30, with the contact layer divided into 10 sections, is analyzed (data as in Section 3.3.2.3). The structure is subjected to a static force

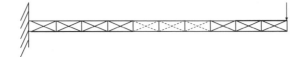

**Figure 3.30**    Double-layer beam analyzed to identify delamination in statics

applied at the free end. The delamination zone extends over three sections, 5, 6 and 7. The beam is equipped with 20 sensors, located in each horizontal element of both the lower and upper beam. The results of identification, obtained after 200 optimization iterations, are shown in Figure 3.31. The assumed zone of delamination has been detected correctly.

In dynamics, the contact layer, shown in Figure 3.32, is divided into 20 sections (data as in Section 3.3.2.4). An impulse sine load (cf. Figure 3.28(b)), with a duration equal to the time period of the fourth eigenfrequency of the beam, is applied as excitation at the free end. The delamination zone extends over four sections, 11, 12, 13 and 14. Contrary to the static case, only four sensors mounted on the upper beam and marked by bold lines (see Figure 3.32) are considered in the identification process. The results reached after 300 iterations are presented in Figure 3.33. The obtained accuracy is very good.

Another example in dynamics focuses on identification of two zones of delamination, including the inner part and the edge of the cantilever beam (see Figure 3.34). The damaged zone extends over sections 15 and 16 and 19 and 20 of the contact layer. This scenario of delamination is harder to detect, as can be seen in Figure 3.35, but both zones of damage have been correctly located. The result can be improved with more sophisticated optimization.

### 3.3.3.3 Experimental Verification of Off-Line Identification in Dynamics

Analogously to the experimental verification carried out for delamination modeling, described in Section 3.3.2.5, similar measurements were collected to check the correctness of delamination identification. The same experimental stand was used. Delamination was applied by removing two screws tightening the two beams in the middle part of the structure, as shown in Figure 3.26(a). Thus the delamination extends over three sections of the contact layer modeled by the VDM. The first stage was to tune the numerical model to an experimental response. It was achieved with the material data given in Section 3.3.2.5. With the well-tuned model, the identification algorithm was run and the results of delamintion identification are shown in Figure 3.36.

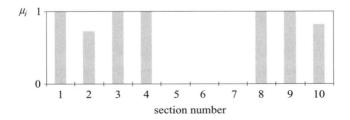

**Figure 3.31**    Results of the search for the inner delamination – static load

**Figure 3.32** Double-layer beam analyzed to identify the inner delamination in dynamics

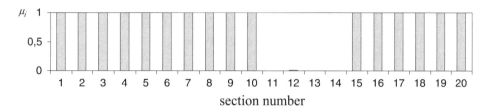

**Figure 3.33** Results of search for the inner delamination – dynamic load

**Figure 3.34** Double-layer beam analyzed to identify the inner and edge delamination in dynamics

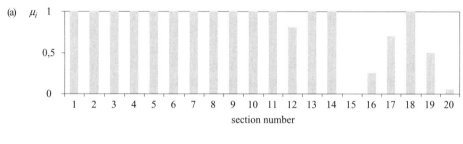

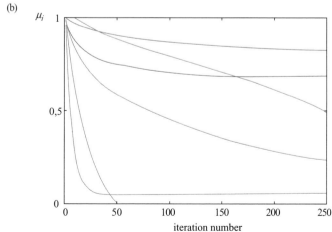

**Figure 3.35** (a) Results of the search for the inner and edge delaminations – dynamic load, (b) variations of $\mu_i$ in 250 iterations

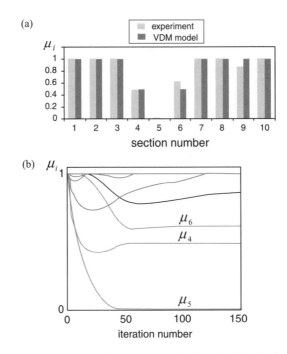

**Figure 3.36** (a) Numerical versus experimental identification of delamination for the double-layer beam, (b) variations of $\mu_i$ in 150 iterations

The agreement between the experiment and the numerical model (very simple and with coarse discretization) is very good. Due to the fact that the double beam was connected with screws, there is always some contact between layers in the vicinity of the screws. That is why the identified stiffness in the assumed delaminated sections close to the tightened screws does not drop to zero but just a half of the initial stiffness. In the middle section of assumed delamination, where both neighboring screws are removed, the ideal zero value was detected.

### 3.3.3.4 Numerical On-Line Identification in Dynamics

On-line identification of delamination is extremely important in some kinds of applications requiring 'allow' or 'not allow' decisions, e.g. assessing the structural health of a helicopter rotor in motion. Intuitively, it is likely that the problem will require more sensors, able to detect a defect quickly in real time.

The concept of on-line identification of delamination assumes a uniformly distributed net of sensors (see Figure 3.37), acting in pairs, attached to the upper and lower surfaces of the double-layer beam structure. Piezoelectric sensors measure voltage on the upper and lower surface, which is proportional to the flexural strain of the beam. Local delamination manifests itself in an apparent difference observed in the responses of collocated (paired) sensors.

Numerical simulation was run to validate the idea. Figure 3.38(a) depicts the time signals, captured by the sensors, with delamination occurring in the midpoint of the analyzed period. The corresponding difference in response for the pairs of sensors is presented in Figure 3.38(b). The numerically assumed delamination extends over sections 13 to 15 and the difference

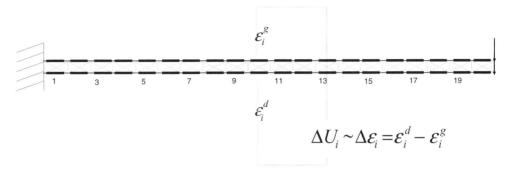

**Figure 3.37**   The net of paired sensors able to detect delamination on-line

in voltage (see Figure 3.38(b)) clearly identifies this zone. Experimental verification of this promising result will be pursued in the future.

### 3.3.4 Conclusions

This section presents an idea of modeling and identification of delamination in a two-layer beam, using the virtual distortion method. A novel concept of the contact layer, consisting of simple truss elements connecting two beam layers, has been proposed. The model has been simplified (Bernoulli beam, no friction) in order to reduce numerical costs. However, it has been shown to follow the experiment faithfully for the problem of delamination modeling (see Figure 3.29).

An approach to performing off-line identification by solving an analytically posed inverse problem has been proposed. With the adopted simplified model of the contact layer between two beams, the identification has proceeded to the expected solutions in a reasonable computational time. The major difference between the static and dynamic approach is that many sensors are needed in statics and only a few in dynamics. The reason is that the number of available data in statics is very limited, while time histories of the monitored quantities in dynamics compensate for the fact of mounting just a few sensors. However, the price to pay for the priviledge of

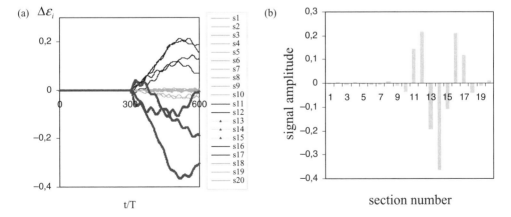

**Figure 3.38**   (a) Time histories and (b) signal amplitudes for on-line identification of delamination

having lots of data in dynamics is a much more time-consuming numerical analysis. Good qualitative identification of two delamination zones (including the edge) has been achieved (see Figure 3.35). Experimental verification of delamination identification in dynamics has also been successful (see Figure 3.36).

A proposition for on-line identification has been put forward. The problem has just been recognized at the numerical level. An experimental verification will be the subject of future research.

## 3.4 Leakage Identification in Water Networks

### 3.4.1 Introduction

The problem of management of water sources is more and more important in the world scale. In particular, the problem of detection and identification of leakages (mostly due to corrosion) in water networks is an important engineering challenge, especially in tropical countries often suffering from the lack of water. On the other hand, the consequences of unpredicted large-scale leakage in the operating water network may be very serious. Therefore, there is a need for an automatic monitoring system able to detect and localize leakages in the early stage of their development. A number of papers addressing the issue have been published in the hydraulics-related journals of the American society of civil engineers (ASCE). The use of a genetic algorithm for solving the inverse transient problem has been proposed in References [45] and [46]. Frequency response, provoked in open-loop piping systems by periodic opening and closing of the valves, has been investigated in Reference [47]. The damping of transient events has been discussed in [48]. The papers [49] and [50] demonstrate that leakages can be not only detected but also reduced by optimal valve control.

The proposed approach is based on continuous observation of the water heads at the nodes of the water network. By having a reliable (verified versus field tests) numerical model of the network and its responses for determined inlet and outlet conditions, any perturbations to the original network response (water head distribution) can be detected. Then, applying the proposed numerical procedure, the inverse problem of the water flow distribution can be solved. The possibility of simultaneous identification of several leakages with different locations and intensities is included in the presented approach.

The presented methodology for leakage identification [51] is an extension of the VDM applications in structural mechanics (cf. Section 3.2). Analogies between truss structures and water networks, both modeled as closed-loop systems, the system theory and graph representation of the truss-like system, have been effectively used.

### 3.4.2 Modeling of Water Networks and Analogies to Truss Structures

Advantage will be taken of the general *system theory* (cf. Reference [52]) in order to build a numerical tool for modeling and analysis of water networks (cf. Reference [53]). To this end, a water network needs to be visualized as an oriented *graph* (a two-loop example is depicted in Figure 3.39) with the direction of water flow indicated by arrows.

The network shown in Figure 3.39 consists of five real branches, connecting four nodes. The water inlet is located at node 1 and the outlet at node 4. There is also a reference node $W_0$ serving as an arbitrary level $H_0$ for measuring water heads at the network's nodes. Four fictitious branches connecting the reference node $W_0$ with the corresponding four network nodes are provided in the analysis. It can be seen that the topology of the water network is

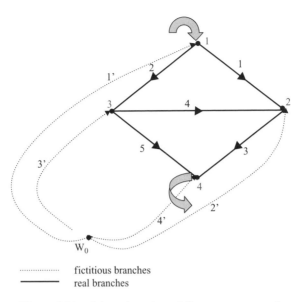

**Figure 3.39**   Oriented graph modeling a water network

quite similar to the topology of a truss structure. The major distinction is the orientation of the graph in water networks, which is obviously disregarded in trusses. The analogy between the structural mechanics and the network analysis was discovered by Cross over 70 years ago. He used the method of successive corrections to find the distribution of flow in networks of pipes as well as in electrical circuits (cf. Reference [54]).

The mathematical tool enabling the direction of water flow to be found is the incidence matrix $\mathbf{N}$, taking the following form for the presented example:

$$
N_{Ki} =
\begin{bmatrix}
1 & 1 & 0 & 0 & 0 \\
-1 & 0 & 1 & -1 & 0 \\
0 & -1 & 0 & 1 & 1 \\
0 & 0 & -1 & 0 & -1
\end{bmatrix}
\tag{38}
$$

The rows of the matrix $N_{Ki}$ correspond to the network's nodes $K$ while its columns correspond to the branches $i$. For instance, column 1, responsible for network branch 1, indicates that the branch connects nodes 1 and 2 and the assumed direction of flow is from node 1 (value $= 1$) to node 2 (value $= -1$). For the remaining unconnected nodes in column 1, the matrix $\mathbf{N}$ takes the values 0.

The governing equations will now be defined for the water network in a steady-state flow. The equilibrium of the system relating the internal flow distribution in the network's branches $\mathbf{Q}$ (m$^3$/s) with the external inlet/outlet $\mathbf{q}$ (m$^3$/s) is expressed as follows:

$$
\mathbf{NQ} = \mathbf{q}
\tag{39}
$$

The analogous relation for truss structures relates internal and external forces via the transposed geometric matrix (cf. Equation (11) in chapter 2). The following continuity equation relates the water heads $\mathbf{H}$ (m) at the network's nodes with the pressure heads $\varepsilon$ (m) in the network's branches:

$$\mathbf{N}^{\mathrm{T}}\mathbf{H} = \varepsilon \tag{40}$$

The analogy for truss structures is the geometrical relation between displacements and strains (cf. Equation (2) in Chapter 2). The constitutive relation for water networks relates the pressure head $\varepsilon$ with the flow $\mathbf{Q}$ in the branches (strain–force relation in trusses, cf. Equations (9) and (10) in Chapter 2):

$$\mathbf{Q}^2 = \mathbf{R}\varepsilon \tag{41}$$

The constitutive relation (41) is nonlinear. The hydraulic compliance $R_i$ is a ratio of the characteristic of a branch $K_i$ (m$^3$/s) (depending upon pipe material, diameter, filtration, etc.) to its length $l_i$ (m) as follows:

$$R_i = \frac{K_i^2}{l_i} \tag{42}$$

Nevertheless, let it temporarily be assumed that the linearity of the relation (41) holds

$$\mathbf{Q} = \mathbf{R}\varepsilon \tag{43}$$

Substituting Equations. (43) and (40) into (39), the following formula, analogous to the linear elastic force–displacement equilibrium condition in mechanics (1) in Chapter 2, can be obtained:

$$\mathbf{N}\mathbf{R}\mathbf{N}^{\mathrm{T}}\mathbf{H} = \mathbf{q} \tag{44}$$

For the water network shown in Figure 3.39, assuming one inlet at node 1 and one outlet at node 4, the set of equations (44) takes the following form:

$$\begin{bmatrix} R_1 + R_2 & -R_1 & -R_2 & 0 \\ -R_1 & R_1 + R_3 + R_4 & -R_4 & -R_3 \\ -R_2 & -R_4 & R_2 + R_4 + R_5 & -R_5 \\ 0 & -R_3 & -R_5 & R_3 + R_5 \end{bmatrix} \begin{bmatrix} H_1 \\ H_2 \\ H_3 \\ H_4 \end{bmatrix} = \begin{bmatrix} q_1 \\ 0 \\ 0 \\ -q_4 \end{bmatrix} \tag{45}$$

It is assumed that the network is supplied only through node 1 with the inlet intensity $q_1$. The flow coefficients $R'_K$ for fictitious branches, varying in the range between 0 and 1, are used to determine the degree of opening of the outlet valves (index $K$ runs through the potential outlet nodes only). The intensity $q_4$ of the single outlet at node 4 can be expressed as

$$q_4 = R'_4(H_4 - H_0) \tag{46}$$

where the flow coefficient $R'_4$ is equal to 1 (with the outlet valve fully open). The coefficients $R'_2$ and $R'_3$ are equal to 0, which means that the outlet valves at nodes 2 and 3 are closed. Note that the inlet/outlet balance must be met (an analogy to the external equilibrium condition for truss structures, which means that external loads have to be equilibrated by reaction forces at supports), i.e. $\sum_K q_K = 0$ (index $K$ runs over all nodes). In this case the condition simply yields $q_1 = q_4$.

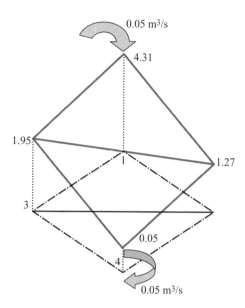

**Figure 3.40**   Water head distribution in the original network

Substituting Equations (46) to (45), assuming the reference value $H_0 = 0$ and rearranging the set of equations, gives the following system:

$$
\begin{bmatrix}
R_1 + R_2 & -R_1 & -R_2 & 0 \\
-R_1 & R_1 + R_3 + R_4 & -R_4 & -R_3 \\
-R_2 & -R_4 & R_2 + R_4 + R_5 & -R_5 \\
0 & -R_3 & -R_5 & R_3 + R_5 + R'_4
\end{bmatrix}
\begin{bmatrix}
H_1 \\ H_2 \\ H_3 \\ H_4
\end{bmatrix}
=
\begin{bmatrix}
q_1 \\ 0 \\ 0 \\ 0
\end{bmatrix}
\tag{47}
$$

For the network depicted in Figure 3.39, assuming the following data: $K_1 = 0.2\,\text{m}^3/\text{s}$, $K_2 = K_3 = K_4 = K_5 = 0.4\,\text{m}^3/\text{s}$, $l_1 = l_2 = l_3 = l_5 = 10.000\,\text{m}$, $l_4 = 14.142\,\text{m}$, $q = 0.050\,\text{m}^3/\text{s}$, and $H_0 = 0.000\,\text{m}$, the following water head distribution is obtained (see Figure 3.40):

$$
\mathbf{H}^{\mathbf{L}} = [4.312,\ 1.279,\ 1.946,\ 0.050]^{\mathrm{T}}
\tag{48}
$$

Calculating the pressure head from Equation (40) gives

$$
\varepsilon^{\mathbf{L}} = [3.033,\ 2.366,\ 1.229,\ 0.667,\ 1.896]^{\mathrm{T}}
\tag{49}
$$

### 3.4.3  VDM-Based Simulation of Parameter Modification

Local modification of a network parameter can be introduced into the system through the *virtual distortion* $\varepsilon^0$, incorporated into the formula (44) by analogy to Equation (13) in Chapter 2:

$$
\mathbf{NR}(\mathbf{N}^{\mathrm{T}}\mathbf{H} - \varepsilon^0) = \mathbf{q}
\tag{50}
$$

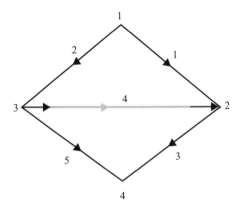

**Figure 3.41**    Distortion simulating water flow (pressure head modification) in branch 4

The virtual distortion $\varepsilon^0$ is of the same character as the pressure head $\varepsilon$ (see Figure 3.41) and is therefore able to modify it. The influence of virtual distortions on the resultant flow redistribution can be calculated making use of the *influence matrix* $D^H_{Kj}$, collecting $j$ responses (columnwise) in terms of water heads $H^{@\varepsilon^0=1}_K$, induced in the network by imposing the unit virtual distortion $\varepsilon^0_j$, generated consecutively in each branch $j$ of the network. Thus each *influence vector* $H^{@\varepsilon^0=1}_K$ can be calculated on the basis of the following equation obtained from Equation (50):

$$\mathbf{NRN^T H}^{@\varepsilon^0=1} = \mathbf{q}^* + \mathbf{NRI} \tag{51}$$

The vector $\mathbf{q}^*$ disregards the external inlet (the flow is now provided by the imposition of virtual distortion), but it accounts for the outlets in the network (cf. Equation (45)) as its physics must not be changed. There are $j$ sets of equations (51) to be solved in order to create the full influence matrix $D^H_{Kj}$. Each time the right-hand side changes as the unit virtual distortion is applied to another branch. In practice this can be realized by applying a pair of water heads corresponding to each branch (cf. Equation (40)) – it is the *compensative charge*.

The parameter modification in the system is accounted for by superposing the *linear response* of the original network and the *residual response* due to imposition of the virtual distortion. Therefore the resultant water head distribution can be expressed as

$$H_K = H^L_K + H^R_K = H^L_K + D^H_{Kj}\varepsilon^0_j \tag{52}$$

and the resultant water flow (cf. Equations (40) and (43)) as

$$Q_i = Q^L_i + Q^R_i = Q^L_i + R_{\underline{i}}N^T_{\underline{i}K}D^H_{Kj}\varepsilon^0_j \tag{53}$$

Coming back to the example shown in Figure 3.39, the unit virtual distortion is generated in branch 4, connecting nodes 2 and 3. The corresponding set of equations (51), with the vector $q^* = [0,\ 0,\ 0,\ -q_4]^T$ moved to the left-hand side (cf. Equation (47)), takes the following form:

$$\begin{bmatrix} R_1 + R_2 & -R_1 & -R_2 & 0 \\ -R_1 & R_1 + R_3 + R_4 & -R_4 & -R_3 \\ -R_2 & -R_4 & R_2 + R_4 + R_5 & -R_5 \\ 0 & -R_3 & -R_5 & R_3 + R_5 + R_4' \end{bmatrix} \begin{bmatrix} H_1^{@\varepsilon^0=1} \\ H_2^{@\varepsilon^0=1} \\ H_3^{@\varepsilon^0=1} \\ H_4^{@\varepsilon^0=1} \end{bmatrix}$$

$$= \begin{bmatrix} 0 \\ -R_4\varepsilon_4^0 \\ R_4\varepsilon_4^0 \\ 0 \end{bmatrix} \tag{54}$$

where $\varepsilon_4^0 = 1$. The resulting distribution of water heads:

$$\mathbf{H}^{@\varepsilon^0=1} = [0.151, \ -0.251, \ 0.251, \ 0.000]^{\mathrm{T}} \tag{55}$$

constitutes the fourth column of the influence matrix $D_{Kj}^{\mathrm{H}}$. Continuing this procedure for virtual distortions generated in other branches, the full influence matrix can be determined as

$$D_{Kj}^{\mathrm{H}} = \begin{bmatrix} 0.243 & 0.757 & 0.393 & 0.151 & 0.607 \\ -0.071 & 0.071 & 0.678 & -0.251 & 0.322 \\ 0.071 & -0.071 & 0.322 & 0.251 & 0.678 \\ 0.000 & 0.000 & 0.000 & 0.000 & 0.000 \end{bmatrix} \tag{56}$$

Note that the matrix $D_{Kj}^{\mathrm{H}}$ has zero values at node 4 (fourth row) where the outlet is assumed. In general the sum of components, corresponding to the outlet nodes in each column of the matrix $D_{Kj}^{\mathrm{H}}$, equals zero. Alternatively, taking into account relation (40) and applying it consecutively to each influence vector $H_K^{@\varepsilon^0=1}$, another influence matrix $D_{ij}^{\varepsilon}$ can be created, collecting the response to unit virtual distortions in terms of the pressure head $\varepsilon_i^{@\varepsilon^0=1}$:

$$D_{ij}^{\varepsilon} = \begin{bmatrix} 0.314 & 0.686 & -0.284 & 0.402 & 0.284 \\ 0.172 & 0.828 & 0.071 & -0.101 & -0.071 \\ -0.071 & 0.071 & 0.678 & -0.251 & 0.322 \\ 0.142 & -0.142 & -0.355 & 0.503 & 0.355 \\ 0.071 & -0.071 & 0.322 & 0.251 & 0.678 \end{bmatrix} \tag{57}$$

The following relations (cf. Equations (8) and (9) in Chapter 2) are valid:

$$\varepsilon_i = \varepsilon_i^{\mathrm{L}} + \varepsilon_i^{\mathrm{R}} = \varepsilon_i^{\mathrm{L}} + D_{ij}^{\varepsilon}\varepsilon_j^0 \tag{58}$$

$$Q_i = Q_i^{\mathrm{L}} + Q_i^{\mathrm{R}} = Q_i^{\mathrm{L}} + R_i D_{ij}^{\varepsilon}\varepsilon_j^0 \tag{59}$$

Note that the influence matrices $D_{Kj}^{\mathrm{H}}$, $D_{ij}^{\varepsilon}$ are related via the transposed incidence matrix $N_{iK}$ (cf. Equation (40)):

$$D_{ij}^{\varepsilon} = N_{iK} D_{Kj}^{\mathrm{H}} \tag{60}$$

### 3.4.3.1 Simulation of Network Parameter Remodeling

First, it will be demonstrated how the virtual distortion generated in branch 4 can simulate the network modification consisting in total blocking of the flow in this branch. To this end, the condition of zero flow should be postulated in the branch under remodeling ($Q_4 = 0$). Making use of the formula (59) or (58) gives, respectively:

$$Q_4 = Q_4^L + R_4 D_{44}^\varepsilon \varepsilon_4^0 = 0 \quad \text{or} \quad \varepsilon_4^L + D_{44}^\varepsilon \varepsilon_4^0 = 0 \tag{61}$$

which leads to

$$\varepsilon_4^0 = -\frac{\varepsilon_4^L}{D_{44}^\varepsilon} = -1.340 \tag{62}$$

Multiplying the network response $H_K^{@\varepsilon^0=1}$ due to imposition of the unit virtual distortion $\varepsilon_4^0 = 1$, i.e. the fourth column $D_{K4}^H$ of the matrix $D_{Kj}^H$ (cf. Equation (56)) by the above-determined value $\varepsilon_4^0 = -1.340$ (cf. Equation (62)), the residual response is determined (see Figure 3.42(b)). Superposing the correction over the water head distribution in the original network (see Figure 3.42(a)), the resultant water head distribution for the modified network $H_K = H_K^L - 1.340 D_{K4}$ is determined (see Figure 3.42(c)). Similarly, other types of network modifications can be simulated through virtual distortions using the once-determined influence matrix $\mathbf{D}^H$ (or $\mathbf{D}^\varepsilon$) and the linear response $\mathbf{H}^L$.

### 3.4.3.2 Simulation of Constitutive Nonlinearity

Nonlinearity of constitutive relations (cf. Equation (41)) can also be simulated through virtual distortions. To this end, it will be assumed that this relation is approximated through a piecewise linear function, e.g. composed of two pieces (see Figure 3.43). The algorithm for nonlinear analysis of water networks is analogous in this case to the progressive collapse analysis of elastoplastic truss structures (cf. Reference [55]), where the sequence of overloaded (i.e. exceeding the yield stress limit due to increasing load intensity) elements should be determined and the corresponding sequence of 'growing' sets of linear equations should be solved.

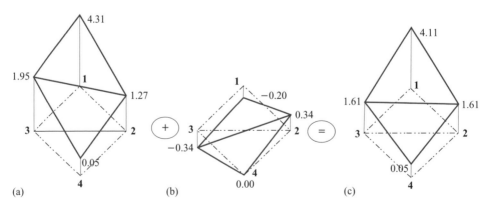

**Figure 3.42**  Pressure head distributions for the (a) original, (b) residual, (c) modified network

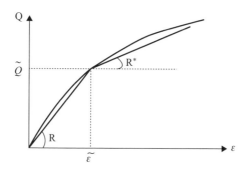

**Figure 3.43**   Piecewise linear approximation of the nonlinear constitutive relation

By analogy to structural mechanics (cf. Equation (22) in Chapter 2), the conditions for simulation of nonlinear behaviour of network branches (see Figure 3.43) take the following form:

$$Q_i - \tilde{Q}_i = R_i^*(\varepsilon_{\underline{i}} - \tilde{\varepsilon}_{\underline{i}}) \tag{63}$$

Denoting $\gamma_i = R_i^*/R_{\underline{i}}$ and substituting Equations (58) and (59) into (63), the following set of linear equations is obtained:

$$(1 - \gamma_{\underline{k}})D_{\underline{k}l}^\varepsilon \beta_l^0 = -(1 - \gamma_{\underline{k}})(\varepsilon_k^L - \tilde{\varepsilon}_k) \tag{64}$$

The set (64) should be solved with respect to the unknown virtual distortions denoted here by $\beta_l^0$ and the indices $k, l$ run over the branches of nonlinear characteristics only.

As an example, a nonlinear analysis will be performed of the network, shown in Figure 3.39. It is assumed that $\varepsilon_k > \tilde{\varepsilon}_k$, e.g. $\tilde{\varepsilon}_k = 0.8 \max(\varepsilon_k^L) = 2.426$ and $\gamma_k = 0.1$. Values of the distortions generated by solving Equation (64) are $\beta_1^0 = 1.212$ and $\beta_2^0 = 0.525$. The values are positive since the pressure heads in branches 1 and 2 are also positive (cf. Equation (49)). The distortion $\beta_k^0$ and the pressure head $\varepsilon_k$, referring to the same network branch, are always consistent in sign. Thus for the network shown in Figure 3.39, the residual water head equals

$$\mathbf{H}^R = [0.692, -0.049, 0.049, 0.000]^T \tag{65}$$

Therefore the water head distribution accounting for the nonlinear constitutive relation is the sum of Equations (48) and (65), and yields

$$\mathbf{H} = [5.004, 1.230, 1.995, 0.050]^T \tag{66}$$

and the corresponding pressure head distribution (cf. Equation (49)) yields

$$\varepsilon = [3.773, 3.010, 1.181, 0.763, 1.944]^T \tag{67}$$

A check can be made to ensure that the relations (50) are satisfied.

### 3.4.4 Leakage Identification

#### 3.4.4.1 Modeling of Leakage by Virtual Distortions

Apart from parameter modifications (see Section 3.4.3), the applied virtual distortion can also simulate leakage from the water network and therefore it can be useful in identification of this type of defect. For example, a leakage from branch 4 of the testing network (see Figure 3.39) can be modeled through the solution of the following problem (cf. Equation (54)):

$$
\begin{bmatrix}
R_1 + R_2 & -R_1 & -R_2 & 0 \\
-R_1 & R_1 + R_3 + R_4 & -R_4 & -R_3 \\
-R_2 & -R_4 & R_2 + R_4 + R_5 & -R_5 \\
0 & -R_3 & -R_5 & R_3 + R_5 + R_4'
\end{bmatrix}
\begin{bmatrix}
H_1^{@\varepsilon^0=1} \\
H_2^{@\varepsilon^0=1} \\
H_3^{@\varepsilon^0=1} \\
H_4^{@\varepsilon^0=1}
\end{bmatrix}
$$
$$
=
\begin{bmatrix}
0 \\
R_4\varepsilon_4^0 \\
R_4\varepsilon_4^0 \\
0
\end{bmatrix}
\tag{68}
$$

where $\varepsilon_4^0 = 1$. Note that the applied distortion (see Figure 3.44) is now supposed to model the leakage of water from branch 4 (not the flow through the branch as in Section 3.4.3).

The modeling is realized by applying the water heads of positive sign to nodes 2 and 3 (contrary to the case of distortion simulating water flow, in which the signs coincide with the incidence matrix $\mathbf{N}$). The resulting distribution of potentials yields:

$$
\mathbf{H}^{\mathbf{d}@\varepsilon^0=1} = [0.730,\ 0.730,\ 0.730,\ 0.023]^{\mathrm{T}}
\tag{69}
$$

Applying the same procedure to other branches, the following influence matrix $\mathbf{D}^{\mathrm{Hd}}$ (cf. Equation (56)), modeling the leakage, can be determined:

$$
D_{Kj}^{\mathrm{Hd}} =
\begin{bmatrix}
0.447 & 2.003 & 0.425 & 0.730 & 0.639 \\
0.276 & 0.748 & 0.710 & 0.730 & 0.354 \\
0.240 & 1.316 & 0.354 & 0.730 & 0.710 \\
0.008 & 0.032 & 0.032 & 0.023 & 0.032
\end{bmatrix}
\tag{70}
$$

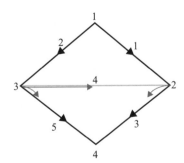

**Figure 3.44**  Distortion simulating leakage in branch 4

Note that with the direction of flow assumed in branch 4 in Figure 3.44 (inward to the element), all values of the matrix $\mathbf{D}^{\mathrm{Hd}}$ are always positive as the flow generated in this way creates a state of superpressure in the network. If the opposite direction of flow (outward of the element) is assumed, the network will exhibit the state of subpressure and all values of the matrix $\mathbf{D}^{\mathrm{Hd}}$ will be negative.

### 3.4.4.2 Leakage Identification – Linear Constitutive Relation

The leakage identification problem will now be formulated for a network governed by the linear constitutive relation (43). Assuming a leakage in branch 4 of the network shown in Figure 3.39, this situation will first be modeled by adding node 5 in the middle of the former branch 4 (see Figure 3.45).

The corresponding flow distribution can be described by the following set of equations (cf. Equation (47)):

$$
\begin{bmatrix}
R_1+R_2 & -R_1 & -R_2 & 0 & 0 \\
-R_1 & R_1+R_3+R_4 & 0 & -R_3 & -R_4 \\
-R_2 & 0 & R_2+R_5+R_6 & -R_5 & -R_6 \\
0 & -R_3 & -R_5 & R_3+R_5+R_4' & 0 \\
0 & -R_4 & -R_6 & 0 & R_4+R_6+R_5'
\end{bmatrix}
\begin{bmatrix}
H_1 \\ H_2 \\ H_3 \\ H_4 \\ H_5
\end{bmatrix}
=
\begin{bmatrix}
q_1 \\ 0 \\ 0 \\ 0 \\ 0
\end{bmatrix}
\tag{71}
$$

where the outlet intensities $q_4$ and $q_5$ incorporated into the system matrix are expressed by:

$$
q_4 = R_4'(H_4 - H_0) \quad \text{and} \quad q_5 = R_5'(H_5 - H_0)
\tag{72}
$$

It has been assumed that the only inlet is applied at node 1, while two outlets at nodes 4 and 5 are determined by the flow coefficient values $R_4' = 1$ and $R_5' = 1$. Assuming all other data as in Section 3.4.2, the set of equations (71) leads to the following solution in terms of water heads (cf. Equation (48)):

$$
\mathbf{H}^{\mathrm{M}} = [3.373,\ 0.340,\ 1.006,\ 0.021]^{\mathrm{T}}
\tag{73}
$$

and the following solution in terms of pressure heads (cf. Equation (49)):

$$
\varepsilon^{\mathrm{M}} = [3.033,\ 2.367,\ 0.319,\ 0.666,\ 0.985]^{\mathrm{T}}
\tag{74}
$$

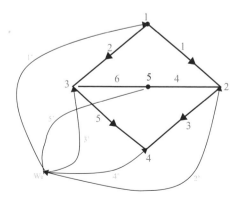

**Figure 3.45** Network subject to one leakage modeled by inclusion of node 5 in the middle of the branch

where only the first four nodes of the network shown in Figure 3.45 were considered (the combined pressure heads in branches 4 and 6) in order to match the layout of the original network, presented in Figure 3.39. In this way it is assumed that the responses supposed to model measurements are known in every node of the original network.

It can be demonstrated that the result (73) can be obtained from the linear water head $H_K^L$ (cf. Equation (48)) by adding the following residual vector $H_K^{Rd} = -1.288 D_{K4}^{Hd}$ (cf. Equation (70)):

$$\mathbf{H}^{Rd} = [-0.939, \ -0.939, \ -0.939, \ -0.029]^T \tag{75}$$

Therefore, the solution of the set of equations (71), reduced to the layout of the original network, can be described in general as the solution of the following optimization problem:

$$\min F = \min \left( \frac{H_K - H_K^M}{H_K^M} \right)^2 \tag{76}$$

subject to

$$H_K = H_K^L + H_K^{Rd} = H_K^L + D_{Kj}^{Hd} \varepsilon_j^0 \geq 0 \tag{77}$$

and

$$\varepsilon_j^0 \leq 0 \tag{78}$$

where

$\mathbf{H}^M$ = water head values measured in experiment or simulated numerically
$\mathbf{H}^L$ = linear response (water heads in original network)
$\mathbf{H}^{Rd}$ = residual response of a malfunctioning network (water heads due to virtual distortions modeling leakage)
$\mathbf{H}$ = sum of the linear and residual responses
$\varepsilon^0$ = virtual distortion modeling leakage (location and intensity)
$\mathbf{D}^{Hd}$ = influence matrix (in water heads) modeling leakage (cf. Equation 70)

The optimization problem posed by Equations (76), (77) and (78) is a classical quadratic programming problem, which may be solved by standard optimization codes, e.g. the routine TOLMIN [56]. The constraints (78) stem from the fact that the matrix $\mathbf{D}^{Hd}$ is built with the sign convention shown in Figure 3.44 (inward to the element), due to which components of the matrix are always positive. Most of the constraints (78) are active in the final solution. Only negative values of the generated distortions $\varepsilon^0$ indicate leakage.

For this small example, the constraints (77) take the following form:

$$H_K = H_K^L + D_{K4}^{Hd} \varepsilon_4^0 \geq 0 \tag{79}$$

### 3.4.4.3 Leakage Identification – Nonlinear Constitutive Relation

In the case of nonlinear constitutive relations, a distinction will be made between the virtual distortion field simulating constitutive nonlinearities $\beta^0$ and leakage $\varepsilon^0$. Consequently, the water head $\mathbf{H}$ will be expressed as

$$H_K = H_K^{\mathrm{L}} + H_K^{\mathrm{R}} + H_K^{\mathrm{Rd}} = H_K^{\mathrm{L}} + D_{Kj}^{\mathrm{H}}\beta_j^0 + D_{Kj}^{\mathrm{Hd}}\varepsilon_j^0 \tag{80}$$

In the numerical algorithm the distortion $\beta^0$ has to be determined (first with $\varepsilon^0 = 0$ as a solution of the following system of equations, including only the network branches of the nonlinear constitutive characteristics (cf. Equation (64)):

$$(1 - \gamma_{\underline{i}})N_{\underline{i}K}D_{Kj}^{\mathrm{H}}\beta_j^0 = -(1 - \gamma_{\underline{i}})(\varepsilon_{\underline{i}}^{\mathrm{L}} - \tilde{\varepsilon}_{\underline{i}}) - (1 - \gamma_{\underline{i}})N_{\underline{i}K}D_{Kj}^{\mathrm{Hd}}\varepsilon_j^0 \tag{81}$$

Having determined $\beta^0$, the primary design variable $\varepsilon^0$, which models leakage in the network, can be calculated by TOLMIN. If the detected leakages have an influence on the distribution of flow in the branches of the nonlinear characteristics, the resultant $\varepsilon^0$ is substituted to Equation (81) and a new $\beta^0$ is found. The iterative process proceeds until convergence is obtained, i.e. the final $\varepsilon^0$ does not imply any variation of $\beta^0$.

#### 3.4.4.4 Sensitivity Analysis

As demonstrated previously, leakages in the water network can be simulated through a virtual distortion state. The original (linear) response $\mathbf{H}^{\mathrm{L}}$ of the network is combined with the residual response $\mathbf{H}^{\mathrm{R}}$, generated by virtual distortions, in order to match the measured response $\mathbf{H}^{\mathrm{M}}$. The components of virtual distortions (design variables) are calculated by the quadratic programming routine TOLMIN and therefore a sensitivity analysis allowing for the calculation of gradients is necessary.

Having an analytical description of water network relations, presented in Section 3.4.2, the gradient of the objective function (76) for linear constitutive relations can be calculated in the following way:

$$\frac{\partial F}{\partial \varepsilon_j^0} = \frac{2}{\left(H_K^{\mathrm{M}}\right)^2}\left(H_K^{\mathrm{L}} + D_{Kj}^{\mathrm{Hd}}\varepsilon_j^0 - H_K^{\mathrm{M}}\right)D_{Kj}^{\mathrm{Hd}} \tag{82}$$

In the case of the nonlinear constitutive relation advantage can be taken of the following chain differentiation (cf. Equation (55) in Chapter 2):

$$\frac{\partial F}{\partial \varepsilon^0} = \frac{\partial F}{\partial \mathbf{H}}\left(\frac{\partial \mathbf{H}}{\partial \varepsilon^0} + \sum_{\mathrm{plastic}}\frac{\partial \mathbf{H}}{\partial \beta^0}\frac{\partial \beta^0}{\partial \varepsilon^0}\right) \tag{83}$$

The components $\partial \mathbf{H}/\partial \varepsilon^0$ and $\partial \mathbf{H}/\partial \beta^0$ are determined from Equation (80), whereas the component $\partial \beta^0/\partial \varepsilon^0$ is calculated from Equation (81).

### 3.4.5 Numerical Examples

#### 3.4.5.1 Example 1

Assume now that the water head distribution $\mathbf{H}^{\mathrm{M}}$ (cf. Equation (73)) has been measured in every node of the water network shown in Figure 3.39, which may be subject to leakages. Location as well as intensity of the leakages needs to be identified, assuming that they may be located in any branch of the network and the constitutive relation is linear. Employing the gradient-based (cf. Equation (82)) TOLMIN routine, the minimization problem of Equations (76), (77) and (78) with five unknowns $\varepsilon_i^0$, $i = 1, ..., 5$, leads to the following solution: $\varepsilon_1^0 = \varepsilon_2^0 = \varepsilon_3^0 = \varepsilon_5^0 = 0$ and $\varepsilon_4^0 = -1.288$. This means that the proposed procedure is able to identify the location (branch 4) as well as the intensity of the leakage (distortion value), making use of the water head distribution $\mathbf{H}^M$ measured in every node of the network.

The distortion value $\varepsilon_4^0 = -1.288$ m, referring to the original network (see Figure 3.39), does not translate directly into the leakage intensity $q_5 = 0.029$ m$^3$/s, referring to node 5 of the network, which is supposed to model the malfunctioning state (see Figure 3.45). However, once the distortion $\varepsilon_4^0$ has been determined, it is known that the leakage occurs in branch 4. By substituting the distortion $\varepsilon_4^0$ to the flow balance equations (50) the unbalanced right-hand side vector $q_K$ is obtained, i.e. $\sum_K q_K \neq 0$. In the case of just one leakage, this indicates immediately how intensive the leakage is. In the case of many simultaneous leakages, the unbalanced vector $q_K$ provides information about the overall loss of water from the network, but the branches in which leakages occur (see Example 2) are exactly known. By modeling the leakage through a node placed in the middle of a damaged branch and solving a system of equations analogous to Equation (71), the leakage intensity in each defective branch can be determined exactly in terms of $q_K$.

Note that there is a subtlety in checking the flow balance equations (50) with the distortion $\varepsilon_4^0$, modeling leakage. When considering the equilibrium of node 3 (see Figure 3.39), the calculated value $\varepsilon_4^0 = -1.288$ must be considered, whereas the equilibrium of node 2 requires the opposite value $\varepsilon_4^0 = 1.288$. This is due to the fact that the assumed flow in branch 4, from node 3 to node 2 (see Figure 3.39), differs in direction from the flow assumed when creating the influence matrix $\mathbf{D}^{Hd}$, modelling leakage (cf. Figures 3.41 and 3.44).

The progress of the optimization process for five unknown virtual distortions is illustrated in Figure 3.46. The nonzero virtual distortion $\varepsilon_4^0$, reached in several iterations, indicates leakage in branch 4. The combination of linear and residual responses producing the response with leakage is demonstrated in Figure 3.47.

For the nonlinear constitutive relation assuming the level $\bar{\varepsilon}_i = 0.8 \max(\varepsilon_i^L) = 2.426$, the linear state $\mathbf{H}^L$ (cf. Equation (48)) can simply be combined with the residual states $\mathbf{H}^R$ (cf. Equation (65)) and $\mathbf{H}^{Rd}$ (cf. Equation (75)) in order to model the leakage in branch 4 of the original network. This example is rather exceptional because the leakage distortion $\varepsilon_4^0$ does not affect pressure heads in branches 1 and 2 (cf. Equations (49) and (74)) which are affected by distortions $\beta_1^0$ and $\beta_2^0$, modeling constitutive nonlinearity. Thus the measured state of water heads (referring to the original layout) in this case yields (cf. Equations (66) and (73))

$$\mathbf{H}^M = [4.065, \ 0.291, \ 1.055, \ 0.021]^T \tag{84}$$

and the state of pressure heads yields (cf. Equations (67), and (74))

$$\varepsilon^M = [3.774, \ 3.010, \ 0.270, \ 0.764, \ 1.034]^T \tag{85}$$

With the previously assumed data, the nonlinear constitutive relation and the measured state (73), the optimization program TOLMIN identifies the leakage in branch 4 ($\varepsilon_4^0 = -1.288$), similarly to the linear case. The flow balance equations (50) are satisfied here as well. The previous comment in this section on the sign of the distortion $\varepsilon_4^0$ for the linear constitutive law is also valid.

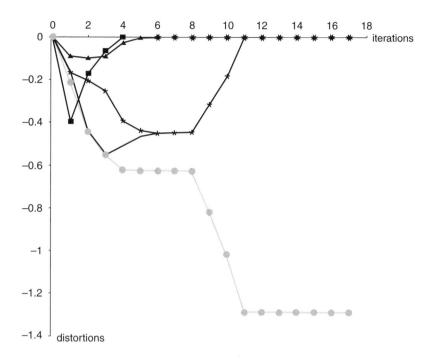

**Figure 3.46**  Development of distortions $\varepsilon^0$ for the two-loop network

### 3.4.5.2 Example 2

Another four-loop network in its original state is shown in Figure 3.48. The inlet is provided at node 1 and the outlet at node 6. Assuming the following data: $K_{1-9} = 0.5\,\mathrm{m^3/s}$, $l_1 = l_2 = l_4 = l_5 = l_6 = l_8 = l_9 = 10.000\,\mathrm{m}$, $l_3 = l_7 = 14.142\,\mathrm{m}$, $q_1 = 0.050\,\mathrm{m^3/s}$, $H_0 = 0.000\,\mathrm{m}$, the following linear response for the original network in terms of the water head is obtained:

$$\mathbf{H}^{\mathrm{L}} = [2.789,\ 1.897,\ 1.681,\ 1.158,\ 0.942,\ 0.050]^{\mathrm{T}} \tag{86}$$

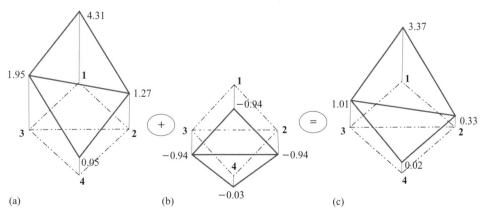

**Figure 3.47**  Water head distributions for (a) the original, (b) locally distorted and (c) damaged two-loop network

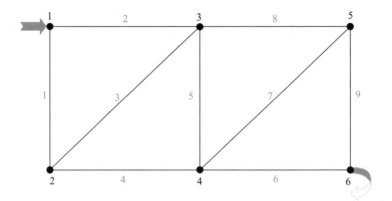

**Figure 3.48**    Four-loop network in the original configuration

and in terms of the pressure head:

$$\varepsilon^{L} = [0.892,\ 1.108,\ 0.216,\ 0.739,\ 0.522,\ 1.108,\ 0.216,\ 0.739,\ 0.892]^{T}. \qquad (87)$$

Assuming the nonlinear constitutive relation ($\tilde{\varepsilon}_i = 0.9$, $\gamma_i = 0.1$) for the original network, it turns out that four branches of the network exhibit nonlinearity ($\beta_1^0 = 0.762$, $\beta_2^0 = 1.038$, $\beta_6^0 = 1.038$, $\beta_9^0 = 0.762$). The solution in terms of the water head yields

$$\mathbf{H} = [4.619,\ 2.871,\ 2.566,\ 2.103,\ 1.797,\ 0.050]^{T} \qquad (88)$$

and in terms of the pressure head yields

$$\varepsilon = [1.747,\ 2.053,\ 0.306,\ 0.769,\ 0.463,\ 2.053,\ 0.306,\ 0.769,\ 1.747]^{T} \qquad (89)$$

In order to model a malfunctioning network with two simultaneous leakages, the configuration shown in Figure 3.49 was adopted. Assuming the flow coefficients $R_7' = 1$ and $R_8' = 1$,

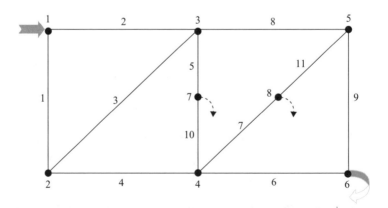

**Figure 3.49**    Four-loop network in the damaged configuration

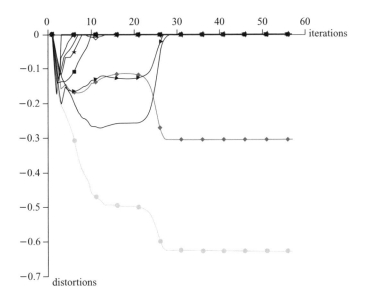

**Figure 3.50**   Development of distortions $\varepsilon^0$ for the four-loop network

the following response in terms of the water head is obtained:

$$\mathbf{H}^{M} = [1.670, \ 0.816, \ 0.525, \ 0.166, \ 0.160, \ 0.008]^{T} \tag{90}$$

and in terms of the pressure head (both referring to the original configuration):

$$\varepsilon^{M} = [0.855, \ 1.145, \ 0.291, \ 0.649, \ 0.358, \ 0.159, \ 0.006, \ 0.364, \ 0.153]^{T}. \tag{91}$$

Based on the measured state (90), TOLMIN generates the following distortions: $\varepsilon_5^0 = -0.628$, $\varepsilon_7^0 = -0.305$, modeling leakages. The development of the distortions in the optimization process is shown in Figure 3.50.

Assuming the nonlinear constitutive relation ($\tilde{\varepsilon}_i = 0.9$, $\gamma_i = 0.1$) for the malfunctioning network, it turns out that only two branches (compared to four branches for the original network) exhibit nonlinearity ($\beta_1^0 = 0.727$, $\beta_2^0 = 1.073$), which is due to the leakages. The measured state in terms of the water head yields

$$\mathbf{H}^{M} = [2.595, \ 0.888, \ 0.503, \ 0.180, \ 0.154, \ 0.008]^{T} \tag{92}$$

and in terms of the pressure head (both referring to original network), it yields

$$\varepsilon^{M} = [1.707, \ 2.093, \ 0.385, \ 0.708, \ 0.323, \ 0.172, \ 0.026, \ 0.349, \ 0.146]^{T} \tag{93}$$

TOLMIN returns two values of the distortions modelling leakage $\varepsilon_5^0 = -0.620$, $\varepsilon_7^0 = -0.312$. They are different from the corresponding distortions for the linear case because: (i) the number of branches exhibiting nonlinear behavior is different for original (four branches) and malfunctioning networks (two branches), and (2) the leakages influence the pressure head state in all branches of the network (cf. Equations (87) and (91)), contrary to Example 1. Nevertheless the relations (50) are always satisfied.

In both examples the computational time was negligible and the objective function value dropped to numerical zero ($1.0 \times 10^{-14}$), so the identification result is extremely accurate.

### 3.4.6 Conclusions

The section proposes a numerical approach for closed-loop water networks based on the virtual distortion method and system theory. At the start of the numerical simulation, the influence matrix is built. Then the analysis of flow distribution in the network reduces to linear combinations of the components of the matrix and virtual distortions (design variables). Unlike the Cross iterative method, the presented approach determines the flow distribution in one step. Various modifications introduced to water networks may be modeled by the presented approach, e.g. network remodeling (change of flow in selected branches). In particular, a state of network malfunctioning due to leakage may be effectively diagnosed on the basis of *in situ* measurements of water heads at network nodes. The presented approach assumes that the analysed network exhibits a steady state of water flow. Transient behavior will be investigated in the future.

It has been demonstrated that having a numerical model of the water network (with the pre-calculated influence matrix $\mathbf{D}^{Hd}$) and knowing the current state of the system, i.e. the nodal water head distribution $\mathbf{H}^M$ measured *in situ*, the locations of possible simultaneous leakages in network branches and their intensities can be determined. The VDM-based sensitivity analysis enables a gradient-based optimization algorithm to be employed to solve the leakage identification problem. As for location of a leakage, with the currently formulated approach just the defective branch can be determined. The authors also have an idea for precise location of leakage along the branch (the subject of future research). Leakage intensity is fully determined by the distortion $\varepsilon^0$ value (cf. the comment in Example 1 in Section 3.4.5). All results presented in the section were obtained for numerically simulated responses $\mathbf{H}^M$, so no uncertainties of measurements were accounted for. The approach has been proved to work for the noise-free data. The experimental verification has just been initiated and will be continued.

It has been assumed that the current water head is measured in every node of the network, but leakage identification is also possible by making measurements in selected nodes only. It seems that a minimum number of measurements should be specified in order to ascertain good convergence of the optimization algorithm (cf. Reference [57]). This will be a crucial issue for large networks, in which the inspection zone must be naturally limited. The investigation of the problem will be the subject of future research.

Linear constitutive relation has been assumed in the first formulation. Then, the nonlinearity of the pressure head/flow constitutive relation has been taken into account through superposition of the following two virtual distortion fields: the first one $\beta^0$, modeling physical nonlinearities and being a solution of the set of equations (64), and the second one $\varepsilon^0$, describing leakages and being a solution of the optimization problem of Equations (76), (77) and (78).

## 3.5 Damage Identification in Electrical Circuits

### 3.5.1 Introduction

In this section, an adaptation of the *virtual distortion method* to the analysis and diagnostics of electrical circuits is presented. Similarly to the case of water networks (Section 3.4), the implementation of VDM concepts is based on the system of analogies with plain truss structures (see Table 2.3). A motivation for introducing the VDM concepts is to model and identify changes of conductance in elements of electrical circuits, which is analogous to modeling and identification of stiffness parameters in truss members (see Section 3.2). The presented approach can be applied to the diagnostics and fault detection in linear time-invariant analog circuits, both in the steady state as well as dynamic cases (direct current (DC) [58], alternating current (AC) and transient analysis in the discrete-time domain, based on numerical integration of the circuit equations). In-depth reviews of various methods commonly used in the diagnostics of analog circuits can be found in References [59] and [60].

For the sake of consistency, the term *(virtual) distortion* is introduced to describe a certain input function, which models modifications of circuit parameters. However, it should not be confused with the same term used to describe the change of circuit response caused by a single-frequency input signal of small magnitude (*distortion analysis*).

### 3.5.2 *Modeling of Electrical Circuits and Analogies to Truss Structures*

In terms of topological compatibility, the form of global governing equations and overall characteristics of behavior, models of truss structures are a close mechanical equivalent of electrical circuits consisting of bulk passive elements and independent sources. On a very basic level, a foundation for defining the mutual equivalence of physical quantities and parameters is the concept of electromechanical inverse analogy [61]. Accordingly, a simple 1DOF mass–spring–damper system (Figure 3.51) has an electrical counterpart in the form of a parallel connection of a resistor, capacitor and inductor. The behavior of both systems is described by integral-differential equations of the same form:

$$f(t) = m\frac{dv(t)}{dt} + cv(t) + k\int v(t)dt \tag{94a}$$

$$i(t) = C\frac{du(t)}{dt} + \frac{1}{R}u(t) + \frac{1}{L}\int u(t)dt \tag{94b}$$

indicating a direct correlation between external force $f(t) \leftrightarrow$ current source $i(t)$ as inputs and velocity $v(t) \leftrightarrow$ voltage $u(t)$ as responses.

However, with regard to truss members and electrical components, especially in the steady state of response, it is more convenient to use the different system of analogies: the state of currents remains compatible with the field of internal forces, but the state of voltages is associated with the state of deformations (not velocities). As a result, a direct analogy is obtained between constitutive relations (Hooke's versus Ohm's law) and global governing equations (equilibrium of forces and continuity of deformations versus Kirchhoff's laws). Excitations by external forces and initial deformations are analogous with the power supply by current and voltage sources. Equivalent nodal quantities are the displacements of truss

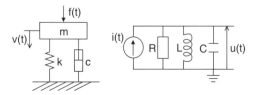

**Figure 3.51**   Electromechanical inverse analogy

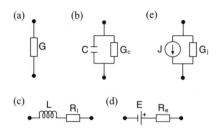

**Figure 3.52**   Models of electrical components

nodes and electrical potentials. There is no mechanical equivalent for inductance and electrical equivalent for mass, but those parameters are not directly related to the VDM procedures to be used.

The following analysis will be carried out with regard to electrical circuits of arbitrary topology and the distribution of elements. Basic passive elements and sources will be treated as independent components, represented by simple two-terminal models of real elements. The following models (Figure 3.52) will be taken into consideration:

(a) Resistor of conductance $G$
(b) Capacitor of capacitance $C$ and parallel conductance $G_c$
(c) Coil of inductance $L$ and serial resistance $R_l$
(d) Voltage source of electromotive force $E$ and inner resistance $R_e$
(e) Current source of intensity $J$ and inner conductance $G_j$

The proposed models of electrical components enable a residuary drop of voltage on a coil, current leakage through a capacitor and changes of nominal values of sources due to their inner resistance to be taken into account. Models of ideal components are obtained when assuming $G_c = R_l = 0$, $G_j$ and $R_e = 0$. In an arbitrary state of the circuit, instantaneous values of current and voltage on element terminals are described by the following relations:

$$\text{Resistor:} \qquad i(t) = -G\,u(t) \tag{95a}$$

$$\text{Capacitor:} \qquad i(t) = -C\frac{du}{dt} - G_c\,u(t) \tag{95b}$$

$$\text{Coil:} \qquad u(t) = -L\frac{di}{dt} - R_l\,i(t) \tag{95c}$$

$$\text{Voltage source:} \qquad u(t) = E(t) - R_e\,i(t) \tag{95d}$$

$$\text{Current source:} \qquad i(t) = J(t) - G_j\,u(t) \tag{95e}$$

Signs in the above relations result from the assumed conventional flow notation, where directions of current and polarity of voltage are opposite (the arrow of voltage points towards higher potential). The topology of circuit connections will be described by making use of an oriented graph, where every passive element of the circuit is represented by an edge of fixed orientation linking two separate nodes of the graph. Orientation of the edge establishes a reference for the direction of current flow and polarity of voltage for the element. To describe mutual connections between nodes and edges of a circuit graph, an incidence matrix $\mathbf{N}$ (cf. Equation (38)) will be used. Entry of the matrix $N_{Kj} = 1$ denotes that the $j$th edge enters the $K$th node, $N_{Kj} = -1$ denotes that the edge leaves the node and $N_{Kj} = 0$ states that the given edge and node are disjoint. The incidence matrix enables the relation between voltages on elements and potentials in nodes to be formulated:

$$\mathbf{u}(t) = \mathbf{N}^T \mathbf{v}(t) \tag{96}$$

as well as a relation for a summation of currents in nodes:

$$\mathbf{N}\,\mathbf{i}(t) = \sum_{node} i(t) \tag{97}$$

The following procedure for assembling the system of circuit equations is based on the modified nodal analysis (MNA) [62]. In a general outline, the MNA is very similar to the finite element method (FEM) formulation for truss structures. Using the assumed system of analogies and FEM-based terminology, models of passive components can be regarded and handled as 'electrical finite elements', sources as 'external loads' and certain predefined configurations of the circuit (grounding, short-circuits) as 'boundary conditions'. A global system of circuit equations is formulated with regard to unknown nodal potentials, in accordance with Kirchhoff's current law.

### 3.5.2.1 DC Analysis

Relation (97), applied to all elements and sources, produces a system of equations for global equilibrium of currents in circuit nodes:

$$\mathbf{N}_r\,\mathbf{i}_r + \mathbf{N}_c\,\mathbf{i}_c + \mathbf{N}_l\,\mathbf{i}_l + \mathbf{N}_e\,\mathbf{i}_e + \mathbf{N}_j\,\mathbf{i}_j = \mathbf{0}$$

Currents can then be substituted by voltages in accordance with relations (95), and subsequently voltages can be expressed as functions of nodal potentials (Equation (96)). Taking into account that in the steady-state DC analysis the ideal capacitor acts as a break ($i_c = 0$) and the ideal coil acts as a short-circuit ($u_l = 0$), the following relation is obtained:

$$\mathbf{G}^*\,\mathbf{v} - \mathbf{N}_l\,\mathbf{i}_l - \mathbf{N}_e\,\mathbf{i}_e = \mathbf{N}_j\,\mathbf{J}\,,$$

where

$$\mathbf{G}^* = \mathbf{N}_r\,\mathbf{G}\,\mathbf{N}_r^T + \mathbf{N}_c\,\mathbf{G}_c\,\mathbf{N}_c^T + \mathbf{N}_j\,\mathbf{G}_j\,\mathbf{N}_j^T$$

In the case of coils and ideal voltage sources, currents cannot be expressed directly as functions of potentials. Hence they are treated as additional unknown variables and relations (95c) and (95d) have to be used as additional equations (boundary conditions). Ultimately, the following system of equations can be obtained:

$$\begin{bmatrix} \mathbf{G}^* & -\mathbf{N}_l & -\mathbf{N}_e \\ \mathbf{N}_l^T & \mathbf{R}_l & \mathbf{0} \\ \mathbf{N}_e^T & \mathbf{0} & \mathbf{R}_e \end{bmatrix} \begin{bmatrix} \mathbf{v} \\ \mathbf{i}_l \\ \mathbf{i}_e \end{bmatrix} = \begin{bmatrix} \mathbf{N}_j\,\mathbf{J} \\ \mathbf{0} \\ \mathbf{E} \end{bmatrix}$$

The system can be expressed in a more compact form as

$$\mathbf{K}\,\mathbf{x} = \mathbf{z} \tag{98}$$

Because of the fact that the number of independent Kirchhoff's current laws is less by one than the number of nodes, at least one node has to be grounded in order to obtain a reference for values of potentials (and nonsingularity of the matrix $\mathbf{K}$). To implement the condition $v_i = 0$, all entries in the $i$th row and column of the matrix $\mathbf{K}$ have to be set to zero, except $K_{ii} = 1$, and all entries of $z_i = 0$. After imposing boundary conditions on the nodes, Equation (98) can be solved using standard routines and algorithms for linear systems of equations.

### 3.5.2.2 AC Analysis

Steady-state AC analysis is based on a symbolic method making use of linear algebra in the complex numbers domain. According to Euler's formula, a harmonic signal of amplitude $U$, angular frequency $\omega$ and phase $\varphi$ can be expressed in the following form:

$$u(t) = U \, \sin(\omega t + \varphi) = \mathrm{Im}(U \, e^{i\varphi} \, e^{i\omega t})$$

The quantity $\underline{U} = U \, e^{i\varphi}$ is defined as a complex amplitude. It presents a temporary state of the signal in a time instant $t=0$, where the modulus and argument of the complex number are associated with the amplitude and phase of the signal. In the steady state of response, the time factor $e^{i\omega t}$ is the same for all considered signals (currents, voltages, potentials); hence it can be neglected and all relations between the signals can be written for complex amplitudes. Constitutive relations for reactive elements are described by the following equations:

$$\underline{I_c} = (i\omega C + G_c) \, \underline{U_c} = \underline{Y_c} \, \underline{U_c}$$
$$\underline{U_l} = (i\omega L + R_l) \, \underline{I_l} = \underline{I_l} / \underline{Y_l}$$

The global system of equations can be written in the following form:

$$\underline{\mathbf{Y}} \, \underline{\mathbf{x}} = \underline{\mathbf{z}} \tag{99}$$

where

$$\underline{\mathbf{Y}} = \begin{bmatrix} \mathbf{Y}^* & -\mathbf{N}_e \\ \mathbf{N}_e^T & \mathbf{R}_e \end{bmatrix}; \qquad \underline{\mathbf{x}} = \begin{bmatrix} \mathbf{v} \\ \mathbf{i}_e \end{bmatrix}; \qquad \underline{\mathbf{z}} = \begin{bmatrix} \mathbf{N}_j \, \mathbf{J} \\ \mathbf{E} \end{bmatrix}$$

and

$$\mathbf{Y}^* = \mathbf{N}_r \, \mathbf{G} \, \mathbf{N}_r^T + \mathbf{N}_c \, \mathbf{Y}_c \, \mathbf{N}_c^T + \mathbf{N}_l \, \mathbf{Y}_l \, \mathbf{N}_l^T + \mathbf{N}_j \, \mathbf{G}_j \, \mathbf{N}_j^T .$$

Boundary conditions on grounded nodes are imposed in the same manner as in the DC case. The solution to the system (99) can be obtained by using standard routines for linear systems of equations in the complex number domain.

### 3.5.2.3 Transient Analysis

A global system of circuit equations for the dynamic analysis is formulated in such a way as to obtain the system of second-order differential equations, which is analogous to the equations of motion describing truss models. The procedure for assembling matrices is also based on the MNA, but it requires that in-between nodes need to be introduced for real coils. Voltage sources are aggregated in the same manner as in the steady-state cases (i.e. by additional boundary conditions with unknown currents through sources). As a result the following system of equations is obtained:

$$\tilde{\mathbf{C}}\,\ddot{\mathbf{y}}(t) + \tilde{\mathbf{G}}\,\dot{\mathbf{y}}(t) + \tilde{\mathbf{L}}\,\mathbf{y}(t) = \mathbf{z}(t) \tag{100}$$

where

$$\dot{\mathbf{y}}(t) = \begin{bmatrix} \mathbf{v}^+(t) \\ \mathbf{i}_e(t) \end{bmatrix}; \qquad \mathbf{z}(t) = \begin{bmatrix} \hat{\mathbf{N}}_j\,\mathbf{J}(t) \\ \mathbf{E}(t) \end{bmatrix};$$

$$\tilde{\mathbf{C}} = \begin{bmatrix} \mathbf{C}^* & 0 \\ 0 & 0 \end{bmatrix}; \qquad \tilde{\mathbf{G}} = \begin{bmatrix} \mathbf{G}^* & -\hat{\mathbf{N}}_e \\ \hat{\mathbf{N}}_e^T & \mathbf{R}_e \end{bmatrix}; \qquad \tilde{\mathbf{L}} = \begin{bmatrix} \mathbf{L}^* & 0 \\ 0 & 0 \end{bmatrix}$$

and

$$\mathbf{G}^* = \hat{\mathbf{N}}_r\,\mathbf{G}\,\hat{\mathbf{N}}_r^T + \hat{\mathbf{N}}_c\,\mathbf{G}_c\,\hat{\mathbf{N}}_c^T + \hat{\mathbf{N}}_j\,\mathbf{G}_j\,\hat{\mathbf{N}}_j^T + \hat{\mathbf{N}}_{rl}\,\frac{1}{\mathbf{R}_l}\,\hat{\mathbf{N}}_{rl}^T;$$

$$\mathbf{C}^* = \hat{\mathbf{N}}_c\,\mathbf{C}\,\hat{\mathbf{N}}_c^T; \qquad \mathbf{L}^* = \hat{\mathbf{N}}_l\,\frac{1}{\mathbf{L}}\,\hat{\mathbf{N}}_l^T$$

The vector of nodal potentials $\mathbf{v}^+(t)$ is assembled for standard and additional nodes. Incidence matrices for all elements besides coils ($\hat{\mathbf{N}}_r$, $\hat{\mathbf{N}}_c$, $\hat{\mathbf{N}}_j$, $\hat{\mathbf{N}}_e$) are expanded by inserting zeros in entries associated with additional nodes. The incidence matrix for the model of a real coil is expanded and transformed separately for the coil ($\hat{\mathbf{N}}_l$) and its resistance ($\hat{\mathbf{N}}_{rl}$). Nodal boundary conditions are imposed in the same manner as in the DC case (all entries in columns and rows of matrices $\tilde{\mathbf{C}}$, $\tilde{\mathbf{G}}$, $\tilde{\mathbf{L}}$ and excitation vector $\mathbf{z}(t)$, associated with the $K$th grounded node, are set to zero, except $G_{KK} = 1$).

The solution to the system (100), for the uniformly defined discrete time domain, is obtained by utilizing the Newmark algorithm. The circuit response to the impulse excitation (a crucial issue for VDM procedures) can be computed by setting nonzero values of the excitation vector only for the first time step of the Newmark algorithm.

### 3.5.3  VDM Formulation

Making use of analogies with truss structures, concepts of the *virtual distortion method* can be easily and intuitively adapted to the analysis of simple linear electrical circuits. Basic notions and definitions of the method can be introduced in the following way:

**Linear response** is an arbitrary circuit response (potentials in nodes, currents or voltages on elements), generated by real sources in the circuit described by the reference set of parameters.

**Modified response** is a response, generated by real sources in the circuit described by the set of modified parameters.

**Virtual distortions** are, conceptually, certain additional input functions that alter the system response in the same way as modifications of structural parameters. They can be interpreted as additional 'virtual' sources, coupled with elements where modifications are modeled.

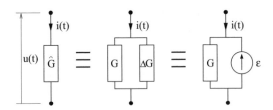

**Figure 3.53**    Modeling of conductance modification by distortion

**Residual response** is a balanced state of response induced by distortions in the initial, passive
    circuit (i.e. with real sources discarded).
**Modeled response** is a superposition of linear and residual responses, assumed to be equal
    with the modified response.
**Influence matrix D**, similarly as for truss structures, is a stationary predefined matrix, which
    stores selected system responses to certain elementary states of distortions imposed on
    individual elements of the circuit.

It is assumed that distortions are associated with modifications of conductance in resistive
elements of the circuit. The remaining parameters and configurations of the circuit are considered to be invariant. Taking into account forms of global circuit equations, the most convenient
way to implement distortions is to treat them as ideal current sources inserted in parallel with
elements where modifications are modeled. In analogy with truss models, inserting a current
source is equivalent to imposing a pair of nodal forces that realize distortion (initial strain) in
the truss member. Figure 3.53 presents the interpretation of distortion as an equivalent to the
modification of conductance.

Relations between voltage and current in the selected resistive element in modified and
modeled circuits can be written as follows:

Modified element: $\qquad -\hat{G}_i\, u_i(t) = i_i(t)$

Modeled element: $\qquad -G_i\, u_i(t) = i_i(t) - \varepsilon_i^0(t)$

where $\hat{G}_i$ depicts the modified value of conductance and $\varepsilon_i^0(t)$ is a distortion modeling that
modification. The form of distortion complies with the type of conducted analysis. It is associated with the constant source in the DC case, the harmonic source in the AC case (of the same
frequency as real sources) and is time-dependent in the transient state. From the assumption
that responses in the modified and modeled elements are compatible, distortion must fulfill the
following condition:

$$\varepsilon_i^0(t) = (G_i - \hat{G}_i)\, u_i(t) = (1 - \mu_i)\, G_i\, u_i(t) \tag{101}$$

where the parameter of conductance modification $\mu_i$ (cf. Equation (17) in Chapter 2) is defined
as

$$\mu_i = \frac{\hat{G}_i}{G_i}$$

Relation (101) between the distortion and modification parameter is not direct because the modeled voltage response $u_i(t)$ generally depends also on other distortions imposed on the circuit (local modifications of conductance have an influence on global responses of the circuit). To obtain a direct relation, the modeled response of the circuit has to be expressed as a superposition of linear and residual responses, where the residual response is calculated as a linear combination of distortions, making use of the influence matrix.

### 3.5.3.1 Steady-State Case

Distortions can be implemented in the model of the circuit in the same way as real current sources. Having a vector of distortions $\varepsilon^0$, which has the same number of entries as the number of resistors, and a corresponding incidence matrix $\mathbf{N}_{\varepsilon^0} = \mathbf{N}_r$, the global vector of excitation for the modeled circuit in the DC case can be expressed in the following way:

$$\hat{\mathbf{z}} = \mathbf{z} + \mathbf{z}(\varepsilon^0) = \begin{bmatrix} \mathbf{N}_j \mathbf{J} \\ 0 \\ \mathbf{E} \end{bmatrix} + \begin{bmatrix} \mathbf{N}_{\varepsilon^0} \, \varepsilon^0 \\ 0 \\ 0 \end{bmatrix}$$

For the AC case, modification of the vector $\mathbf{z}$ can be introduced similarly. Inserting the modified vector $\hat{\mathbf{z}}$ into the main system of Equations (98), a solution is obtained as a sum of the linear response (generated by real sources) and the residual response (induced by distortions):

$$\mathbf{x} = \mathbf{K}^{-1}\mathbf{z} + \mathbf{K}^{-1}\mathbf{z}(\varepsilon^0) = \mathbf{x}^L + \mathbf{x}^R$$

Limiting the response to nodal potentials only, the following relation can be written:

$$\mathbf{v} = \mathbf{v}^L + \mathbf{D}^v \varepsilon^0 \tag{102}$$

Matrix $\mathbf{D}^v$ will be called a *potential influence matrix*. An entry $D^v_{Kj}$ stores the value of potential induced in the $K$th node by the *unit virtual distortion* imposed on the $j$th element. In the DC case, unit distortion means a constant current source of unit intensity, and a harmonic source of unit amplitude and zero phase in the AC case.

The potential linear response and potential influence matrix are the only quantities that have to be calculated from the system of circuit equations. Other system responses (voltages and currents) can be calculated from potentials using Equation (96) and constitutive relations (with initial values of parameters). They can also be expressed as a combination of the linear and residual responses. The modeled voltage response is described by the relation:

$$u_i = u_i^L + D^u_{ij}\varepsilon^0_j \tag{103}$$

where the voltage influence matrix is calculated as

$$\mathbf{D}^u = \mathbf{N}_r^T \mathbf{D}^v \tag{104}$$

The direct relation between distortions $\varepsilon^0$ and parameters of conductance modifications $\mu$ can now be derived by inserting (103) into (101). As a result, the following system of equations can be obtained:

$$A^0_{ij}\,\varepsilon^0_j = b_i \tag{105}$$

where

$$A^0_{ij} = \delta_{ij} - (1 - \mu_{\underline{i}})\,G_{\underline{i}}\,D^u_{\underline{i}j}$$
$$b_i = (1 - \mu_{\underline{i}})\,G_{\underline{i}}\,u^L_{\underline{i}}$$

The system of equations (105) enables the global vector of distortions, corresponding to a given vector of modifications, to be calculated (with a linear voltage response and voltage influence matrix pre-computed). To reduce the dimension of the system, it can be assembled only for modified elements of the circuit. If $\mu_i = 1$ (no modification of conductance) then $A^0_{ij} = \delta_{ij}$, $b_i = 0$ and immediately $\varepsilon^0_i = 0$.

### 3.5.3.2 Dynamic Case

According to Equation (101) in the transient state of the circuit, modifications of conductance are modeled by time-dependent distortions. Analogously to the truss structures in dynamics (see Section 2.5), the analysis will be conducted in a homogeneously discretized time domain. Distortions will be approximated by the sequence of impulses occurring in the defined time instants and the overall modeled response of the circuit will be calculated as a superposition of impulse responses, making use of the *impulse influence matrix*. The modeled voltage response can be described by the following relation:

$$u_i(t) = u^L_i(t) + \sum_{\tau=0}^{t} D^u_{ij}(t - \tau)\,\varepsilon^0_j(\tau) \tag{106}$$

$\mathbf{D}^u(t)$ is a three-dimensional matrix, where the component $D^u_{ij}(\tau)$ stores the value of voltage in the $i$th element in the time instant $t = \tau$, induced by the unit impulse distortion imposed on the $j$th element in the time instant $t = 0$. Columns of the impulse influence matrix can be calculated making use of the procedure for calculation of the impulse response.

A direct relation between distortions and modification parameters in a transient state can be derived by inserting Equation (106) into (101). For every discrete-time instant, the following system of equations can be obtained:

$$A^0_{ij}\,\varepsilon^0_j(t) = b_i(t) \tag{107}$$

where

$$A^0_{ij} = \delta_{ij} - (1 - \mu_{\underline{i}})\,G_{\underline{i}}\,D^u_{\underline{i}j}(0)$$
$$b_i(0) = (1 - \mu_{\underline{i}})\,G_{\underline{i}}\,u^L_{\underline{i}}(0)$$
$$b_i(t > 0) = (1 - \mu_{\underline{i}})\,G_{\underline{i}}\left[ u^L_{\underline{i}} + \sum_{\tau=0}^{t-1} D^u_{\underline{i}j}(t - \tau)\varepsilon^0_j(\tau) \right]$$

The system of equations (107) must be solved sequentially for every time instant starting from $t = 0$. The main matrix $\mathbf{A}^0$ is time-invariant; only vector $\mathbf{b}$ needs to be updated using values of distortions calculated in the previous time steps. Similarly to the steady-state cases, the system of equations can be assembled only for the modified elements of the circuit.

### 3.5.3.3 Sensitivity Analysis

Making use of relations between the modeled system response, distortions and modification parameters, the virtual distortion method enables the gradient of circuit responses (sensitivity analysis) to be calculated directly and accurately. In the steady states, the modeled response of the circuit can be described as follows:

$$\mathbf{f} = \mathbf{f}^{\mathrm{L}} + \mathbf{D}^f \, \varepsilon^0 \tag{108}$$

The vector $\mathbf{f}$ stores arbitrary, various circuit responses (potentials, voltages or currents) and the general influence matrix $\mathbf{D}^f$ is assembled for those selected responses. The partial derivative of the modeled response $f_i$ with respect to the modification parameter $\mu_k$ can be calculated using the chain rule:

$$\frac{\partial f_i}{\partial \mu_k} = \frac{\partial f_i}{\partial \varepsilon_j^0} \frac{\partial \varepsilon_j^0}{\partial \mu_k} \tag{109}$$

The modelled response is a linear function of distortions. Hence its partial derivative with respect to distortion equals to the corresponding entry of the influence matrix. The problem then reduces to finding the components of the *gradient of distortions*:

$$\frac{\partial f_i}{\partial \mu_k} = D_{ij}^f \frac{\partial \varepsilon_j^0}{\partial \mu_k} \tag{110}$$

After differentiation of both sides of relation (105) with respect to the selected modification parameter, the following system of equations can be obtained:

$$A_{ij}^0 \frac{\partial \varepsilon_j^0}{\partial \mu_k} = B_{ik} \tag{111}$$

where the matrix $A_{ij}^0$ is the same matrix as in the Equation (105) used to calculate distortions and $B_{ik}$ is a diagonal matrix, dependent on the modeled voltage response:

$$B_{ik} = -\delta_{ik} \, G_{\underline{i}} \, u_{\underline{i}}$$

To find the gradient of selected circuit responses with respect to the modification parameter $\mu_k$, in the given space of parameters $\boldsymbol{\mu}$, the following procedure needs to be applied:

*Initial computations*:

- Calculate the influence matrix for the selected responses $D_{ij}^f$.
- Calculate the voltage influence matrix $D_{ij}^u$.

***For the given vector μ :***

1. Calculate the modeled voltage responses $u_i$.
2. Solve the system of equations (111) to find the gradient of distortions.
3. Calculate the gradient of responses from Equation (110).

A similar procedure for obtaining the gradient of circuit responses can be applied in the case of transient analysis. An arbitrary modeled response can be described by the relation

$$f_i(t) = f_i^L(t) + \sum_{\tau=0}^{t} D_{ij}^f(t - \tau)\,\varepsilon_j^0(\tau) \tag{112}$$

Using the chain rule and knowing that the partial derivative of the response with respect to the selected distortion is equal to an appropriate component of the impulse influence matrix, the gradient of the circuit response with respect to the selected modification parameter can be described as follows:

$$\frac{\partial f_i(t)}{\partial \mu_k} = \sum_{\tau=0}^{t} D_{ij}^f(t - \tau)\,\frac{\partial \varepsilon_j^0(\tau)}{\partial \mu_k} \tag{113}$$

As a result of differentiation of Equation (107), for every time instant, the following system of equations is obtained:

$$A_{ij}^0 \frac{\partial \varepsilon_j^0(t)}{\partial \mu_k} = B_{ik}(t) \tag{114}$$

where

$$A_{ij}^0 = \delta_{ij} - (1 - \mu_{\underline{i}})\, G_{\underline{i}}\, D_{\underline{i}j}^u(0)$$
$$B_{ik}(0) = -\delta_{ik}\, G_{\underline{i}}\, u_{\underline{i}}^L(0)$$
$$B_{ik}(t > 0) = -\delta_{ik}\, G_{\underline{i}}\, u_{\underline{i}}(t) + (1 - \mu_{\underline{i}})\, G_{\underline{i}} \sum_{\tau=0}^{t-1} D_{\underline{i}j}^u(t - \tau)\frac{\partial \varepsilon_j^0(\tau)}{\partial \mu_k}$$

System (114) must be calculated sequentially for every time instant.

### 3.5.4 Defect Identification

In this section, the focus is on the defect identification problem, formulated as a search for modifications of conductance in resistive elements of electrical circuits. Number, distribution and magnitude of defects are assumed arbitrarily, in the range from short-circuits ($G \rightarrow \infty$) to breaks ($G \rightarrow 0$). Input data include the numerical model of the circuit and a certain set of responses (reference points), obtained from the modified configuration of the circuit. Using the VDM methodology, the problem of defect identification can be formulated as a search for the equivalent state of distortions, which generate the same responses in reference points as those obtained from the modified circuit. Distortions are sought over a certain set of possible locations, which supposedly contains all unknown locations of defects. In the most general case, the set of distortion locations includes all resistive elements of the circuit.

### 3.5.4.1 Defect Identification in the Steady State

In the steady-state cases, the problem can be solved by transforming Equation (108), in which modeled responses are substituted by the responses from reference points:

$$\mathbf{D}^f \varepsilon^0 = \mathbf{f}^{\mathrm{ref}} - \mathbf{f}^{\mathrm{L}} \tag{115}$$

Obviously, to obtain a unique solution, the influence matrix $\mathbf{D}^f$, assembled for the selected reference points and distortion locations, needs to be square and nonsingular. This means that the number of independent reference responses cannot be lower than the assumed number of distortion locations. The mutual independence of responses is determined by distortion locations and Kirchhoff's laws. In a circuit consisting of $n$ elements and $K$ nodes, there are at most $(K-1)$ independent current Kirchhoff's laws and $(n - K + 1)$ voltage Kirchhoff's laws. When all resistive elements are considered as possible defect locations, in order to obtain a nonsingular influence matrix, the set of reference points must include a current reference in every independent loop of the circuit and voltage reference in all but one node. With every element excluded from the set of distortion locations, the number of unknowns is reduced and one reference response becomes redundant.

An alternative approach uses iterative methods based on gradient optimization. Although more computationally demanding, they enable constraints to be imposed. The evolution of optimized parameters can be actively controlled, ensuring more reliable results (when reference responses are perturbed by sloppy measurement or noise). In some cases a reduction in the number of necessary reference responses is possible. The proposed procedure, valid for both the DC and AC cases, is based on the steepest-descent method, with the objective function defined as the least square problem and distortions chosen as optimization variables.

The vector of distance functions $\mathbf{d}$ describes differences between the modeled responses (defined by the actual state of distortions) and responses obtained from the reference points:

$$\mathbf{d} = \mathbf{f}(\varepsilon^0) - \mathbf{f}^{\mathrm{ref}} \tag{116}$$

Components of the gradient of distance functions with respect to distortions (complex derivatives in the AC case) are equal to entries of the influence matrix:

$$\nabla \mathbf{d} = \frac{\partial d_i}{\partial \varepsilon_j^0} = D_{ij}^f \tag{117}$$

The objective function $g$ is defined as

$$g = d_i d_i = (\mathbf{d})^{\mathrm{H}} \mathbf{d} \tag{118}$$

where $(\cdot)^{\mathrm{H}}$ denotes the conjugate transpose. Function $g$ is a real-valued function of real or complex arguments. With respect to the objective function, the direction of steepest descent $\mathbf{h}$ is defined as

$$\mathbf{h} = 2 \, [\nabla \mathbf{d}]^{\mathrm{H}} \mathbf{d} \tag{119}$$

In the DC case (real distortions), $\mathbf{h}$ corresponds to the gradient of objective function $\nabla g$. In the AC case (complex distortions), $\mathbf{h}$ is equal to the directional derivative and formally cannot be associated with the gradient of $g$, since the objective function takes only real values and has no derivatives with respect to complex distortions. In every iteration of the steepest-descent method, distortions are updated according to the following formula:

$$\varepsilon^{0(p+1)} = \varepsilon^{0(p)} - \lambda^{(p)} \mathbf{h} \qquad (120)$$

where $p$ denotes the number of iteration steps and $\lambda^{(p)}$ is a nonnegative factor that normalizes and scales the value of distortion updating. An approximation procedure for $\lambda^{(p)}$ is

$$\lambda^{(p)} = \frac{\|[\nabla \mathbf{d}]^H \mathbf{d}\|^2}{2 \|\nabla \mathbf{d} [\nabla \mathbf{d}]^H \mathbf{d}\|^2} \qquad (121)$$

Constraints are usually defined with respect to the modification parameter $\mu$. Constraints include physical conditions ($\mu_k > 0$) but can also be defined in relation to the specific type of defects. In the case of breaks $\mu_k \in [0; 1]$ and in the case of short-circuits $\mu_k \geqslant 1$. To impose constraints, a corresponding vector of modification parameters needs to be calculated after every distortion updating.

The gradient-based optimization method also usually demands that the number of distortion locations should not be larger than the number of independent reference responses. This ensures that a generation of certain false configurations of distortions during the optimization process is restrained. Generally, when the mentioned condition is not fulfilled, the vector of distortions obtained from the optimization procedure can consist of three components:

$$\varepsilon^0 = \varepsilon^0(\mu) + \varepsilon^{0(i=0)} + \varepsilon^{0(u=0)} \qquad (122)$$

where $\varepsilon^0(\mu)$ represents the appropriate solution associated with the modeled modifications. The remaining components are called impotent states of distortions because they generate only a voltage ($\varepsilon^{0(i=0)}$) or a current ($\varepsilon^{0(u=0)}$) response. Impotent states occur only in the specific configurations: $\varepsilon^{0(u=0)}$ in all elements of a loop, while $\varepsilon^{0(i=0)}$ in all elements connected with a node.

### 3.5.4.2 Defect Identification in Dynamics

The procedure of defect identification in the transient state is also based on the steepest-descent method, but the gradient of the objective function will be calculated with respect to the modification parameters $\mu$. The vector of the distance functions is defined for all discrete time instants:

$$\mathbf{d}(t) = \mathbf{f}(\mu, t) - \mathbf{f}^{\text{ref}}(t) \qquad (123)$$

The objective function $g$ is defined as a sum of squares of all distance functions:

$$g = \sum_t d_i(t) d_i(t) \qquad (124)$$

The gradient of the objective function with respect to the modification parameters $\mu_k$ can be obtained from the following relation:

$$\nabla \mathbf{g} = \frac{\partial g}{\partial \mu_k} = 2 \sum_t d_i(t) \frac{\partial f_i(t)}{\partial \mu_k} \tag{125}$$

The procedure for calculation of the gradient of response with respect to the modification parameters was presented in the previous section. In every iteration of the optimization procedure, the vector of the modification parameters is updated according to the following formula:

$$\boldsymbol{\mu}^{(p+1)} = \boldsymbol{\mu}^{(p)} - \lambda^{(p)} \nabla \mathbf{g} \tag{126}$$

where $\lambda^{(p)}$ is a non negative factor. The full procedure of defect identification in dynamics is presented below:

*Initial computations*:

- Linear responses and the impulse influence matrix for reference points are to be used for updating the modeled responses and calculation of the gradient of the objective function.
- Linear voltage responses and the voltage impulse influence matrix for all elements included in the set of distortion locations are to be used for updating the distortions and calculation of the gradient of the distortions.

*In every iteration*, for the actual vector of modifications $\boldsymbol{\mu}^{(p)}$, compute:

- the vector of the distortions using Equation (107);
- the modeled response in reference points using Equation (112);
- the distance functions using Equation (123);
- the gradient of the distortions using Equation (114);
- the gradient of the reference response using Equation (113);
- the gradient of the objective function using Equation (125);
- the updated value of the modification vector $\boldsymbol{\mu}^{(p+1)}$ using Equation (126).

### 3.5.5 Numerical Example

The considered circuit of the symmetrical hexagonal topology is shown schematically in Figure 3.54. It consists of 12 uniform resistive elements of resistances 1 k$\Omega$ and two ideal capacitors of capacitances $C_1 = 10$ μF and $C_2 = 100$ μF. The circuit is supplied by an ideal current source of intensity 1 mA in the DC case and amplitude 1 mA, zero phase and frequency 50 Hz in the AC case.

#### 3.5.5.1 Modeling Conductance Modifications by Distortions

Changes of conductance in resistors $R_2$ and $R_6$ are now introduced, described by the modification parameters $\mu_2 = 0.1$ and $\mu_6 = 0.2$. To find the equivalent state of distortions, the linear voltage responses and voltage influence matrices need to be calculated first, using the circuit

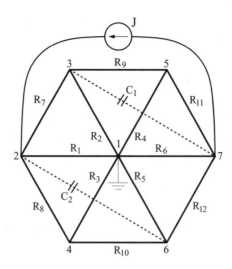

**Figure 3.54**    Numerical example of an electrical circuit

equations (98) or (99). In the DC case, the following results are obtained:

$$\begin{bmatrix} u_2^L \\ u_6^L \end{bmatrix} = \begin{bmatrix} -0.1 \\ 0.4 \end{bmatrix}; \qquad \mathbf{D}^u = \begin{bmatrix} 450 & 75 \\ 75 & 450 \end{bmatrix}$$

In the AC case,

$$\begin{bmatrix} u_2^L \\ u_6^L \end{bmatrix} = \begin{bmatrix} -0.12019 + 0.06525i \\ -0.17981 - 0.06525i \end{bmatrix}$$

$$\mathbf{D}^u = \begin{bmatrix} 293.3097 - 64.7071i & 231.6903 + 64.7071i \\ 231.6903 + 64.7071i & 293.3097 - 64.7071i \end{bmatrix}$$

The columns of influence matrices are calculated as voltage responses to unit distortions, imposed consecutively in elements $R_2$ and $R_6$. The values of distortions, modeling the introduced modifications, are calculated from Equation (105):

$$\text{DC}: \qquad \begin{bmatrix} \varepsilon_2^0 \\ \varepsilon_6^0 \end{bmatrix} = \begin{bmatrix} 0.0956 \times 10^{-3} \\ 0.4910 \times 10^{-3} \end{bmatrix}$$

$$\text{AC}: \qquad \begin{bmatrix} \varepsilon_2^0 \\ \varepsilon_6^0 \end{bmatrix} = \begin{bmatrix} (0.2226 + 0.0656i) \times 10^{-3} \\ (0.2338 - 0.0531i) \times 10^{-3} \end{bmatrix}$$

An arbitrary response of the modified circuit can now be quickly calculated as a superposition of the linear and residual responses.

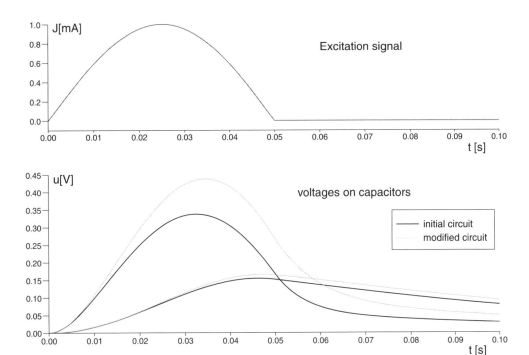

**Figure 3.55**   Responses of the analysed circuit

### 3.5.5.2 Defect Identification

Assume that the set of possible distortion locations (i.e. the potentially defective elements) includes elements $R_1$ to $R_6$. The previously introduced two modifications (see Section 3.5.5.1) – strictly speaking their location and magnitude – are now considered unknown. The voltages on both capacitors are chosen as the reference responses. Figure 3.55 presents the reference responses in the initial and modified circuits for the dynamic excitation in the form of a half sine wave. The time of analysis is 100 ms, divided into 100 time steps. The distance functions are defined for all discrete-time instants. Figure 3.56 presents the evolution of the modification parameters during the optimization process.

### 3.5.6 Conclusions

It has been demonstrated that the concept of virtual distortions, due to the system of analogies, can be easily adapted and implemented in the steady-state and dynamic analyses of simple electrical circuits. The distortions, introduced to model modifications of conductance, can be interpreted and handled as additional ideal current sources, incorporated into the elements where modifications occur. Such an approach enables global responses of a circuit to be calculated as a superposition of actual responses generated by the real sources and virtual responses generated by the distortions, using the pre-computed influence matrices. Another important advantage is the analytical calculation of gradients of circuit responses. Thus, the VDM provides effective algorithms for the circuit reanalysis, sensitivity analysis and identification of conductance modifications.

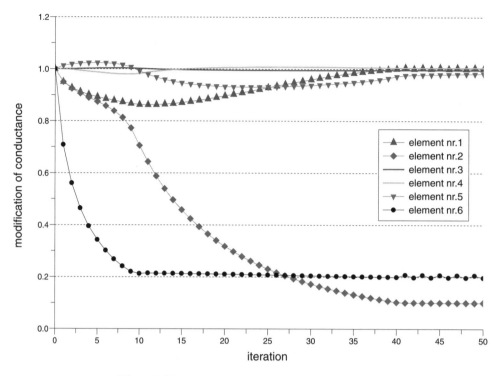

**Figure 3.56**   Evolution of the modification parameters

In this section, only simple models of RLC elements and independent sources have been taken into account for simplicity of presentation. However, the scope of application can be extended to include other types of models and components. On the one hand, formulation of the circuit equations by the MNA allows for aggregation of other linear, electrical components like dependent sources, ideal transformers, transmission lines or ideal operational amplifiers. On the other hand, the VDM formulation of plasticity allows for implementation of nonlinear characteristics, approximated by piecewise linear sections. Using an analogous approach (introducing plastic-like distortions in electrical circuits), implementation of simple models of switches or diodes can be accomplished.

## References

1. H. Sohn, C.R. Farrar, F.M. Hemez, D.D. Shunk, D.W. Stinemates and B.R. Nadler, A review of structural health monitoring literature: 1996–2001, Technical report, Los Alamos National Laboratory Report LA-13976-MS, 2003.
2. G. Gautschi, *Piezoelectic Sensorics*, Springer, Berlin, 2002.
3. E. Udd, *Fibre Optic Smart Structures*, John Wiley & Sons, Inc., New York, 1995.
4. B. Peeters, J. Maeck and G. De Roeck, Vibration-based damage detection in civil engineering: Excitation sources and temperature effects, *Smart Materials and Structures*, **10**, 2001, 518–527.
5. A.V. Oppenheim and R.W. Schafer, *Discrete-Time Signal Processing*, PTR Prentice-Hall, Englewood Cliffs, New Jersey, 1999.
6. S. Mallat, *A Wavelet Tour of Signal Processing*, Academic Press, San Diego, 1999.
7. A. Rytter, *Vibration Based Inspection of Civil Engineering Structures*, PhD thesis, Department of Building Technology and Structural Engineering, Aalborg University, Denmark, 1993.

8. H. Natke and C. Cempel, *Model-Aided Diagnosis of Mechanical Systems*, Springer, Berlin, 1997.

9. http://www.ndt.net/.

10. P. Kołakowski, Structural health monitoring – a review with the emphasis on low-frequency methods. *Engineering Transactions*, IPPT-PAN, **55**(3), 2007, 239–275.

11. M.I. Friswell and J.E. Mottershead, *Finite Element Model Updating in Structural Dynamics*, Kluwer Academic Publishers, 1995.

12. J.N. Juang, *Applied System Identification*, PTR Prentice-Hall, Englewood Cliffs, New Jersey, 1994.

13. W.J. Staszewski, Structural health monitoring using guided ultrasonic waves, in *Advances in Smart Technologies in Structural Engineering (eds C.A. Mota Soares and J. Holnicki-Szulc)*, Springer, Berlin, 2003, pp. 117–162.

14. J.L. Rose, *Ultrasonic Waves in Solid Media*, Cambridge University Press, Cambridge, 1999.

15. K. Worden, G. Manson and C. Surace, Aspects of novelty detection, in *Proceedings of the 7th International Conference on Damage Assessment of Structures DAMAS'07*, Torino, Italy, 25–27 June 2007, pp. 3–16.

16. A. Deraemaeker and A. Preumont, Vibration-based damage detection using large array sensors and spatial filters, *Mechanical Systems and Signal Processing*, **20**, 2006, 1615–1630.

17. J. Maeck and G. De Roeck, Damage assessment using vibration analysis on the z24 bridge, *Mechanical Systems and Signal Processing*, **71**(1), 2003, 133–142.

18. C. Biemans, W.J. Staszewski, C. Boller and G.R. Tomlinson, Crack detection in metallic structures using broadband excitation of acousto-ultrasonics. *Journal of Intelligent Material Systems and Structures*, **12**(8), 2001, 589–597.

19. B. Peeters, *System Identification and Damage Detection in Civil Engineering*, PhD thesis, Katholieke Univeriteit Leuven, Belgium, 2000.

20. S. Vanlanduit, R. Pintelon, T. De Troyer and P. Guillaume, Operational modal analysis of time varying systems using an exogenous multi-sine excitation, in *Proceedings of the International Conference on Noise and Vibration Engineering ISMA2006*, Leuven, Belgium, 18–20 September 2006, pp. 3047–3060.

21. J. Maeck and G. De Roeck, Dynamic bending and torsion stiffness derivation from modal curvatures and torsion rates, *Journal of Sound and Vibration*, **225**(1), 1999, 153–170.

22. C.P. Fritzen and K. Bohle, Damage identification using a modal kinetic energy criterion and 'output-only' modal data – application to the Z24-bridge, in *Proceedings of the 2nd European Workshop on Structural Health Monitoring*, Munich, Germany, 7–9 July 2004, pp. 185–194.

23. E.J. Williams and A. Messina, Applications of the multiple damage location assurance criterion, in *Proceedings of the International Conference on Damage Assessment of Structures DAMAS'99*, Dublin, Ireland, 1999, pp. 256–264.

24. J.-N. Juang, *Applied System Identification*, PTR Prentice-Hall, Englewood Cliffs, New Jersey, 1994.

25. B. Peeters and G. De Roeck, Reference-based stochastic subspace identification for output-only modal analysis, *Mechanical Systems and Signal Processing*, **13**(6), 1999, 855–878.

26. R.J. Allemang and D.L. Brown, A complete review of the complex mode indicator function (CMIF) with applications, in *Proceedings of the International Conference on Noise and Vibration Engineering ISMA2006*, Leuven, Belgium, 18–20 September 2006, pp. 3209–3246.

27. B. Peeters, H. Van der Auweraer, P. Guillaume and J. Leuridan, The PolyMAX frequency-domain method: a new standard for modal parameter estimation, *Shock and Vibration*, **11**, 2004, 395–409.

28. W. Heylen, S. Lammens and P. Sas, *Modal Analysis Theory and Testing*, Celestijnenlaan 300B, B-3001, Katholieke Univeriteit Leuven, Leuven, Belgium, 1997.

29. P. Kołakowski, T.G. Zieliński and J. Holnicki-Szulc, Damage identification by the dynamic virtual distortion method, *Journal of Intelligent Material Systems and Structures*, **15**(6), 2004, 479–493.

30. A. Świercz, P. Kołakowski and J. Holnicki-Szulc, Damage identification in skeletal structures using the virtual distortion method in frequency domain, *Mechanical Systems and Signal Processing*, 2008, (to appear).

31. *EU GROWTH Project – Smart Structural Diagnostics using Piezo-Generated Elastic Waves, Piezo-diagnostics*, GRD1-2001-40589, 2002–2005, http://mice.udg.es/piezo/.

32. Y. Zhou, L. Tong and G.P. Steven, Vibration-based model-dependent damage (delamination) identification and health monitoring for composite structures – a review, *Journal of Sound and Vibration*, **230**(2), 2000, 357–378.

33. M. Meo and E. Thieulot, Delamination modelling in a double cantilever beam, *Composite Structures*, **71**, 2005, 429–434.

34. L. Iannucci, Dynamic delamination modelling using interface elements, *Computers and Structures*, **84**, 2006, 1029–1048.

35. R. De Borst and J.C.J. Remmers, Computational modelling of delamination, *Composites Science and Technology*, **66**, 2006, 713–722.

36. H.Y. Kim and W. Hwang, Effect of debonding on natural frequencies and frequency response functions of honeycomb sandwich beams, *Composite Structures*, **55**, 2002, 51–62.

37. H.C.H. Li, M. Weis, I. Herszberg and A.P. Mouritz, Damage detection in a fibre reinforced composite beam using random dechap3.3:crement signatures, *Composite Structures*, **66**, 2004, 159–167.

38. A.J. Zak, Non-linear vibration of a delaminated composite beam, *Key Engineering Materials*, **293–294**, 2005, 607–614.

39. C. Bois, P. Herzog and C. Hochard, Monitoring a delamination in a laminated composite beam using *in-situ* measurements and parametric identification, *Journal of Sound and Vibration*, **299**, 2007, 786–805.

40. S.I. Ishak, G.R. Liu, H.M. Shang and S.P. Lim, Locating and sizing of delamination in composite laminates using computational and experimental methods, *Composites Part B: Engineering*, **32**, 2001, 287–298.

41. E. Schnack, S. Dimitrov and T.A. Langhoff, Computational identification on interface delaminations in layered composites based on surface measurements, *Archive of Applied Mechanics*, **76**, 2006, 747–758.

42. N. Ramanujam, T. Nakamura and M. Urago, Identification of embedded interlaminar flaw using inverse analysis, *International Journal of Fracture*, **132**, 2005, 153–173.

43. A. Orłowska, P. Kołakowski and J. Holnicki-Szulc, Modelling and identification of delamination in double-layer beams by the virtual distortion method, *Computers and Structures*, 2008 (to appear).

44. W. Ostachowicz and A.J. Zak, Vibration of a laminated beam with a delamination including contact effects, *Shock and Vibration*, **11**, 2004, 157–171.

45. R.W. Gentry, D. Larsen and S. Ivey, Efficacy of genetic algorithm to investigate small scale aquitard leakage, *Journal of Hydraulic Engineering, ASCE*, **129**, 2003, 527–535.

46. J.P. Vitkovsky, A.R. Simpson and M.F. Lambert, Leak detection and calibration using transients and genetic algorithms, *Journal of Water Resources Planning and Management, ASCE*, **126**, 2000, 262–265.

47. W. Mpesha, S.L. Gassman and M.H. Chaudhry, Leak detection in pipes by frequency response method, *Journal of Hydraulic Engineering, ASCE*, **127**, 2001, 134–147.

48. X.-J. Wang, M.F. Lambert, A.R. Simpson, J.A. Liggett and J.P. Vitkovsky, Leak detection in pipelines using the damping of fluid transients, *Journal of Hydraulic Engineering, ASCE*, **128**, 2002, 697–711.

49. K. Vairavamoorthy and J. Lumbers, Leakage reduction in water distribution systems: optimal valve control, *Journal of Hydraulic Engineering, ASCE*, **124**, 1998, 1146–1154.

50. T. Tucciarelli, A. Criminisi and D. Termini, Leak analysis in pipeline systems by means of optimal valve regulation, *Journal of Hydraulic Engineering, ASCE*, **125**, 1999, 277–285.

51. J. Holnicki-Szulc, P. Kołakowski and N. Nasher, Leakage detection in water networks, *Journal of Intelligent Material Systems and Structures*, **16**, 2005, 207–219.

52. N.C. Lind, Analysis of structures by system theory, *Journal of Structural Division, ASCE* 88 ST2, April, 1962.

53. L.A. Rossman, EPANET – User's Manual, Water Supply and Water Resources Division, National Risk Management Research Laboratory. Technical report EPA/600/R-00/057, National Risk Management Research Laboratory, Cincinnati, Ohio, 2000.

54. H. Cross, Analysis of flow in networks of conduits or conductors, Technical report 286, University of Illinois Engineering Experiment Station Bulletin, 1936.

55. J. Holnicki-Szulc and J.T. Gierliński, *Structural Analysis, Design and Control by the Virtual Distortion Method*, John Wiley & Sons, Ltd. Chichester, 1995.
56. M.J.D. Powell, TOLMIN: a FORTRAN package for linearly constrained optimisation calculations, Report damtp/1989/na2, Technical report, University of Cambridge, Cambridge, 1989.
57. J.P. Vitkovsky, J.A. Liggett, A.R. Simpson and M.F. Lambert, Optimal measurement site locations for inverse transient analysis in pipe networks, *Journal of Water Resources Planning and Management, ASCE*, **129**, 2003, 480–492.
58. M. Kokot and J. Holnicki-Szulc, Defect identification in electrical circuits via virtual distortion method. Part I: steady-state case, *Journal of Intelligent Material Systems and Structures*, 2008 (to appear).
59. Ruey-Wen Liu (ed.), *Selected Papers on Analog Fault Diagnosis*, IEEE Press, 1987.
60. T. Ozawa (ed.), *Analog Methods for Computer-Aided Circuit Analysis and Diagnosis*, Marcel Dekker, 1988.
61. P. Gardonio and M.J. Brennan, On the origins and development of mobility and impedance methods in structural dynamics, *Journal of Sound and Vibration*, **249**, 2002, 557–573.
62. L.M. Wedepohl and L. Jackson, Modified nodal analysis: an essential addition to electrical circuit theory and analysis, *Engineering Science and Educational Journal*, **11**, 2002, 84–92.

# 4

# Dynamic Load Monitoring

Łukasz Jankowski, Krzysztof Sekuła, Bartłomiej D. Błachowski,
Marcin Wikło, and Jan Holnicki-Szulc

Load identification constitutes an important type of engineering problem. Together with the problems considered in Chapter 3, it belongs to the category of inverse problems, since its objective is to determine the reason (load) on the basis of the result (measured response). Identification of a static or a quasi-static load is usually easy to perform. However, the difficulties increase drastically in the case of a dynamic excitation.

This chapter discusses three general approaches to the problem of determination of a dynamic load acting on a structure on the basis of a locally measured response. The approaches differ in strictness of the time restrictions and are respectively aimed at real-time, on-line and off-line load identification. A potential application area for *real-time* techniques are *adaptive impact absorption* (AIA) systems, to be considered in the next chapter. Basic impact parameters (e.g. mass and velocity of the impacting object) identified in real-time, i.e. during the load, within its first milliseconds, can be used to trigger an AIA system for optimum mitigation of impact effects on the structure. However, due to rigorous time requirements, such real-time systems can be used only for well-localized loads. On the contrary, *off-line* techniques are based on post-accident analysis of a stored local structural response, and are intended for black-box type monitoring systems for efficient reconstruction of the scenario of a sudden load (impact, collision, etc.). Systems intended for *on-line* load identification are aimed at bridging the gap between real-time and off-line systems: they can be used for continuous identification and monitoring of distributed, nonlocalized loads, performed on-line, but not in real-time.

## 4.1 Real-Time Dynamic Load Identification

In dynamic load identification many difficulties tend to appear, especially when impact loads are considered. This is caused by the short duration of the phenomenon and by relatively high load values. This section focuses on the real-time impact load identification using locally installed sensors and a falling mass as the impact excitation. The main motivation is the potential possibility of applications to adaptive impact absorption (AIA) systems. The basic prerequisite for their operation is the real-time identification of the actual impact scenario, quick enough to control the energy dissipation process [1]. Contrary to the off-line identification, which usually makes use of a numerical model of the structure and requires a long computational time, the real-time impact load identification has to be performed in a few milliseconds. This is the reason why it should be based on a locally operating simple hardware device rather than on sophisticated software and a distributed sensor system. Nevertheless, taking into account the always quicker computer analysis, some methods originally used to be applicable off-line will be applicable in real-time in the near future.

Most of the research done so far does not discuss the operation time of the considered methods. Therefore, the identification techniques presented in this section are very strongly focused on minimizing the time required to apply them and to obtain the results. It is crucial to fulfill the real-time condition as the objective of the analysed methods, in order to make them applicable to AIA systems and quick enough to predict the parameters of the impact ahead of its destructive effects. A short response time is also crucial in many other safety systems: e.g. if an airbag device is considered, the impact parameters must be obtained immediately for activation. Real-time dynamic load identification techniques can also be used in dynamic scales (weigh-in motion systems), which should recognize the weights of passing vehicles without delay.

In the following, three general techniques for load identification are discussed [2]. The first is a pattern recognition concept and can be called the *solution map approach*. Conversely, the two next methods are based on well-known mechanical principles and can be called *analytical model-based approaches*. All have been developed and verified by means of a laboratory drop testing stand. The analysis considers the accuracy of the algorithms, complexity of the instrumentation and the operation time of each approach, which has been counted from the very beginning of the impact up to the moment when identification of the unknown impact parameters was feasible.

### 4.1.1 Impact Load Characteristics

Determination of the impact load requires identification of the mass and the parameters of motion of the object, which impacts the structure. Figure 4.1 shows schematically how the impacting mass and velocity affect the impacted structure. The curves mark different impact scenarios with the same impact kinetic energy. Two characteristic areas can be determined in the mass–velocity domain:

- *Fast dynamics* (area 1) corresponds to small mass impacting with high velocity and results in local structural deformations and damages in the vicinity of the impact point.
- *Slow dynamics* (area 2) corresponds to heavy mass impacting with low velocity, which causes quasi-static structural deformation. In the case of the cantilever shown in Figure 4.1, the overloaded parts are in the vicinity of the supports.

It is assumed further on that impact load identification means identification of two parameters: (a) mass of the falling object and (b) impact velocity (or kinetic energy). Real-time determination of these two parameters allows further development of the impact scenario to be predicted and optimum structural adaptation to be triggered.

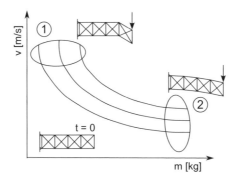

**Figure 4.1** Influence of falling mass and impact velocity on structural response

## 4.1.2 Solution Map Approach

The objective of identification is to determine the impact parameters (impacting mass and its initial velocity) that cause the measured local structural response. Therefore, it can be based on a pre-fetched map

$$
\begin{aligned}
Y_1 &= Y_1(x_1, x_2, \ldots, x_n) \\
Y_2 &= Y_2(x_1, x_2, \ldots, x_n) \\
&\ldots \\
Y_m &= Y_m(x_1, x_2, \ldots, x_n)
\end{aligned}
\tag{1}
$$

of structural responses $Y_1, Y_2, \ldots, Y_m$ (amplitude of excitation, period, etc.) to various impact scenarios defined by the parameters $x_1, x_2, \ldots, x_n$ (mass, velocity, acceleration, amplitude, etc.). The number $n$ of parameters to be identified is limited by the number $m$ of responses: $n \leq m$. The map can be generated via a calibrated numerical model of the structure or by performing experimental tests. In general, one of the main concerns is that the inverse problem can have no uniquely defined solution [3].

The proposed approach leads to an optimization problem, since the best solution for the considered measurement can be determined from the existing solution map. It can be obtained by minimization of the discrepancy between the actually measured and the stored responses $Y_i$. With the normalized least squares discrepancy measure, the objective function to be minimized takes the form

$$
L(x_1, x_2, \ldots, x_n) = \sum_{i=1}^{m} \left[ \frac{Y_i^{\mathrm{M}} - Y_i(x_1, x_2, \ldots, x_n)}{Y_i^{\mathrm{M}}} \right]^2
$$

where $Y_i^{\mathrm{M}}$ denotes the values actually measured.

## 4.1.3 Approach Based on Force and Acceleration

Consider the structure impacted by a falling mass (Figure 4.2). The equation of motion for the falling mass during the impact is

$$
F(t) = mg - m\ddot{u}
$$

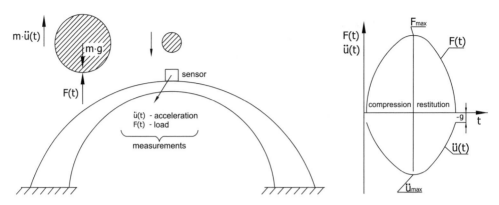

**Figure 4.2**  Force-acceleration approach: (left) schema of structure and falling mass; (right) signals measured during contact of falling mass with structure

Hence the mass can be determined as

$$m = \frac{F(t)}{g - \ddot{u}} \qquad (2)$$

Therefore, the *mass* identification is possible if measurements from two sensors (acceleration and force) are available. The other parameter required for the full impact identification is the *velocity*, which can be measured directly by an additional velocity detector. Nevertheless, it is always reasonable to build the sensor system as simple as possible and to use the minimum number of detectors. In the considered case the velocity can be determined indirectly on the basis of the measured acceleration as

$$\dot{u}(t) = \int_{t_0}^{t} \ddot{u}(\tau) \, d\tau$$

Simplicity of the procedures is the main advantage of this approach. On the other hand, the disadvantage of the method may be the necessity of locating the accelerometer on the impacting body, which limits the range of possible applications.

### 4.1.4 Approaches Based on Conservation of Momentum

Impact identification can also be based on the principle of conservation of momentum and the restitution coefficient, provided the involved structures and objects can be treated as rigid bodies.

Consider a system with two degrees of freedom (see Figure 4.3). Assume at the very beginning of the impact the masses $m_A$, $m_B$ and the velocities $V_{A1}$, $V_{B1}$, and after the impact the velocities $V_{A2}$, $V_{B2}$. Assume the conditions $V_{A1} > V_{B1}$ and $V_{A2} < V_{B2}$ to be satisfied. The equation of momentum can be written in the form:

$$m_A V_{A2} + m_B V_{B2} = m_A V_{A1} + m_B V_{B1}. \qquad (3)$$

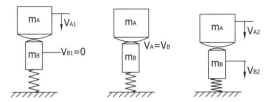

**Figure 4.3**  Different phases of impact: (left) compression phase; (middle) middle phase; (right) rebound phase

In the classical impact theory, one of the basic parameters proposed originally by Newton is the restitution coefficient, which is formulated by the proportion of the relative velocities of colliding rigid bodies:

$$k = \frac{V_{B2} - V_{A2}}{V_{A1} - V_{B1}} \tag{4}$$

where in the case of in-line collisions the actual values of velocities are used, and if the impact is oblique, the velocity vectors perpendicular to the plane of impact should be used instead. The coefficient of restitution $k$ was originally considered to be a material constant independent of the impact characteristics. Later it was proved that this presumption was incorrect [4]. The main quantity affecting the value of the restitution coefficient is the relative velocity of the bodies in the instance of the impact. Therefore, the coefficient of restitution is assumed here to be velocity-dependent. If the initial velocity $V_{B1}$ of the impacted body is assumed to be zero, then Equations (3) and (4) take simpler forms:

$$m_A V_{A2} + m_B V_{B2} = m_A V_{A1}$$
$$k(V_{A1}) = \frac{V_{B2} - V_{A2}}{V_{A1}} \tag{5}$$

which yield the following formula for the *mass* of the impacting body:

$$m_A = \frac{m_B V_{B2}}{V_{A1} - V_{B2} + V_{A1} k(V_{A1})} \tag{6}$$

To apply Equation (6), it is necessary to know the value of the mass of the impacted body $m_B$, which is usually acceptable, because the properties of the AIA system are known in advance. A more difficult problem is the uncertainty of the restitution coefficient, which in most cases should be obtained experimentally. The velocities $V_{A1}$ and $V_{B2}$ have to be directly measured or identified by a modified solution map approach.

The momentum approach can also be formulated in another way, on the basis of the impulse parameter $P$. Let $t_1$ and $t_2$ denote the pre-impact and post-impact time instants. If the impacting body maintains a constant mass $m_A$, then

$$P = \int_{t_1}^{t_2} F(t)\,dt = m_A(V_{A2} - V_{A1})$$

and hence the impacting mass $m_A$ can be determined as

$$m_A = \frac{P}{V_{A2} - V_{A1}} \qquad (7)$$

The denominator can be determined with an accelerometer:

$$\Delta v = V_{A2} - V_{A1} = \int_{t_1}^{t_2} \ddot{u}(t)\,dt \qquad (8)$$

Integration of signals from an accelerometer and a force sensor, by Equations (7) and (8), allows the impacting mass $m_A$ to be determined. A modified version of the solution map approach can be used to obtain the impact initial velocity, as described in the experimental part below.

### 4.1.5 Experimental Test Stand

The assembled experimental drop test stand is shown in Figure 4.4. A gas spring was used as the impact absorption system. The main parts of the set-up are the pneumatic cylinder (1) mounted in the vertical position (diameter 63 mm, maximum stroke 250 mm, atmospheric pressure at full elongation), the frame (2) and the carriage (3). The lift mechanism includes an electromagnet (4) used for releasing the impacting mass (5), which is guided by the rail

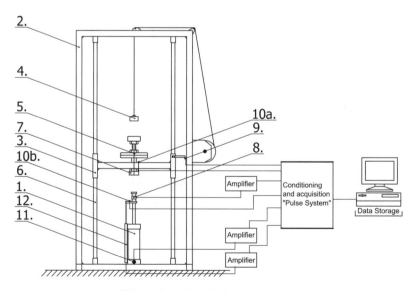

**Figure 4.4**  Experimental test stand

system (6) embedded in the frame. The mass is impacting on to the pneumatic cylinder via a rubber bumper (7).

The data acquisition system included all necessary conditioning systems and amplifiers, and served to acquire real-time measurements of all significant signals:

- force signal from the piezoelectric sensor (8) fixed to the piston rod of the pneumatic cylinder, in order to measure the actual impact history;
- signal from the optical switch (9), which acts as a trigger and allows the vertical velocity of the impacting mass to be determined just before the impact;
- deceleration of the falling mass (10a);
- acceleration (10b) of the piston rod of the pneumatic cylinder;
- pressure in the cylinder, measured with a fast pressure sensor (11);
- displacement of the piston by a linear variable differential transducer (LVDT) sensor (12).

The drop test stand enables a wide variety of impact scenarios to be simulated, defined by the mass and the velocity of the impacting body in the instant of impact. The used mass range was 7.2–50 kg, while the impact velocity was dependent on the drop height, which was confined to the range 0–0.5 m. Three different examples of direct measurements of impact forces are presented in Figure 4.5. The kinetic impact energy in all three scenarios was approximately 30 N m.

In all observed impact phenomena it is possible to distinguish between two phases of the process, as shown in Figure 4.5. The first phase (phase A) is characterized by rebounds between the falling mass and the piston. The number of rebounds, the characteristic successive reduction of the amplitudes and the time intervals are similar for the whole range of initial conditions (mass and velocity) used in the experiment. The total duration of the first impact

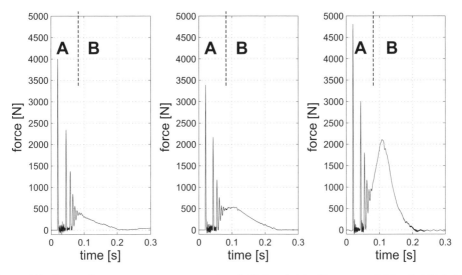

**Figure 4.5** Impact force measurements, examples: (left) drop height 40 cm, mass 7.2 kg; (middle) drop height 25 cm, mass 12.2 kg; (right) drop height 15 cm, mass 22.2 kg. A, first impact phase; B, second impact phase

**Figure 4.6** High-speed snapshots of the impact process (drop height 20 cm, mass 27.2 kg) at 0 ms, 10 ms, 20 ms, 30 ms and 40 ms

phase was approximately 50 ms, while the duration of an average rebound (contact peak) was approximately 5 ms. The tests were filmed with a high-speed camera. Figure 4.6 shows a sequence of frames with 10 ms interval, starting from the beginning of the impact, taken for the mass 27.2 kg falling from 20 cm. The rebound between the falling mass and the piston rod can be clearly seen. A qualitative analysis leads to the conclusion that the impact force measured in the first impact phase shows rather low sensitivity to the size of the falling mass and much stronger sensitivity to its velocity. This relationship is presented in Figure 4.7.

In the second impact phase (phase B), a simultaneous joint movement of the falling mass and the piston can be observed. In this phase the measured force showed much stronger sensitivity to the value of the falling mass. The maximum peak force occurred at the maximum compression of the gas spring and was clearly distinguishable for higher mass values. The duration time of

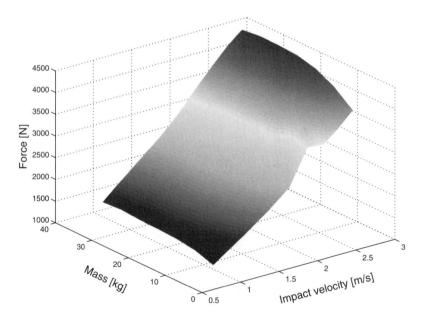

**Figure 4.7** Influence of mass and velocity on the first force peak in the first phase of impact

the second impact phase was 120–150 ms, while the duration of the whole phenomenon for the analysed structure was 170–200 ms.

## 4.1.6 Experimental Verification

The impact identification approaches proposed in the previous subsections were verified experimentally with the described laboratory test stand. Each of the approaches was tested with a wide variety of 56 different impact scenarios admissible by the stand:

- seven masses spaced in the range 7.2–37.2 kg;
- eight drop heights equally spaced in the range 0.05–0.40 m, which corresponds to the impact velocities in the range of approximately 1–2.8 m/s.

### 4.1.6.1 Solution Map Approach

The objective was the maximum simplicity of the data acquisition set-up, i.e. to apply an algorithm able to operate on the basis of measurements from one sensor only. Moreover, an important issue was to use a sensor that was not directly fixed to the falling mass. It was decided to use the force sensor. Two impact parameters (mass $m$ and velocity $v$ of the falling body) had to be identified using the single force sensor. Therefore, in order to construct the solution map, at least two characteristic quantities had to be extracted from only one measured force history. Figure 4.8 shows an example of force measurement

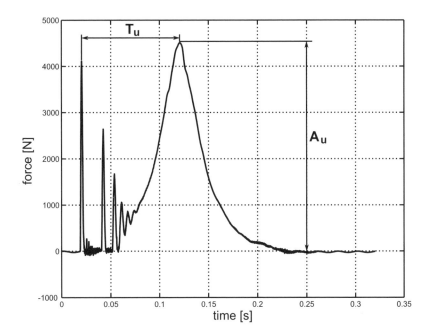

**Figure 4.8** Measured impact force evolution and parameters used to construct the solution map

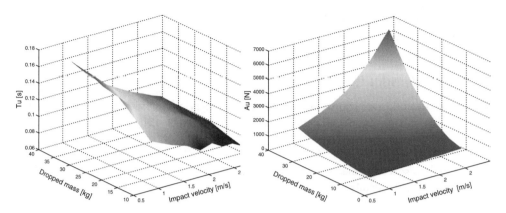

**Figure 4.9** Sensitivities of parameters used in the solution map approach to impact mass $m$ and velocity $v$: (left) $T_u$; (right) $A_u$

during an impact, together with the two chosen parameters ($T_u$, $A_u$) used to construct the solution map:

- $T_u$, the time interval between the maximum force value in the first impact phase and the maximum force value measured in the second impact phase;
- $A_u$, the maximum force value in the second impact phase.

The choice of the parameters was motivated by practical aspects. First of all, chosen values must be characteristic in the impact process. This means that the parameters should be unambiguously determined in real-time during the impact phenomena. Moreover, in order to ensure the uniqueness of the identification, it was important that the chosen parameters revealed a strong sensitivity to the impact parameters $m$ and $v$. Figure 4.9 depicts the values of the chosen parameters $A_u$ and $T_u$ in the impact mass–velocity domain.

Given the solution map, impact identification amounts to minimizing the error function defined as

$$f(m, v) = \left[\frac{A_u^M - A_u(m, v)}{A_u^M}\right]^2 + \left[\frac{T_u^M - T_u(m, v)}{T_u^M}\right]^2$$

where $m$ and $v$ are the impact parameters to be identified, $A_u(m, v)$ and $T_u(m, v)$ are obtained from the solution map for the specific values of $m$ and $v$, while $A_u^M$ and $T_u^M$ denote the actually measured values. The identification problem can be presented graphically. Constant values of $A_u$ and $T_u$ can be represented in Figure 4.9 by contour lines in the mass–velocity space. The solution map approach leads to the two contour lines, which correspond to the actually measured values $T_u^M$ and $A_u^M$. Their intersection point defines the identified impact parameters, which are shown schematically in Figure 4.10.

A practical verification of this approach was performed in two steps: (1) building the solution map and (2) testing the impact identification process within the solution map range of the impact parameters $m$ and $v$. The test results are presented in Figure 4.11, which compares the exact and the identified values of impact parameters. Additionally, the accuracy of identification was

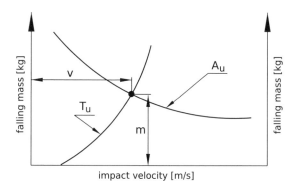

**Figure 4.10**  Graphical interpretation of the solution map approach

assessed using the following measure based on standard deviation from the exact value $\bar{x}$:

$$\sigma = \sqrt{\frac{1}{n-1}\sum_{i=1}^{n}(x_i - \bar{x})^2} \tag{9}$$

which was computed separately for each group of impact scenarios defined by a constant value of the impact velocity (to assess the accuracy of mass identification) or mass (for the accuracy of velocity identification). The results are presented in Figure 4.12.

An important advantage of the solution map approach is the possibility of application in a relatively wide range of technical systems. An important factor is the simplicity, which is evident if the data acquisition set-up is considered: only one sensor was necessary. Moreover, a relatively precise determination of impact parameters is possible. In the experiment, the average accuracy of the determined parameters was $\pm 5\,\%$.

Nevertheless, the approach has also disadvantages. The method includes a lengthy process of performing tests to determine the solution map. Moreover, the impact parameters have been identified after a relatively long time. In the presented system, the mass and velocity of the falling body could be obtained only in the second phase (B) of the impact, not earlier than 100–110 ms and after the maximum peak of the measured force.

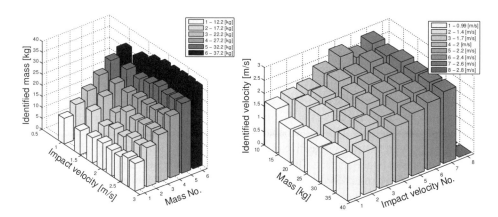

**Figure 4.11**  Impact identification (solution map approach): (left) mass; (right) velocity

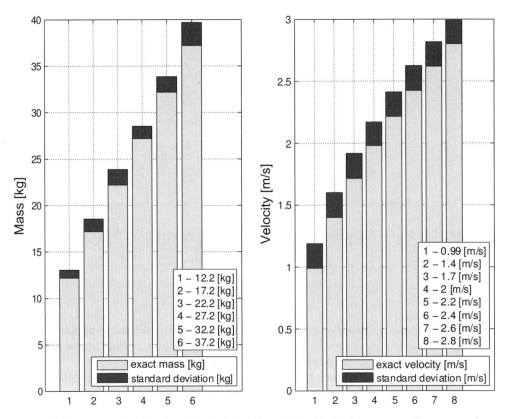

**Figure 4.12**  Impact identification. Standard deviation of identified values compared to exact values (solution map approach): (left) mass; (right) velocity

### 4.1.6.2 Approach Based on Force and Acceleration

The objective was to *minimize the impact identification time* towards the real-time. Equation (2) is used, and it is assumed that measurements from two sensors (force and acceleration) are available.

*Mass* identification was tested in two cases: with the accelerometer attached (1) to the falling body and (2) to the piston rod. Examples of accelerations measured in these two locations are shown in Figure 4.13, along with the measured force. The figure shows the very first impulse due to the first contact between the falling mass and the piston rod (the beginning of phase A of the impact).

The procedure was tested for the whole variety of impacting masses used in the experiments. Chosen results of identification in the first phase of impact are presented in Figure 4.14. The identified mass $m$ shows a very volatile behaviour. This is mainly due to the intrinsic feature of accelerometers, which are very sensitive to external factors like, for example, vibration of the testing stand caused by the impact. The graph shows temporary values of the identified mass, the mean value in the considered time interval and the accurate value of the falling mass. If accelerations of the piston rod are measured (Figure 4.14, left), the identified mass value is almost independent of the actual falling mass and its velocity. This behaviour turned out to be typical for the first impact phase, which is characterized by no joint movement and several

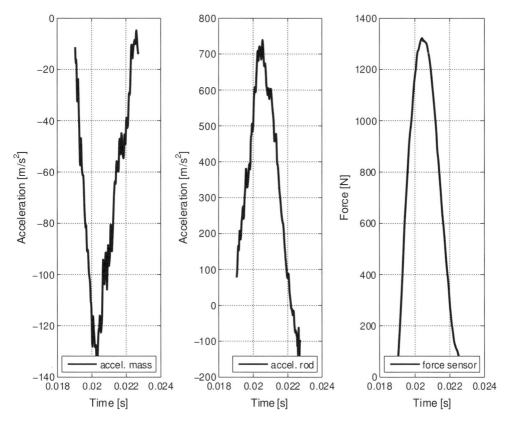

**Figure 4.13**   Examples of measured signals in the first impulse (phase A of impact): (left) falling mass acceleration; (middle) piston rod acceleration; (right) force

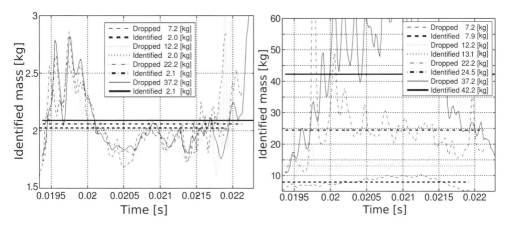

**Figure 4.14**   Mass identification in the first phase of the impact (force and acceleration approach). Acceleration measured on: (left) piston rod of the gas spring; (right) falling mass

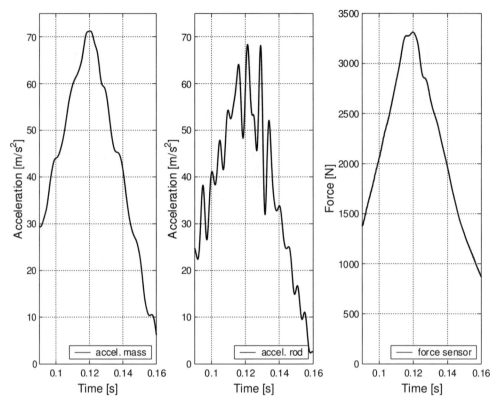

**Figure 4.15**  Examples of measured signals in the second impact phase: (left) falling mass acceleration; (middle) piston rod acceleration; (right) force

rebounds between the piston rod and the falling mass. In fact, when the accelerations are measured on the piston rod, only the mass of the piston rod can be estimated in the first impact phase. More practical results were obtained when the deceleration of the falling body was measured (see Figure 4.14, right). A reliable mass value ($\pm 10\,\%$ accuracy) could be computed already after 5 ms from the very beginning of the impact.

The mass identification based on Equation (2) has also been tested with the signals measured in the second phase (B) of the impact. Examples of measurements are shown in Figure 4.15; notice the different time scale. A common joint movement of the piston rod and the falling mass occurs, which is confirmed by similarity of the accelerations measured on the piston rod and on the falling mass. The results of the mass identification are presented in Figure 4.16. In the second impact phase the mass identification is feasible and the location of the accelerometer does not have any crucial importance: the values obtained for both locations are relatively similar, but slightly more stable results were obtained for the accelerometer mounted on the falling mass. The accuracy was approximately $\pm 5\,\%$.

The impact velocity identification was tested, based on differentiation (integration) of the measured displacement (acceleration) of the piston rod (see the comparison in Figure 4.17). Both led to similar results, although the differentiated displacement was noisier and required filtering, while the integration smoothed down the noise effect. Therefore, integration of the acceleration is more justified for the analysed structure, especially taking into account the simplicity of the data acquisition set-up. However, identification of the velocity of the falling

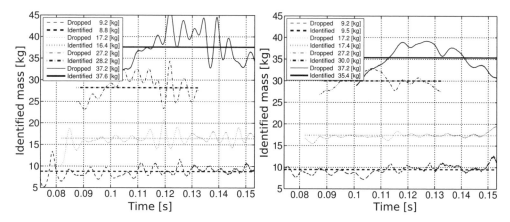

**Figure 4.16** Mass identification in the second phase of the impact (force and acceleration approach). Acceleration measured on: (left) piston rod of the gas spring; (right) falling mass

mass (instead of the piston rod) is feasible only when joint movement of the body and the piston occurs. In the case of the analysed structure it happened first 80 ms from the very beginning of the impact process. Notice that the identification is not possible with the accelerometer sensor located on the falling mass, since for integration the initial condition is required. In the case of the piston rod it is zero, while in the case of the falling body it is exactly the impact velocity being sought.

### 4.1.6.3 Approaches Based on The Conservation of Momentum

Verification of the momentum conservation approach for *mass identification* was performed on the basis of Equation (6), which uses the restitution coefficient, and Equation (7), which is based directly on the principle of conservation of momentum during the first contact impulse.

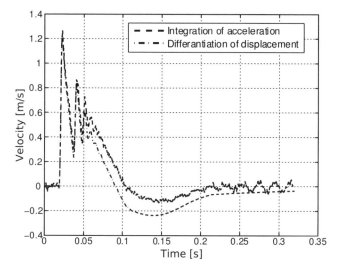

**Figure 4.17** Velocity identification: integrated accelerations and differentiated displacements of piston rod – an example

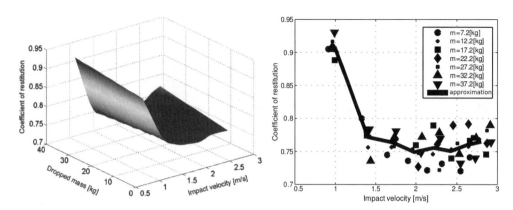

**Figure 4.18**   Restitution coefficient in terms of the impact: (left) mass and velocity; (right) velocity

In the first approach the task is to identify the impact parameters by means of the coefficient of restitution (Equation (5)), which should be determined previously using the measurements at the beginning of the impact process (phase A, first rebound). Experimental values of the restitution coefficient are presented in Figure 4.18. For the whole variety of the tested impacts, the restitution coefficient stayed within the range 0.70–0.95 and showed poor sensitivity to the value of the impacting mass. The effect of the velocity was more significant: impact velocities higher than 1.5 m/s correspond to lower values of the restitution coefficient. Literature data [4–6] generally confirm the negative correlation between the restitution coefficient and impact velocities.

Equation (6) was used for mass identification. The impact velocity $V_{A1}$ of the falling mass was directly measured by means of the photo switch sensor and was used to determine the corresponding value of the restitution coefficient $k(V_{A1})$ by approximation of the previously measured data. In order to test the utility of the pressure sensor, the post-impact velocity of the impacted body (piston rod) $V_{B2}$ was determined by differentiation of the displacement, based on the pressure measurement. An adiabatic compression process with the ratio 1.3 was assumed. The results of mass identification are presented in Figure 4.19. Figure 4.20 (left) compares the exact mass values to the corresponding standard deviation (Equation (9)), while the right-hand figure lists the percentage accuracy of identification:

$$p = \frac{1}{n} \sum_{i=1}^{n} \frac{|x_i - \bar{x}|}{\bar{x}} 100\,\%  \tag{10}$$

Relatively low accuracy (worse than 10 % as an average) is the main disadvantage of the method. Moreover, the approach requires preliminary determination of the restitution coefficient, which can lead to additional uncertainty.

The disadvantages encourage the testing of another method, proposed on the basis of the impulse definition (Equation (7)). The identification used force measurements and accelera-tion measured on the falling mass; only the data concerning the first contact impulse were considered. The results are presented in Figure 4.21. The standard deviation and accuracy of identification are shown in Figure 4.22, and are significantly better than in the case of the method based on the restitution coefficient: ±5 % accuracy as an average. A change in velocity of the impacting mass was computed here via integration of the acceleration measured on the

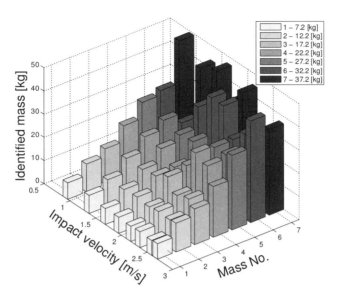

**Figure 4.19** Mass identification by Equation (6) (momentum conservation and the restitution coefficient approach)

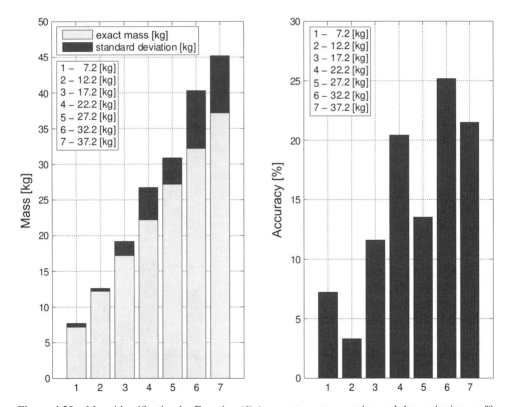

**Figure 4.20** Mass identification by Equation (6) (momentum conservation and the restitution coefficient approach): (left) standard deviation compared to exact mass values; (right) percentage accuracy of identification

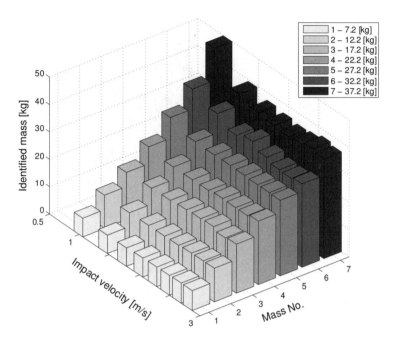

**Figure 4.21**  Mass identification by Equation (7) (momentum conservation and the impulse approach)

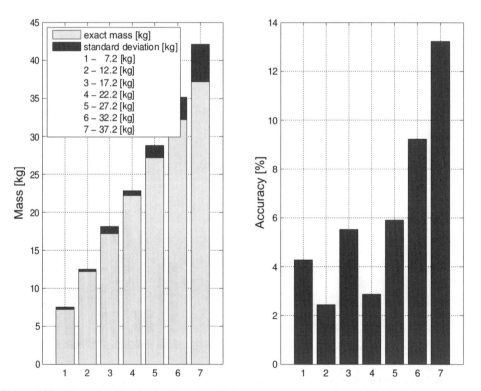

**Figure 4.22**  Mass identification by Equation (7) (momentum conservation and the impulse approach):
(left) standard deviation compared to exact mass values; (right) percentage accuracy of identification

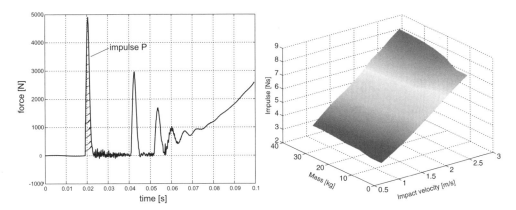

**Figure 4.23**    First impulse for impact velocity identification: (left) definition; (right) sensitivity to impact parameters

falling mass, in accordance with Equation (8). To test the velocity identification a modified solution map approach was used, based on the previously identified mass and the value of the impulse due to the first contact, which reveals strong sensitivity to the impact velocity (see Figure 4.23). The percentage accuracy of identification is shown in Figure 4.24. In all cases it is better than ±5 % and ±3 % as an average. The main disadvantage of the method is the necessity to have preliminary determination of the solution map for the impulse values.

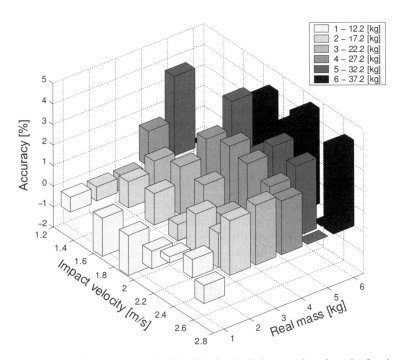

**Figure 4.24**    Accuracy of impact velocity identification (solution map based on the first impulse and previously identified mass)

**Table 4.1**  Comparison and summary of impact load identification procedures ($F$ denotes force; $a_A$ accelerometer (falling body); $a_B$ accelerometer (piston rod); $p$ pressure; $b$ photo switch)

|  | Solution map | Force and acceleration | | Momentum conservation | |
|---|---|---|---|---|---|
| Sensors | $F$ | $F, a_B$ | $F, a_A$ | $p, b$ | $F, a_A$ |
| Solution maps | 2 | 1 | — | 1 | 1 |
| Operation time [ms] | 100–110 | 80 | 5/80 | 5 | 5 |
| Accuracy (mass) | 5 % | 5 % | 10 % / 5 % | 15 % | 5 % |
| Accuracy (velocity) | 5 % | 10 % | — | Measured | 3 % |

In general, the identification of both impact parameters based on the characteristics of the first impulse is possible relatively quickly, already 5 ms from the very beginning of the impact process. Therefore, the approach can operate in real-time.

### 4.1.7  Comparison of Approaches

A real-time impact mass and velocity estimation is extremely difficult with only one sensor. Much better results can be obtained with two sensors; in the tested case good results were obtained with a combination of an accelerometer and a force sensor. However, the location of the accelerometer is very important: fixing the accelerometer to the falling body allows precise mass identification in a short time after the impact, while with the accelerometer placed on the impacted body the identification is far more difficult and only feasible after a longer period. Therefore, mass determination of the impacting body is possible in a short time and with high accuracy only when one of the sensors is fixed to the impacting object or when the parameters of its movement are directly measured.

For velocity identification on the basis of acceleration, the initial condition is needed. Thus in the case of the analysed structure, the velocity identification was feasible only when a joint movement of the piston rod and the falling mass was observed.

The tested algorithms are summarized and compared in Table 4.1.

## 4.2  Observer Technique for On-Line Load Monitoring

The concept of the observer was introduced by Luenberger in the beginning of 1970s [7]. Originally it was devoted to feedback control systems, since first multivariable control strategies had assumed that the whole state vector was available for feedback [8]. However, from the practical point of view, it was not always possible to measure all the components of the state. The state observer, which allows the state vector to be estimated based on only a few measurements, encountered great interest among the control engineering community [9,10]. Another important issue connected with real-life measurements was noise corrupting the measured signal. With the aid of the theory of stochastic processes, this problem was solved by Rudolf Kalman and nowadays his stochastic estimator is frequently called the Kalman filter [11].

Currently, a lot of attention is paid to simultaneous reconstruction of both the state and the inputs in linear and nonlinear systems [12–15]. In the case of a mechanical system, the state can be represented by displacements and velocities, while the inputs can be external forces.

## 4.2.1 State-Space Representation of Mechanical Systems

Consider the standard linear dynamical system in the state-space form:

$$\begin{cases} \dot{\mathbf{x}}(t) = \mathbf{A}\mathbf{x}(t) + \mathbf{B}\mathbf{u}(t), & \mathbf{x}(0) = \mathbf{x}_0 \\ \mathbf{y}(t) = \mathbf{C}\mathbf{x}(t) \end{cases} \tag{11}$$

where $\mathbf{x}(t) \in \mathbb{R}^n$, $\mathbf{u}(t) \in \mathbb{R}^m$, $\mathbf{y}(t) \in \mathbb{R}^p$ and $\mathbf{A}$, $\mathbf{B}$, $\mathbf{C}$ are matrices of proper dimensions. In the above equations $\mathbf{x}(t)$ is called the *state* vector, $\mathbf{u}(t)$ the *input* vector and $\mathbf{y}(t)$ the *output* vector. Second-order linear models obtained by the finite element method (FEM) are usually of the form

$$\mathbf{M}\ddot{\mathbf{q}}(t) + \mathbf{D}\dot{\mathbf{q}}(t) + \mathbf{K}\mathbf{q}(t) = \mathbf{B}_0\mathbf{u}(t)$$

where $\mathbf{M}$, $\mathbf{D}$ and $\mathbf{K}$ are respectively the mass, damping and stiffness matrices, and can be transformed into the state-space form by the substitution

$$\mathbf{x}(t) = \begin{bmatrix} \mathbf{q}(t) \\ \dot{\mathbf{q}}(t) \end{bmatrix}, \qquad \mathbf{A} = \begin{bmatrix} \mathbf{0} & \mathbf{I} \\ -\mathbf{M}^{-1}\mathbf{K} & -\mathbf{M}^{-1}\mathbf{D} \end{bmatrix}, \qquad \mathbf{B} = \begin{bmatrix} \mathbf{0} \\ \mathbf{M}^{-1}\mathbf{B}_0 \end{bmatrix}$$

where $\mathbf{I}$ is the identity matrix. The form of matrix $\mathbf{C}$ depends on sensor allocation.

## 4.2.2 State Estimation and Observability

As mentioned earlier, at the beginning the observers were used to estimate the state vector only. The state estimate was denoted by $\hat{\mathbf{x}}(t)$. The design of the state observer consists in designing a dynamical system, in which the state estimate $\hat{\mathbf{x}}(t)$ converges asymptotically to the actual state $\mathbf{x}(t)$ of the observed system. It was found by Luenberger [7] that a dynamical system that satisfies this assumption has the following form:

$$\dot{\hat{\mathbf{x}}}(t) = \mathbf{F}\hat{\mathbf{x}}(t) + \mathbf{G}\mathbf{u}(t) + \mathbf{H}\mathbf{y}(t), \qquad \hat{\mathbf{x}}(0) = \mathbf{0}$$

where $\hat{\mathbf{x}}$ is the state estimate of the same dimension as the state vector, $\mathbf{u}(t)$ is the input and $\mathbf{y}(t)$ is the measurement. The matrices $\mathbf{F}$, $\mathbf{G}$ and $\mathbf{H}$ have to be determined separately.

Let the vector $\mathbf{e}(t)$ denote the estimation error:

$$\mathbf{e}(t) = \hat{\mathbf{x}}(t) - \mathbf{x}(t)$$

The error dynamics is governed by the equation:

$$\dot{\mathbf{e}}(t) = \mathbf{F}\hat{\mathbf{x}}(t) + \mathbf{G}\mathbf{u}(t) + \mathbf{H}\mathbf{C}\mathbf{x}(t) - \mathbf{A}\mathbf{x}(t) - \mathbf{B}\mathbf{u}(t)$$

The terms depending on $\mathbf{u}(t)$ can be eliminated by choosing $\mathbf{G} = \mathbf{B}$ so that, by further derivation,

$$\dot{\mathbf{e}}(t) = \mathbf{F}\mathbf{e}(t) + (\mathbf{H}\mathbf{C} - \mathbf{A} + \mathbf{F})\mathbf{x}(t)$$

If the matrix $\mathbf{F}$ is chosen such that

$$\mathbf{F} = \mathbf{A} - \mathbf{HC}$$

then the error dynamics assumes the desired form of an autonomous dynamical system

$$\dot{\mathbf{e}}(t) = (\mathbf{A} - \mathbf{HC})\mathbf{e}(t), \qquad \mathbf{e}(0) = -\mathbf{x}_0$$

and thus

$$\mathbf{e}(t) = -e^{(\mathbf{A} - \mathbf{HC})t}\mathbf{x}_0$$

Therefore, if the matrix $\mathbf{H}$ is chosen in such a way that $\mathbf{F}$ is a *Hurwitz matrix*, i.e. real parts of its eigenvalues are negative,

$$\Re\lambda_i(\mathbf{A} - \mathbf{HC}) < 0$$

then the state estimate $\hat{\mathbf{x}}(t)$ converges asymptotically to the actual state $\mathbf{x}(t)$. It is well known [10] that the eigenvalues of $\mathbf{F}$ can be arbitrarily placed if and only if

$$\operatorname{rank}\mathbf{O} = n$$

where $\mathbf{O} = \begin{bmatrix} \mathbf{C}^{\mathrm{T}} & (\mathbf{CA})^{\mathrm{T}} & (\mathbf{CA}^2)^{\mathrm{T}} & \cdots & (\mathbf{CA}^{n-1})^{\mathrm{T}} \end{bmatrix}^{\mathrm{T}}$ is the *observability matrix* and $n$ is the dimension of the state vector $\mathbf{x}(t)$. This general condition of observability has a very simple mechanical interpretation. As an example, a simply supported beam was chosen with one linear sensor located in the middle (see Figure 4.25). This system is not observable, because none of its antisymmetric modes is measured by the linear sensor, no matter whether it measures displacement, velocity or acceleration.

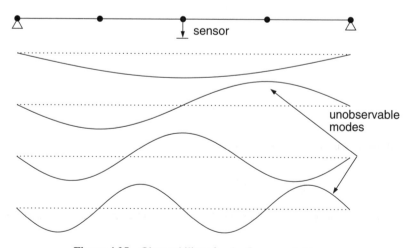

**Figure 4.25**   Observability of a simply supported beam

## 4.2.3 Model-Based Input Estimation

The idea of a state observer can be extended to the case in which both the state and the input are estimated simultaneously. While the most general case with no assumptions about the input is considered in the next subsection, here a special, important case is examined, in which the input is assumed to satisfy the following equation:

$$u^{(s)}(t) + l_{s-1}u^{(s-1)}(t) + l_{s-2}u^{(s-2)}(t) + \cdots + l_1\dot{u}(t) + l_0u(t) = 0 \tag{12}$$

where $u^{(i)}(t)$ denotes the $i$th derivative of $u(t)$ with respect to time. This approach, although not the most general, can be useful in many practical applications. One of them is wind engineering, where – having measured the spectrum of wind fluctuations – a wind load model can be created using the Davenport filter [16].

Equation (12) can be transformed to the state-space form by the following sequence of substitutions:

$$\eta_1 = u, \quad \eta_2 = \dot{u} = \dot{\eta}_1, \quad \eta_3 = \ddot{u} = \dot{\eta}_2, \quad \ldots, \quad \eta_s = u^{(s-1)} = \dot{\eta}_{s-1}$$

so that $\dot{\eta}_s = u^{(s)} = -\sum_{i=1}^{s} l_i\eta_i$. By collecting all the states $\eta_i$ into one state vector $\eta(t) = [\eta_1(t) \quad \eta_2(t) \quad \eta_3(t) \quad \cdots \quad \eta_s(t)]^T$, a set of first-order differential equations is obtained, which can be written in matrix form as

$$\begin{cases} \dot{\eta}(t) = \Lambda\eta(t) \\ u(t) = \Gamma\eta(t) \end{cases}$$

The above state-space representation of input can be combined with the state-space representation of the dynamics of the structure (Equation (11)) to obtain one augmented system of the form

$$\underbrace{\begin{bmatrix} \dot{x}(t) \\ \dot{\eta}(t) \end{bmatrix}}_{\dot{\xi}(t)} = \underbrace{\begin{bmatrix} A & B\Gamma \\ 0 & \Lambda \end{bmatrix}}_{\Theta} \underbrace{\begin{bmatrix} x(t) \\ \eta(t) \end{bmatrix}}_{\xi(t)}$$

$$y = \underbrace{[C \quad 0]}_{\Psi} \underbrace{\begin{bmatrix} x(t) \\ \eta(t) \end{bmatrix}}_{\xi(t)} \tag{13}$$

The last step in input identification is to design the observer for the augmented system

$$\dot{\hat{\xi}}(t) = F\hat{\xi}(t) + Hy(t), \qquad \hat{\xi}(0) = 0$$

where $F = \Theta - H\Psi$. Again, to obtain asymptotic convergence, the system has to be observable, and $H$ should be chosen in such a way that $F$ is a Hurwitz matrix.

## 4.2.4 Unknown Input Observer

The input observer introduced in the last section estimates the load acting on a structure under the assumption that the loading obeys an *a priori* known model. However, a broad class of problems exists in which such a model is not known or does not exist. In this case, in order to estimate on-line the load acting on a mechanical system, the unknown input technique of Hou and Müller [17] can be used. In their approach, the only required information is the model of the system (mass and stiffness matrices) and measurements of a part of the state vector (displacements or velocities).

Consider a dynamical system described by Equation (11). Additionally assume that $m < p$, which means that the number of sensors exceeds the number of forces to be identified.

The first step of the present method is to decompose the system of Equation (11) into two subsystems, so that the unknown input $\mathbf{u}(t)$ acts on one of them only. By the singular value decomposition (SVD), the matrix $\mathbf{B}$ can be restated in the following form:

$$\mathbf{B} = \mathbf{U}\mathbf{\Sigma}\mathbf{V}^{\mathrm{T}} \tag{14}$$

where $\mathbf{U} \in \mathbb{R}^{n \times n}$ and $\mathbf{V} \in \mathbb{R}^{m \times m}$ are unitary matrices, which satisfy the following condition:

$$\mathbf{U}\mathbf{U}^{\mathrm{T}} = \mathbf{I}, \qquad \mathbf{V}\mathbf{V}^{\mathrm{T}} = \mathbf{I} \tag{15}$$

Matrix $\mathbf{\Sigma}$ is a rectangular matrix $(n \times m)$ of the form

$$\mathbf{\Sigma} = \begin{bmatrix} \mathbf{\Sigma}_0 \\ \mathbf{0} \end{bmatrix} = \begin{bmatrix} \sigma_1 & 0 & \cdots & 0 \\ 0 & \sigma_2 & \cdots & 0 \\ \vdots & \vdots & \ddots & \vdots \\ 0 & 0 & \cdots & \sigma_m \\ 0 & 0 & \cdots & 0 \\ \vdots & \vdots & \ddots & \vdots \\ 0 & 0 & \cdots & 0 \end{bmatrix}$$

where matrix $\mathbf{\Sigma}_0$ has singular values of $\mathbf{B}$ on the main diagonal. By substituting Equation (14) into (11),

$$\dot{\mathbf{x}}(t) = \mathbf{A}\mathbf{x}(t) + \mathbf{U}\mathbf{\Sigma}\mathbf{V}^T\mathbf{u}(t)$$

Next, multiplying on the left by $\mathbf{U}^{\mathrm{T}}$ and using the orthogonality (Equation (15)),

$$\mathbf{U}^{\mathrm{T}}\dot{\mathbf{x}}(t) = \mathbf{U}^{\mathrm{T}}\mathbf{A}\mathbf{x}(t) + \mathbf{\Sigma}\mathbf{V}^T\mathbf{u}(t) \tag{16}$$

Linear transformations of the variables,

$$\mathbf{x} := \mathbf{U}\bar{\mathbf{x}}, \qquad \mathbf{u} := \mathbf{V}\bar{\mathbf{u}} \tag{17}$$

yields the two following coupled subsystems:

$$\begin{cases} \dot{\bar{\mathbf{x}}}_1(t) = \bar{\mathbf{A}}_{11}\bar{\mathbf{x}}_1(t) + \bar{\mathbf{A}}_{12}\bar{\mathbf{x}}_2(t) + \mathbf{\Sigma}_0\bar{\mathbf{u}}(t) \\ \dot{\bar{\mathbf{x}}}_2(t) = \bar{\mathbf{A}}_{21}\bar{\mathbf{x}}_1(t) + \bar{\mathbf{A}}_{22}\bar{\mathbf{x}}_2(t) \end{cases} \tag{18}$$

where

$$\bar{\mathbf{A}} = \begin{bmatrix} \bar{\mathbf{A}}_{11} & \bar{\mathbf{A}}_{12} \\ \bar{\mathbf{A}}_{21} & \bar{\mathbf{A}}_{22} \end{bmatrix} = \mathbf{U}^T \mathbf{A} \mathbf{U}$$

The new state vector $\bar{\mathbf{x}}(t)$ substituted into the measurement equation yields

$$\mathbf{y}(t) = \mathbf{C} \mathbf{U} \bar{\mathbf{x}}(t) = \bar{\mathbf{C}} \bar{\mathbf{x}}(t) = \bar{\mathbf{C}}_1 \bar{\mathbf{x}}_1(t) + \bar{\mathbf{C}}_2 \bar{\mathbf{x}}_2(t) \tag{19}$$

Similar to matrix $\mathbf{B}$, matrix $\bar{\mathbf{C}}_1$ can also be decomposed:

$$\bar{\mathbf{C}}_1 = \mathbf{U}_c \mathbf{\Sigma}_c \mathbf{V}_c^T \tag{20}$$

where $\mathbf{U}_c \in \mathbb{R}^{p \times p}$ and $\mathbf{V}_c \in \mathbb{R}^{m \times m}$. The matrix $\mathbf{\Sigma}_c \in \mathbb{R}^{p \times m}$ can be presented as

$$\mathbf{\Sigma}_c = \begin{bmatrix} \mathbf{\Sigma}_{c0} \\ \mathbf{0} \end{bmatrix}$$

Equation (19) can be restated by Equation (20) as

$$\mathbf{y}(t) = \mathbf{U}_c \mathbf{\Sigma}_c \mathbf{V}_c^T \bar{\mathbf{x}}_1(t) + \bar{\mathbf{C}}_2 \bar{\mathbf{x}}_2(t). \tag{21}$$

The measurement variable $\mathbf{y}$ can be linearly transformed into $\bar{\mathbf{y}}$ and decomposed into two parts, $\bar{\mathbf{y}}_1$ and $\bar{\mathbf{y}}_2$, of lengths respectively $m$ and $p - m$:

$$\mathbf{y} := \mathbf{U}_c \bar{\mathbf{y}} = \begin{bmatrix} \mathbf{U}_{c1} & \mathbf{U}_{c2} \end{bmatrix} \begin{bmatrix} \bar{\mathbf{y}}_1 \\ \bar{\mathbf{y}}_2 \end{bmatrix}$$

After substituting the transformation into Equation (21), it is easily seen that $\bar{\mathbf{y}}_2$ does not depend on the state $\bar{\mathbf{x}}_1$:

$$\begin{cases} \bar{\mathbf{y}}_1(t) = \tilde{\mathbf{C}}_{11} \bar{\mathbf{x}}_1(t) + \tilde{\mathbf{C}}_{12} \bar{\mathbf{x}}_2(t) \\ \bar{\mathbf{y}}_2(t) = \qquad\qquad \tilde{\mathbf{C}}_{22} \bar{\mathbf{x}}_2(t) \end{cases} \tag{22}$$

where $\tilde{\mathbf{C}}_{11} = \mathbf{\Sigma}_{c0} \mathbf{V}_c^T$, $\tilde{\mathbf{C}}_{12} = \mathbf{U}_{c1}^T \bar{\mathbf{C}}_2$ and $\tilde{\mathbf{C}}_{22} = \mathbf{U}_{c2}^T \bar{\mathbf{C}}_2$.
  The state $\bar{\mathbf{x}}_1$ can be determined from Equations (22) to be

$$\bar{\mathbf{x}}_1 = \tilde{\mathbf{C}}_{11}^{-1}(\bar{\mathbf{y}}_1 - \tilde{\mathbf{C}}_{12} \bar{\mathbf{x}}_2) \tag{23}$$

which, substituted into the second equation of (18) and combined with the second equation of (22), forms the following dynamical system:

$$\begin{cases} \dot{\bar{\mathbf{x}}}_2(t) = \bar{\mathbf{A}}_{21} \tilde{\mathbf{C}}_{11}^{-1} \bar{\mathbf{y}}_1(t) + \left[ \bar{\mathbf{A}}_{22} - \bar{\mathbf{A}}_{21} \tilde{\mathbf{C}}_{11}^{-1} \tilde{\mathbf{C}}_{12} \right] \bar{\mathbf{x}}_2(t) \\ \bar{\mathbf{y}}_2(t) = \tilde{\mathbf{C}}_{22} \bar{\mathbf{x}}_2(t) \end{cases}$$

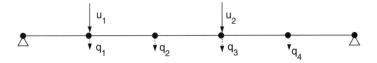

**Figure 4.26** Elastic structure used in the numerical examples

which does not depend on $\mathbf{u}(t)$; hence the classical state observer can be constructed to estimate the state $\bar{\mathbf{x}}_2$:

$$\dot{\hat{\bar{\mathbf{x}}}}_2(t) = \mathbf{F}\hat{\bar{\mathbf{x}}}_2(t) + \mathbf{G}\bar{\mathbf{y}}_1(t) + \mathbf{H}\bar{\mathbf{y}}_2(t)$$

Finally, the unknown input is determined from the first equation of (18) and the transformations of Equation (17) as

$$\hat{\mathbf{u}}(t) = \mathbf{V}\boldsymbol{\Sigma}_0^{-1}\left[\dot{\hat{\bar{\mathbf{x}}}}_1(t) - \bar{\mathbf{A}}_{11}\hat{\bar{\mathbf{x}}}_1(t) - \bar{\mathbf{A}}_{12}\hat{\bar{\mathbf{x}}}_2(t)\right] \tag{24}$$

A drawback of the method is the necessity to compute the derivative of the measurements $\dot{\bar{\mathbf{y}}}_1(t)$, which is required by Equations (24) and 23 to determine the input estimate. However, using a small enough sampling time, satisfactory results can be obtained. A solution of the problem of unknown input estimates, which avoids the differentiation, can be found in Reference [14].

### 4.2.5 Numerical Examples

To illustrate the applicability of both methods, i.e. the model-based and the unknown input observers, two numerical examples are presented. The objective is to find the load acting on the beam structure shown in Figure 4.26. The beam is divided into five finite elements and is simply supported at its ends. To reduce the number of state variables, rotational and horizontal degrees of freedom were removed. In both examples input forces act on the first and third degrees of freedom. The data used in examples are: length $L = 1$ m, cross-sectional area $A = 0.0001$ m$^2$, moment of inertia $I = A^2/12$, Young's modulus $E = 206$ GPa and material density $\rho = 7850$ kg/m$^3$. Forces acting on the beam were assumed to be

$$u_1(t) = \begin{cases} 1, & 0 < t < 1/3 \text{ s,} \\ -0.5, & 1/3 \leq t < 2/3 \text{ s,} \\ 2, & 2/3 \leq t \leq 1 \text{ s,} \end{cases} \qquad u_2(t) = \begin{cases} \sin(25\,t), & 0 < t < 1/2 \text{ s} \\ 3\sin(25\,t), & 1/2 \leq t \leq 1 \text{ s} \end{cases}$$

#### 4.2.5.1 Model-Based Input Estimation

In the first example the model-based input observer was used. It can be easily verified that forces acting on the beam satisfy the following equations:

$$\ddot{u}_1(t) = 0 \qquad \text{and} \qquad \ddot{u}_2(t) + 25^2 u_2(t) = 0$$

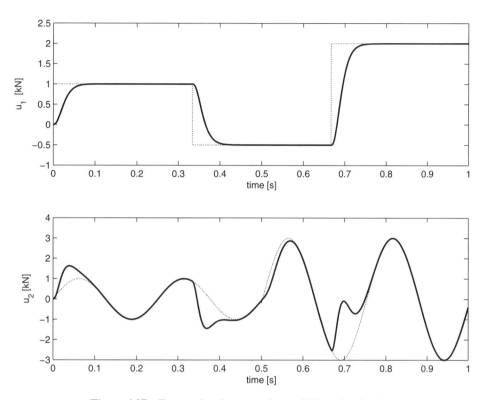

**Figure 4.27**   Force estimation example: model-based estimation

Matrix **C** was chosen to satisfy the observability condition of the augmented system of Equation (13):

$$\mathbf{C} = \begin{bmatrix} 1 & 0 & 0 & 0 & 0 & 0 & 0 & 0 \\ 0 & 1 & 0 & 0 & 0 & 0 & 0 & 0 \\ 0 & 0 & 1 & 0 & 0 & 0 & 0 & 0 \end{bmatrix} \tag{25}$$

Eigenvalues of the observer were chosen to be

$$\lambda(\mathbf{\Theta} - \mathbf{H\Psi}) = [-235.6, -111.0, -110.1, -108.8, -107.4, -101.0, \\ -106.2, -102.0, -105.0, -103.9, -103.0]^{\mathrm{T}}$$

Results obtained for the model-based observer are shown in Figure 4.27.

### 4.2.5.2  Unknown Input Estimation

In the second example the unknown input observer was used and

$$\mathbf{C} = \begin{bmatrix} 1 & -1 & 0 & 0 & 0 & 1 & 0 & 0 \\ 0 & 1 & -1 & 0 & 0 & 0 & 1 & 0 \\ 0 & 0 & 1 & -1 & 0 & 0 & 0 & 1 \end{bmatrix}$$

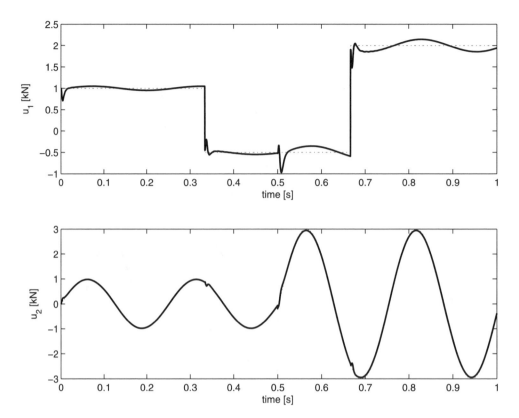

**Figure 4.28**  Force estimation example: unknown input estimation

Eigenvalues of the observer are evenly distributed between $-250$ and $-1500$ with step $250$. In Figure 4.28 the estimates of unknown forces are presented, computed with the sampling time of 0.03 ms. The undesired oscillations in the final stage of the identification process for $u_1$ are caused by the numerical differentiation of measurement. By choosing a smaller sampling time, this phenomenon can be significantly attenuated.

## 4.3  Off-Line Identification of Dynamic Loads

In contrast to the preceding sections, this section considers a methodology for *off-line* reconstruction of spatial and temporal characteristics of dynamic loads. The motivation is the need for a general analysis technique for efficient reconstruction of the scenario of a sudden load (e.g. impact, collision, etc.), which could be used in black-box type systems. The methodology is based on off-line analysis of the local structural response (strain, acceleration, etc.) and requires a dedicated sensor system to be distributed in the structure in order to measure and store the response. In the elastoplastic case the virtual distortion method (VDM) [18] is used, which is restricted to the small deformation case. The section expands and builds upon the approach briefly outlined in Reference [19].

There is an ongoing research effort in the field (see References [20] to [22] for relatively recent reviews). However, the structure is assumed to be linear and the generality of the considered load is usually strictly limited to a single pointwise load, with the location known in advance [23] or determined in an additional non-linear optimization [24]. Moreover, the identification is usually additionally simplified by assuming stationarity of the load. Authors [20, 25] do consider a moving force, but it is assumed to have a constant velocity. A number of papers deal with single pointwise impact loads only and disregard all load characteristics (magnitude, evolution, duration, etc.) besides the location [26, 27]. Papers that do consider multiple independent loads, as Reference [28], still assume a superfluous number of sensors.

Compared to other researches, the approach discussed here is aimed at the fully general case and provides more insight into the identification process by distinguishing between the reconstructible and unreconstructible load components. In the so-called *underestimated* case, the approach allows the use of a limited number of sensors to identify general dynamic loads of unknown locations, including simultaneous multiple impacts, freely moving and diffuse loads. However, this is at the cost of the uniqueness of identification, which can be attained only with additional heuristic assumptions. In this way an equivalent load is identified, which is observationally indistinguishable from the real load and optimum in a given sense. The identification is formulated analytically as a complex optimization problem: find the equivalent impact scenario that:

(1) minimizes the potentially pre-conditioned mean-square distance between simulated and measured dynamic responses in sensor locations;
(2) is optimum according to given heuristic conditions.

Since the reconstruction quality is directly related to the number and location of sensors, two complementary criteria of correct sensor location are proposed, combined in a compound criterion and illustrated in a numerical example. Measurement noise, inevitable in real-world structures, is fully accounted for. In the case of elastoplastic structures gradients of structural response are derived, which allow any general gradient-based optimization approach to be applied to identify the load.

## 4.3.1 Response to Dynamic Loading

A prerequisite for the load reconstruction methodology discussed here is the transfer function of the system being considered, i.e. the structural response to local impulse excitations and plastic distortions. It can be either generated from a numerical model or – at least partially – measured experimentally, the latter being potentially more practical in the case of real-world complex structures.

### 4.3.1.1 Linear Systems

Let the system being considered be linear and spatially discretized. Provided the system transfer function is known and no excitation prior to time $-\Delta T$ has appeared (i.e. zero initial conditions), the response of the system in a given sensor location can be expressed by means of a convolution integral as follows:

$$\varepsilon_i(t) = \int_{-\Delta T}^{t} \tilde{\mathbf{D}}_i^{\mathrm{f}}(t - \tau)^{\mathrm{T}} \mathbf{f}(\tau) \, \mathrm{d}\tau \tag{26}$$

where $\varepsilon_i$ is the response of the system in the $i$th sensor location, vector $\mathbf{f}$ denotes the loading evolution in all potentially load-exposed degrees of freedom (DOFs), while $\tilde{\mathbf{D}}_i^{\mathrm{f}}$ is the vector of the system transfer functions $\tilde{D}_{iN}^{\mathrm{f}}$ relating the response in the $i$th sensor location to local impulse loading in the $N$th potentially load-exposed DOF.

In Equation (26) $\Delta T$ is the maximum system response time (i.e. the maximum propagation time of an elastic wave between a loading-exposed DOF and a sensor). Due to the intended limited number of sensors, the considered system is rarely collocated; thus $\Delta T > 0$. The measurements and the reconstruction process may be triggered by a strong excitation (e.g. an impact), which is picked up delayed at most $\Delta T$; hence the time shift is necessary.

In real-world applications continuous responses are rarely known and so Equation (26) should be discretized with respect to time. With the simplest quadrature rule it takes the following form:

$$\varepsilon_i(t) = \sum_{\tau=-\Delta T}^{t} \mathbf{D}_i^{\mathrm{f}}(t-\tau)^{\mathrm{T}}\mathbf{f}(\tau) \tag{27}$$

where $\mathbf{D}_i^{\mathrm{f}}$ is the discretized and accordingly rescaled system transfer function $\mathbf{D}_i^{\mathrm{f}} = \Delta t \tilde{\mathbf{D}}_i^{\mathrm{f}}$. Equation (27), rewritten for each sensor location $i$ and each measurement time instance $t$, can be stated in the form of a general linear equation:

$$\varepsilon = \mathbf{D}^{\mathrm{f}}\mathbf{f} \tag{28}$$

where $\varepsilon$ is the vector of system responses in all sensor locations $i$ and *measurement* time instances $t = 0, \ldots, T$, loading vector $\mathbf{f}$ represents the loading forces in all loading-exposed DOFs and in all *loading* time instances $\tau = -\Delta T, \ldots T$, while $\mathbf{D}^{\mathrm{f}}$ is the system transfer matrix compound of discretized $\mathbf{D}_i^{\mathrm{f}}$.

### 4.3.1.2 Elastoplastic Systems

The description of system dynamics stated above can be extended to include the elastoplastic system behavior by combining the computationally effective virtual distortion method (VDM) [18] with the return-mapping algorithm of Simo and Hughes [29]. The small deformation restriction still applies and the extension is obviously at the cost of the linearity. For notational simplicity the concept is presented here for trusses only and $\varepsilon_i(t)$ denotes in this subsection the strain in the $i$th truss element. Nevertheless, with inessential modifications (which result, however, in much notational overhead), the concept is readily applicable to other types of structures and linear sensors.

Equation (26) in the elastoplastic case has to take into account the effect of the plastic distortions of the truss elements,

$$\varepsilon_i(t) = \sum_{\tau=-\Delta T}^{t} \mathbf{D}_i^{\mathrm{f}}(t-\tau)^{\mathrm{T}}\mathbf{f}(\tau) + \sum_{\tau=-\Delta T}^{t} \mathbf{D}_i^{\varepsilon}(t-\tau)^{\mathrm{T}}\beta^0(\tau) \tag{29}$$

where vector $\beta^0$ contains the discretized plastic distortions of all truss elements and $\mathbf{D}_i^{\varepsilon}$ is the vector of the discrete system transfer functions $D_{ij}^{\varepsilon}$, which relate the response in the $i$th sensor location (i.e. the strain in the $i$th element according to the convention of this subsection) to the unit plastic distortion of the $j$th truss element. Equation (29), being very similar to Equation (26), seems to be linear, but it obviously cannot be the case here. One of the reasons is that the plastic distortions $\beta^0$ are nonlinearly dependent on the loading $\mathbf{f}$.

Only isotropic hardening plasticity is considered in the following as a relatively basic example requiring only one internal hardening variable $\Psi_i(t)$, which denotes the *total plastic strain*:

$$
\begin{aligned}
\Psi_i(t + \Delta t) &= \Psi_i(t) + |\Delta\beta_i^0(t + \Delta t)| \\
\Delta\beta_i^0(t + \Delta t) &= \beta_i^0(t + \Delta t) - \beta_i^0(t)
\end{aligned}
\tag{30}
$$

The yield criterion can be expressed in the case of isotropic hardening via the total plastic strain as follows:

$$
|\sigma_i(t + \Delta t)| = \sigma_i^\star + \frac{\gamma_i}{1 - \gamma_i} E_i \Psi_i(t + \Delta t)
\tag{31}
$$

where $\sigma_i^\star$, $\gamma_i$ and $E_i$ are respectively the initial plastic flow stress, the hardening coefficient and Young's modulus of the $i$th truss element. Other plasticity models can be relatively easily obtained by increasing the number of internal variables (see Reference [29]).

To determine the set of elements yielding in each time step and the actual amounts of plastic flows, the *trial strain* $\varepsilon_i^{\mathrm{tr}}(t + \Delta t)$ and the *trial stress* $\sigma_i^{\mathrm{tr}}(t + \Delta t)$ have to be computed by *freezing the plastic flow* and performing a purely elastic step, which amounts to a temporary assumption $\beta^0(t + \Delta t) := \beta^0(t)$ in Equation (29):

$$
\begin{aligned}
\varepsilon_i^{\mathrm{tr}}(t + \Delta t) &= \varepsilon_i(t + \Delta t) - \mathbf{D}_i^\varepsilon(0)^{\mathrm{T}} \Delta\beta^0(t + \Delta t) \\
\sigma_i^{\mathrm{tr}}(t + \Delta t) &= E_i \left[ \varepsilon_i^{\mathrm{tr}}(t + \Delta t) - \beta_i^0(t) \right]
\end{aligned}
\tag{32}
$$

The actual stress $\sigma_i(t + \Delta t)$ in the $i$th element at time $t + \Delta t$ can be expressed in two ways:

1. One is in terms of Young's modulus and the actual values of the strain and the plastic distortion as

$$
\sigma_i(t + \Delta t) = E_i \left[ \varepsilon_i(t + \Delta t) - \beta_i^0(t + \Delta t) \right]
$$

which by Equations (30) and (32) can be transformed to

$$
\begin{aligned}
\sigma_i(t + \Delta t) &= E_i \left[ \varepsilon_i^{\mathrm{tr}}(t + \Delta t) + \mathbf{D}_i^\varepsilon(0)^{\mathrm{T}} \Delta\beta^0(t + \Delta t) - \beta_i^0(t) - \Delta\beta_i^0(t + \Delta t) \right] \\
&= \sigma_i^{\mathrm{tr}}(t + \Delta t) + E_i \sum_j \left[ D_{ij}^\varepsilon(0) - \delta_{ij} \right] \Delta\beta_j^0(t + \Delta t)
\end{aligned}
\tag{33}
$$

where $\delta_{ij}$ is Kronecker's delta.
2. Provided the element is yielding, the stress $\sigma_i(t + \Delta t)$ can also be obtained by multiplying the yield criterion of Equation (31) by $\mathrm{sgn}\,\sigma_i(t + \Delta t)$ and noticing that for isotropic hardening the trial stress $\sigma_i^{\mathrm{tr}}$, the actual stress $\sigma_i$ and the plastic flow $\Delta\beta_i^0$ are all of the same sign in each time step:

$$
\sigma_i(t + \Delta t) = \sigma_i^\star \,\mathrm{sgn}\,\sigma_i^{\mathrm{tr}}(t + \Delta t) + \frac{\gamma_i}{1 - \gamma_i} E_i \left[ \Psi_i(t)\,\mathrm{sgn}\,\sigma_i^{\mathrm{tr}}(t + \Delta t) + \Delta\beta_i^0(t + \Delta t) \right]
\tag{34}
$$

Equations (33) and (34) combined together yield the following linear set of equations for the plastic flow $\Delta\beta_i^0(t + \Delta t)$:

$$E_i \sum_{j \in \mathfrak{B}_{t+\Delta t}} \left[ D_{ij}^\varepsilon(0) - \delta_{ij} \left( 1 + \frac{\gamma_i}{1 - \gamma_i} \right) \right] \Delta\beta_j^0(t + \Delta t)$$

$$= -\sigma_i^{tr}(t + \Delta t) + \left[ \sigma_i^* + \frac{\gamma_i}{1 - \gamma_i} E_i \Psi_i(t) \right] \text{sgn}\, \sigma_i^{tr}(t + \Delta t) \qquad (35)$$

which are valid only for the actually yielding elements $\{i, j\} \in \mathfrak{B}_{t+\Delta t}$. The extent of the plastic zone $\mathfrak{B}_{t+\Delta t}$ has to be found iteratively by applying the yield criterion of Equation (31) to the trial stresses $\sigma_i^{tr}(t + \Delta t)$ and, if necessary, to the stresses computed consecutively by Equations (35) and (32).

### 4.3.2 Load Reconstruction

Load reconstruction amounts basically to deconvolution: compare the measured $\varepsilon^M$ and the modeled $\varepsilon$ system responses and obtain the excitation by solving the resulting system of equations. For a linear system it leads either to a system of several Volterra integral equations of the first kind (the continuous time case) (see Equation (26))

$$\varepsilon_i^M(t) = \int_{-\Delta T}^t \tilde{\mathbf{D}}_i^f(t - \tau)^T \mathbf{f}(\tau) \qquad (36)$$

or, in the discrete time case, to a large system of linear equations, (see Equation (28))

$$\varepsilon^M = \mathbf{D}^f \mathbf{f} \qquad (37)$$

An elastoplastic system (see Equation (29)) yields the following set of nonlinear equations:

$$\varepsilon_i^M(t) = \sum_{\tau=-\Delta T}^t \mathbf{D}_i^f(t - \tau)^T \mathbf{f}(\tau) + \sum_{\tau=-\Delta T}^t \mathbf{D}_i^\varepsilon(t - \tau)^T \beta^0(\tau) \qquad (38)$$

In all cases, the unknown is the loading vector $\mathbf{f}$.

Equation (37) is a large linear system, where the vectors $\varepsilon^M$ and $\mathbf{f}$ contain respectively the measured discretized system response and the discretized loading forces, while $\mathbf{D}^f$ is the system transfer matrix. By a proper reordering of the vectors $\varepsilon^M$ and $\mathbf{f}$, the matrix $\mathbf{D}^f$ can take the rearranged form of a large block matrix composed of Toeplitz matrices, which relate the discretized sensor responses to the unit excitations in all potentially load-exposed DOFs.

Note that in the intended practical situations the linear system of Equation (37) is usually underdetermined; i.e. there are significantly fewer equations than unknowns being sought. The reason is twofold: (1) there are significantly fewer sensors than potentially load-exposed DOFs; (2) time intervals of different lengths (measurement $T$ and reconstruction $T + \Delta T$) are discretized with the same time step $\Delta t$. More precisely, let $A$ be the number of sensors, $N$ the number of potentially load-exposed DOFs, $T$ the length of the reconstruction time interval and $\Delta t$ the time discretization. Then in the linear system of Equation (37) there are $AT/\Delta t$ equations (related to the measurements) and $N(T + \Delta T)/\Delta t$ unknowns (related to the loading being reconstructed). Moreover, even with sufficiently many sensors and equations, a specific topology of the mechanical system being modeled can decrease the rank of the matrix $\mathbf{D}^f$ and make it underdetermined.

Both the continuous Equation (36) and the discrete Equation (37) tend to be ill-conditioned. This is mainly due to the inherent ill-conditioning of compact integral operators of the first kind, which cannot have a bounded inverse [30]. A seemingly contradictory behaviour is the result: the finer the time discretization $\Delta t$, the more ill-conditioned Equation (37) is. Moreover, the ill-conditioning of Equation (37) may also arise due to a small or neglected time shift $\Delta T$, which results in the block matrix $\mathbf{D}^f$ including almost triangular Toeplitz matrices with very small values near the diagonal. Therefore, a regularization technique is usually essential [21, 30–32].

Note also that the linear system of Equation (37) can be easily preconditioned by simple substitutions $\varepsilon^M \leftarrow \mathbf{M}\varepsilon^M$ and $\mathbf{D}^f \leftarrow \mathbf{M}\mathbf{D}^f$ to take effectively the form $\mathbf{M}\varepsilon^M = \mathbf{M}\mathbf{D}^f\mathbf{f}$, provided matrix $\mathbf{M}$ is of appropriate dimensions. Preconditioning may be desired in order to weight the responses of different sensors or to speed up the optimisation-based reconstruction process [33]. For reasons of notational simplicity, no preconditioning (or an identity preconditioner matrix $\mathbf{M} = \mathbf{I}$) is assumed further on.

Real-world measurements are always discrete; hence the continuous time case is dropped from the following as impractical. The next two parts deal with the over- and underdetermined discrete linear cases, while the third part considers the elastoplastic case.

### 4.3.2.1 Overdetermined Linear Systems

If the system of Equation (37) is overdetermined, a unique generalized load can be found to minimize its residuum. The direct way to find it would be to use the pseudoinverse matrix $\mathbf{D}^{f+}$, which can be obtained, for example, via the singular value decomposition (SVD) of the matrix $\mathbf{D}^f$. Moreover, the use of the SVD would allow direct truncating too-small singular values, which is a common regularization technique. However, as matrix $\mathbf{D}^f$ is usually very large, a quicker and less memory-consuming way may be to use iterative methods to minimize the residuum of Equation (37), possibly coupled with the Tikhonov regularization term [21, 31, 32].

*Objective Function*
The system of Equation (37) can be solved via the least squares approach [34] using the following objective function:

$$h(\mathbf{f}) = \left\| \varepsilon^M - \mathbf{D}^f\mathbf{f} \right\|^2 + \delta \left\| \mathbf{B}\mathbf{f} \right\|^2 \tag{39}$$

where $\|\mathbf{B}\mathbf{f}\|^2$ is a Tikhonov regularization term and $\delta \geq 0$ may be assigned a specific numerical value, for example by means of the L-curve technique [21, 30–32]. Taking into account Equations (27) and (28), the objective function can thus be rewritten as

$$h(\mathbf{f}) = \sum_i \sum_{t=0}^{T} \left[ \varepsilon_i^M(t) - \varepsilon_i(t) \right]^2 + \delta \left\| \mathbf{B}\mathbf{f} \right\|^2 \tag{40}$$

while its derivatives can be expressed as

$$\frac{\partial h(\mathbf{f})}{\partial f_N(t)} = \delta \frac{\partial \|\mathbf{Bf}\|^2}{\partial f_N(t)} - 2 \sum_i \sum_{\tau=\max(0,t)}^{T} \left[\varepsilon_i^M(\tau) - \varepsilon_i(\tau)\right] D_{iN}^f(\tau - t),$$ (41)

where $f_N(t)$ and $D_{iN}^f$ are respectively elements of the vectors $\mathbf{f}(t)$ and $\mathbf{D}_i^f$, and

$$\frac{\partial \|\mathbf{Bf}\|^2}{\partial f_N(t)} = 2 \left[\mathbf{B}^T \mathbf{Bf}\right]_{N,t}$$ (42)

Note that the formulas (40) and (41), as well as Equation (44) below, make extensive use of the special form of the system matrix $\mathbf{D}^f$ in order to spare the numerical costs by one order of magnitude. However, they hold for the assumed identity preconditioner matrix $\mathbf{M} = \mathbf{I}$ only, and in other cases have to be replaced by their general counterparts.

The objective function (39) is quadratic and convex; hence it can be exactly expanded around a given loading vector $f$ as

$$h(\mathbf{f} + \mathbf{d}) = h(\mathbf{f}) + \nabla h(\mathbf{f})^T \mathbf{d} + \tfrac{1}{2} \mathbf{d}^T \mathbf{H} \mathbf{d}$$ (43)

where $H = \nabla^2 h$ is the positive semidefinite Hessian of $h$. Equation (43) compared with Equation (40) yields the two following basic formulas:

$$\nabla h(\mathbf{f})^T \mathbf{d} = 2\delta \mathbf{f}^T \mathbf{B}^T \mathbf{B} \mathbf{f} - 2 \sum_i \sum_{t=0}^{T} \varepsilon_i^{(d)} \left[\varepsilon_i^M(t) - \varepsilon_i^{(f)}(t)\right]^2$$

$$\mathbf{d}^T \mathbf{H} \mathbf{p} = 2\delta \mathbf{d}^T \mathbf{B}^T \mathbf{B} \mathbf{p} + 2 \sum_i \sum_{t=0}^{T} \varepsilon_i^{(d)}(t) \, \varepsilon_i^{(p)}(t)$$ (44)

where $\varepsilon_i^{(d)}$ denotes the response in the $i$th sensor location to the excitation $\mathbf{d}$.

### Line Optimization

The iterative optimization method used here consists of a series of line optimizations. Each one amounts to finding the minimum of $h$ at a given loading $\mathbf{f}$ along a given direction $\mathbf{d}$, i.e. the minimum of $h(\mathbf{f} + s\,\mathbf{d})$ with respect to $s$. Due to Equation (43) this is a convex quadratic function with the minimum

$$s_{\min} = -\frac{\nabla h(\mathbf{f})^T \mathbf{d}}{\mathbf{d}^T \mathbf{H} \mathbf{d}}$$ (45)

which is directly calculable by Equation (44).

### Conjugate Gradient Method

Given a loading $\mathbf{f}$, the minimum along each optimization direction $\mathbf{d}$ can be directly calculated by Equation (45). However, first the optimization direction $\mathbf{d}$ has to be chosen.

Equations (41) and (42) allow the steepest descent direction $-\nabla h(\mathbf{f})$ to be calculated directly. The objective function is unbounded quadratic, thus perfectly suited for the conjugate gradient

**Table 4.2**  Optimization algorithm for the overdetermined case

Initial calculations:

| | |
|---|---|
| 1. Initialize | $\mathbf{f}^{(0)} := \mathbf{0}, \quad \varepsilon^{(\mathbf{f}^{(0)})} := \mathbf{0}, \quad n := 0$ |

The loop:

| | |
|---|---|
| 2. Calculate | $\mathbf{d}^{(n)} := -\nabla h(\mathbf{f}^{(n)}), \quad \varepsilon^{(\mathbf{d}^{(n)})}$ |
| 3. Conjugate direction | for $(k = 0;\ k < n, \text{++}k)$ |
| (3a) | $\eta := -\mathbf{d}^{(n)\mathrm{T}}\mathbf{H}\mathbf{d}^{(k)}$ |
| (3b) | $\mathbf{d}^{(n)} := \mathbf{f}^{(n)} + \eta\,\mathbf{d}^{(k)}$ |
| (3c) | $\varepsilon^{(\mathbf{d}^{(n)})} := \varepsilon^{(\mathbf{d}^{(n)})} + \eta\,\varepsilon^{(\mathbf{d}^{(k)})}$ |
| 4. Normalize | |
| (4a) | $D := \sqrt{\mathbf{d}^{(n)\mathrm{T}}\mathbf{H}\mathbf{d}^{(n)}}$ |
| (4b) | $\mathbf{d}^{(n)} := \mathbf{d}^{(n)}/D$ |
| (4c) | $\varepsilon^{(\mathbf{d}^{(n)})} := \varepsilon^{(\mathbf{d}^{(n)})}/D$ |
| 5. Line minimum | $s := -\nabla h(\mathbf{f}^{(n)})^{T}\mathbf{d}^{(n)}$ |
| 6. Store | $\mathbf{d}^{(n)}, \quad \varepsilon^{(\mathbf{d}^{(n)})}$ |
| 7. Update | |
| (7a) | $\mathbf{f}^{(n+1)} := \mathbf{f}^{(n)} + s\,\mathbf{d}^{(n)}$ |
| (7b) | $\varepsilon^{(\mathbf{f}^{(n+1)})} := \varepsilon^{(\mathbf{f}^{(n)})} + s\,\varepsilon^{(\mathbf{d}^{(n)})}$ |
| (7c) | $n := n + 1$ |

method: choosing in the $(n + 1)$th optimization step the direction $\mathbf{d}^{(n+1)}$ *conjugate* to all previous optimisation directions $\mathbf{d}^{(0)}, \ldots, \mathbf{d}^{(n)}$ yields, by Equation (45), directly the minimum in the subspace spanned by all the considered directions span $(\mathbf{d}^{(0)}, \ldots, \mathbf{d}^{(n)})$. Therefore, starting with the steepest descent direction and making use of the conjugacy criterion $\mathbf{d}^{(k)\mathrm{T}}\mathbf{H}\mathbf{d}^{(l)} = 0$,

$$\mathbf{d}^{(n+1)} = -\nabla h(\mathbf{f}^{(n+1)}) + \sum_{k=0}^{n} \eta_{n+1,k}\mathbf{d}^{(k)}, \quad \text{where } \eta_{n+1,k} = \frac{\nabla h(\mathbf{f}^{(n+1)})^{\mathrm{T}}\mathbf{H}\mathbf{d}^{(k)}}{\mathbf{d}^{(k)\mathrm{T}}\mathbf{H}\mathbf{d}^{(k)}} \quad (46)$$

which is basically the Gram–Schmidt orthogonalization scheme. The calculations start with the steepest descent direction; hence in exact arithmetic $\eta_{n+1,k} = 0$ holds for $k = 0, 1, \ldots, n - 1$, which is a useful property of the conjugated gradient method [33]. However, the limited accuracy of the floating-point arithmetic makes all the correction terms necessary.

### The Algorithm

The algorithm for the overdetermined case is presented in Table 4.2. The most expensive operations are the calculations of the gradient and of the corresponding response (step 2). Moreover, at large step numbers $n$, it is quicker to calculate the response $\varepsilon^{(\mathbf{d}^{(n)})}$ directly by Equation (27) than to superpose the stored responses (step 3c).

#### 4.3.2.2 Underdetermined Linear Systems

All known research, (see, for example, References [20] to [22] for reviews and analysis), deals with the overdetermined case only. However, in real-world applications the number of sensors is limited by practical reasons. Therefore, in the overdetermined case the generality of the load being reconstructed must be heavily limited. As mentioned in the introduction, the load has usually been assumed to be a single stationary (or moving at a constant velocity) pointwise force, while its location is assumed to be known in advance or determined by a second-stage nonlinear optimization: freely moving, multiple or diffuse loads are excluded. The approach presented in this subsection addresses the general underdetermined case directly, although at the cost of the reconstruction uniqueness. This allows all general loading patterns to be taken into account.

Generally, with an underdetermined system (Equation (37)), the unknown loading $\mathbf{f}$ can be reconstructed in two ways, which differ in accuracy and numerical costs per single reconstruction (time, memory, etc.). The information lost in measurement in an underdetermined system can be completed by heuristic assumptions only; hence they play an important role in both of the following ways:

- The more accurate approach requires a numerically costly (but one-time only) singular value decomposition (SVD) of the matrix $\mathbf{D}^f$. Thereupon, given a measured response vector $\varepsilon^M$, two complementary components of the corresponding load can be relatively quickly reconstructed: the reconstructible component (based on the measurements $\varepsilon^M$) and the unreconstructible component (based on the heuristic assumptions).
- The less accurate approach makes no distinction between both loading components and reconstructs them simultaneously. The system (37) is first transformed to a larger overdetermined system using heuristic (and in fact regularizing) assumptions and then solved with the iterative approach presented above for overdetermined systems. The numerically costly singular value decomposition of the obtained augmented system is hence not required, but at the expense of a higher numerical cost per single reconstruction. This approach is generally less accurate, since the heuristic assumptions influence both reconstructible and unreconstructible components of the loading being reconstructed, while the former component can – and thus should – be reconstructed on the basis of the measurements $\varepsilon^M$ only. The solution of the augmented system by the SVD is possible, but prohibitive in terms of the numerical costs due to its size.

*Load decomposition*
The matrix $\mathbf{D}^f$ of Equation (37) has a singular value decomposition (SVD):

$$\mathbf{D}^f = \mathbf{U}\boldsymbol{\Sigma}\mathbf{V}^T, \tag{47}$$

where $\mathbf{U}$ and $\mathbf{V}$ are square unitary matrices, i.e. $\mathbf{U}^T\mathbf{U} = \mathbf{U}\mathbf{U}^T = \mathbf{I}$ and $\mathbf{V}^T\mathbf{V} = \mathbf{V}\mathbf{V}^T = \mathbf{I}$. Their dimensions equal respectively the number of equations and the number of unknowns. The matrix $\boldsymbol{\Sigma}$ is a rectangular diagonal matrix of appropriate dimensions whose diagonal values are called singular values of $\mathbf{D}^f$ and are customary ordered nonincreasing. The SVD (47) is unique up to the permutation of the singular values [34].

The system (37) is usually very ill-conditioned, which is indicated by its singular values (diagonal elements of $\boldsymbol{\Sigma}$) spanning across several orders of magnitude. A regularization is then a must, which, given the SVD (47), may be performed in a straightforward way by zeroing too small singular values, i.e. the values below the threshold level defined by the expected relative measurement accuracy $\delta \geq 0$ (see References [21], [31] and [32]). In this way the modified diagonal and system matrices $\Sigma^\delta$, $\mathbf{D}^{f\delta}$ are obtained and Equation (47) takes the following regularized form

$$\mathbf{D}^{f\delta} = \mathbf{U}\,\Sigma^\delta \mathbf{V}^{\mathrm{T}} \tag{48}$$

Note that with $\delta = 0$ all singular values are preserved: $\mathbf{D}^{f0} = \mathbf{D}^f$.

Let $\mathbf{F}$ be the linear space of all possible loadings $\mathbf{f}$. Let $\mathbf{V}_1^\delta$ and $\mathbf{V}_2^\delta$ denote the two matrices composing together the matrix $\mathbf{V} = \begin{bmatrix} \mathbf{V}_1^\delta & \mathbf{V}_2^\delta \end{bmatrix}$, where the number of columns of $\mathbf{V}_1^\delta$ equals the number of positive singular values of $\mathbf{D}^{f\delta}$, i.e. the number of positive diagonal values of $\Sigma^\delta$. The matrix $\mathbf{V}$ is unitary; thus the columns of $\mathbf{V}_1^\delta$ and $\mathbf{V}_2^\delta$ are mutually orthonormal vectors, constitute an orthonormal basis in $\mathbf{F}$ and hence span two orthogonal and complementary linear subspaces $\mathbf{F}_1^\delta$ and $\mathbf{F}_2^\delta$:

$$\mathbf{F} = \mathbf{F}_1^\delta \times \mathbf{F}_2^\delta, \qquad \mathbf{F}_1^\delta = \operatorname{span} \mathbf{V}_1^\delta, \qquad \mathbf{F}_2^\delta = \operatorname{span} \mathbf{V}_2^\delta \tag{49}$$

Due to Equation (48), $\mathbf{F}_2^\delta = \ker \mathbf{D}^{f\delta}$, i.e.

$$\mathbf{D}^{f\delta}\mathbf{V}_2^\delta = \mathbf{0} \tag{50}$$

and hence the regularized system transfer matrix $\mathbf{D}^{f\delta}$ is a linear measurement operator, which effectively: (1) transforms $\mathbf{F}$ orthonormally, (2) projects it on to $\mathbf{F}_1^\delta$, losing a part of the load information, (3) rescales along the basis directions by $\Sigma^\delta$ and finally (4) transforms again orthonormally via $\mathbf{U}$. Therefore, with respect to $\mathbf{D}^{f\delta}$, $\mathbf{F}_1^\delta$ is the reconstructible subspace and $\mathbf{F}_2^\delta$ is the unreconstructible subspace of $\mathbf{F}$. In other words, given the relative measurement accuracy $\delta \geq 0$, each load $\mathbf{f}$ can be uniquely decomposed into a sum of two orthogonal components:

$$\begin{aligned} \mathbf{f} = \mathbf{V}\,\mathbf{V}^{\mathrm{T}}\mathbf{f} &= \mathbf{V}_1^\delta {\mathbf{V}_1^\delta}^{\mathrm{T}}\mathbf{f} + \mathbf{V}_2^\delta {\mathbf{V}_2^\delta}^{\mathrm{T}}\mathbf{f} \\ &= \mathbf{V}_1^\delta\,\mathbf{m}_1 + \mathbf{V}_2^\delta\,\mathbf{m}_2 = \mathbf{f}_{\mathrm{R}}^\delta + \mathbf{V}_2^\delta\,\mathbf{m}_2 \end{aligned} \tag{51}$$

where the first component $\mathbf{f}_{\mathrm{R}}^\delta = \mathbf{V}_1^\delta {\mathbf{V}_1^\delta}^{\mathrm{T}}\mathbf{f} = \mathbf{V}_1^\delta\,\mathbf{m}_1$ is a linear combination of the columns of $\mathbf{V}_1^\delta$, and hence fully reconstructible from the noisy measurements $\varepsilon^{\mathrm{M}} = \mathbf{D}^f\mathbf{f}$, while the second component $\mathbf{V}_2^\delta\,\mathbf{m}_2$ is a linear combination of the columns of $\mathbf{V}_2^\delta$, and hence unreconstructible, since all respective information is lost in the noisy measurement process represented by the linear operator $\mathbf{D}^{f\delta}$ (due to Equation (50) $\mathbf{D}^{f\delta}\mathbf{V}_2^\delta\,\mathbf{m}_2 = \mathbf{0}$) and thus not retained in the noisy measurements above the required relative degree of accuracy $\delta \geq 0$.

### Reconstructible Load Component

Given the noisy measurements $\varepsilon^{\mathrm{M}}$ and the regularized system matrix $\mathbf{D}^{f\delta}$, the unique corresponding reconstructible load component $\mathbf{f}_{\mathrm{R}}^\delta = \mathbf{V}_1^\delta\mathbf{m}_1^\delta$ can be found either

- directly by the standard pseudoinverse matrix $\mathbf{D}^{f\delta\,+}$ of the regularized system matrix $\mathbf{D}^{f\delta}$,

$$\mathbf{f}_{\mathrm{R}}^\delta = \mathbf{D}^{f\delta\,+}\varepsilon^{\mathrm{M}} = \mathbf{V}\,\Sigma^{\delta\,+}\mathbf{U}^{\mathrm{T}}\varepsilon^{\mathrm{M}} \tag{52}$$

where the diagonal matrix $\boldsymbol{\Sigma}^{\delta+}$ is obtained from $\boldsymbol{\Sigma}^{\delta}$ by transposition and replacement of all nonzero elements by their reciprocals, or
- by solving the system $\varepsilon^{M} = \mathbf{D}^{f\delta} \mathbf{V}_1^{\delta} \mathbf{m}_1^{\delta}$, which can be done, for example, by the conjugate gradient technique described above for the overdetermined systems, used to minimize the residual

$$h_1(\mathbf{m}_1) := \left\| \varepsilon^{M} - \mathbf{D}^{f\delta} \mathbf{V}_1^{\delta} \mathbf{m}_1 \right\|^2 \tag{53}$$

Since $\mathbf{D}^{f\delta}$ is already regularized, no Tikhonov regularization term is necessary. Note that the regularization of the system matrix perturbs its structure; hence, instead of the optimized formulas (40), (41) and (44), their general counterparts have to be used.

### Unreconstructible Load Component

Assume, given the noisy measurements $\varepsilon^{M}$ and the relative accuracy $\delta \geq 0$, that the corresponding regularized reconstructible load component $\mathbf{f}_R^{\delta}$ has already been calculated. According to Equation (50), any linear combination of columns of $\mathbf{V}_2^{\delta}$ added to $\mathbf{f}_R^{\delta}$ does not change the (noisy) system response $\varepsilon^{M}$. Therefore, all loadings of the form

$$\mathbf{f} = \mathbf{f}_R^{\delta} + \mathbf{V}_2^{\delta} \mathbf{m}_2 \tag{54}$$

where $\mathbf{m}_2$ is a vector of arbitrary coefficients, are measurably indistinguishable and thus admissible regularized solutions to Equation (37). The choice of a particular vector of coefficients $\mathbf{m}_2$ and the corresponding loading $\mathbf{f}$ must hence be based on additional criteria, which are intrinsically heuristic and formulate *a priori* assumptions concerning anticipated characteristics of the loading. For example to obtain smooth admissible loading distributions, $\mathbf{m}_2$ can be chosen so that the corresponding loading of Equation (54) minimizes an *a priori* given objective function

$$h_2(\mathbf{m}_2) := \left\| \mathbf{B} \left( \mathbf{f}_R^{\delta} + \mathbf{V}_2^{\delta} \mathbf{m}_2 \right) \right\|^2 \tag{55}$$

where $\mathbf{B}$ is the first derivative with respect to time and/or space. Notice that:

1. Due to Equation (49), $\mathbf{f}_R^{\delta}$ is perpendicular to span $\mathbf{V}_2^{\delta}$. Hence, if $\mathbf{B} = \mathbf{I}$, then $h_2(\mathbf{m}_2)$ is minimized by $\mathbf{m}_2 = \mathbf{0}$ and $\mathbf{f}_R^{\delta}$ is the optimum loading itself.
2. As $h_2(\mathbf{m}_2)$ is a quadratic function, the optimum loading

$$\mathbf{f}^{\delta} = \mathbf{f}_R^{\delta} + \mathbf{V}_2^{\delta} \arg\min h_2(\mathbf{m}_2) \tag{56}$$

depends linearly on $\mathbf{f}_R^{\delta}$ and, due to Equation (52), on the measurements $\varepsilon^{M}$:

$$\mathbf{f}^{\delta} = \left[ \mathbf{I} - \mathbf{V}_2^{\delta} \left( \mathbf{V}_2^{\delta\mathrm{T}} \mathbf{B}^{\mathrm{T}} \mathbf{B} \, \mathbf{V}_2^{\delta} \right)^{-1} \mathbf{V}_2^{\delta\mathrm{T}} \mathbf{B}^{\mathrm{T}} \mathbf{B} \right] \mathbf{f}_R^{\delta} \tag{57}$$

A scrupulous analysis would reveal that in Equation (54) and (55), upper bounds should be put on the moduli of the elements of $m_2$ that correspond to the columns of $\mathbf{V}_2^{\delta}$ not belonging to $\mathbf{V}_2^{0}$. These columns represent unit loadings that influence the measurement $\varepsilon^{M}$, although below the noise threshold $\delta$.

### Single-Stage Load Reconstruction

Instead of separate successive reconstructions of the reconstructible and unreconstructible loading components, the loading can be reconstructed in one stage only, but at the cost of accuracy. The objective functions $h_1(\mathbf{m}_1)$ of Equation (53) and $h_2(\mathbf{m}_2)$ of Equation (55) resemble the components of the general objective function $h(\mathbf{f})$ in Equation (39), which is used in the overdetermined case. Hence, instead of successive separate optimizations of $h_1(\mathbf{m}_1)$ and $h_2(\mathbf{m}_2)$, both functions may be optimized simultaneously, as in $h(\mathbf{f})$, weighted by an appropriate coefficient $\delta > 0$:

$$h_\delta(\mathbf{f}) := \left\| \varepsilon^M - \mathbf{D}^f \mathbf{f} \right\|^2 + \delta \left\| \mathbf{B} \mathbf{f} \right\|^2 \tag{58}$$

As the second term also plays the role of a regularization term, there is no need to use the regularized system matrix $\mathbf{D}^{f\delta}$, and hence no need for numerically costly (although one-time only) singular value decomposition. The minimization of the compound objective function $h_\delta$ can be performed relatively quickly by the conjugate gradient technique described earlier.

This one-stage approach makes no distinction between the reconstructible and the unreconstructible loading components and retrieves them simultaneously. Therefore, it is generally less accurate, since the heuristic assumptions influence both components of the reconstructed loading, while the two-stage approach described in the preceding subsection properly reconstructs the reconstructible component on the basis of the measurements $\varepsilon^M$ only.

### 4.3.2.3 Elastoplastic Systems

The approaches described before rely heavily on the linearity of the system equation (37). The elastoplastic case of Equation (38) has to be treated separately. In general, three cases are possible:

1. *Strongly overdetermined* system, possible in the case of a very limited loading area. The number of sensors exceeds or equals the total number of potentially loading-exposed DOFs and potentially plastified truss elements. The approach described before for overdetermined linear systems is straightforwardly applicable (unless the system is singular), with both loads $f_N(t)$ and plastic distortions $\beta_i^0(t)$ treated as independent unknowns.
2. *Overdetermined* system. The number of sensors exceeds or equals the number of potentially loading-exposed DOFs. The unique evolution of the load can be reconstructed from the measurements, unless the system is singular. However, the system is not linear and thus the approaches of the preceding sections are not applicable.
3. *Underdetermined* system. The number of potentially loading-exposed DOFs exceeds the number of sensors. In general, the gradient-based optimization approach presented below reconstructs a nonunique evolution of loading, which is observationally indistinguishable from the actual loading.

In the overdetermined elastoplastic case, Equation (38) can be, in general, uniquely solved by minimizing the objective function (40) with any gradient-based optimization algorithm. However, the modeled system response $\varepsilon(t)$ is no longer a linear function of the loading $\mathbf{f}(t)$ and the derivatives, instead of in Equation (41), take the following form:

$$\frac{\partial h(\mathbf{f})}{\partial f_N(t)} = \delta \frac{\partial \left\| \mathbf{B} \mathbf{f} \right\|^2}{\partial f_N(t)} - 2 \sum_i \sum_{\substack{\tau = \\ \max(0,t)}}^{T} \left[ \varepsilon_i^M(\tau) - \varepsilon_i(\tau) \right] \frac{\partial \varepsilon_i(\tau)}{\partial f_N(t)} \tag{59}$$

The first derivative is computable by Equation (42), while the second, for $t \leq \tau$, by Equation (29) and (30), is

$$\frac{\partial \varepsilon_i(\tau)}{\partial f_N(t)} = D_{iN}^{\mathrm{f}}(\tau - t) + \sum_{\kappa=t}^{\tau} \sum_{j} D_{ij}^{\varepsilon}(\tau - \kappa) \sum_{\nu=t}^{\kappa} \frac{\partial \Delta \beta_j^0(\nu)}{\partial f_N(t)}$$

The derivatives of the plastic flow $\Delta \beta_j^0(\nu)$ with respect to the load $f_N(t)$ have to be computed prior to the derivatives of the objective function (59) by iteratively solving the following linear sets of equations, which are obtained by differentiating Equation (35) and making use of Equation (30) to compute the derivatives of the total plastic strain:

$$E_i \sum_{j \in \mathcal{B}_t} \left[ D_{ij}^{\varepsilon}(0) - \delta_{ij} \left( 1 + \frac{\gamma_i}{1 - \gamma_i} \right) \right] \frac{\partial \Delta \beta_j^0(t)}{\partial f_N(t)} = -E_i D_{iN}^{\mathrm{f}}(0)$$

$$E_i \sum_{j \in \mathcal{B}_{\nu + \Delta t}} \left[ D_{ij}^{\varepsilon}(0) - \delta_{ij} \left( 1 + \frac{\gamma_i}{1 - \gamma_i} \right) \right] \frac{\partial \Delta \beta_j^0(\nu + \Delta t)}{\partial f_N(t)}$$

$$= -\frac{\partial \sigma_i^{\mathrm{tr}}(\nu + \Delta t)}{\partial f_N(t)} + \frac{\gamma_i}{1 - \gamma_i} E_i \sum_{\iota=t}^{\nu} \frac{\partial \Delta \beta_i^0(\iota)}{\partial f_N(t)} \, \mathrm{sgn} \, \Delta \beta_i^0(\iota) \, \mathrm{sgn} \, \sigma_i^{\mathrm{tr}}(\nu + \Delta t)$$

where, by Equation (32) and (29), the derivative of the trial stress is

$$\frac{\partial \sigma_i^{\mathrm{tr}}(\nu + \Delta t)}{\partial f_N(t)} = E_i D_{iN}^{\mathrm{f}}(\nu + \Delta t - t)$$

$$+ E_i \sum_{\kappa=t}^{\nu + \Delta t} \sum_{j} \left[ D_{ij}^{\varepsilon}(\nu + \Delta t - \kappa) - \delta_{\kappa \nu} \delta_{ij} \right] \sum_{\iota=t}^{\min(\kappa, \nu)} \frac{\partial \Delta \beta_j^0(\iota)}{\partial f_N(t)}$$

where $\delta_{\kappa \nu}$ and $\delta_{ij}$ denote Kronecker's delta.

### 4.3.3 Optimum Sensor Location

In the load reconstruction strategies described above, the system transfer matrix $\mathbf{D}^{\mathrm{f}}$ is crucial. However, it depends on the location of the sensors. This is a problem on its own, especially in underdetermined systems with the number of potentially loading exposed DOFs exceeding considerably the number of available sensors. There is not much theoretical investigation into the problem. This section proposes two optimality criteria based on conditioning and accuracy, which tend to be negatively correlated, and proposes a third, compound criterion. However, only linear systems are considered. The problem of the optimal sensor location in the case of elastoplastic systems requires further study.

A sensor location is denoted by $\pi$ and can be represented by a nonempty subset of $\{1, 2, \ldots, A_{\max}\}$, where $A_{\max}$ is the number of all possible locations of a single sensor in the structure.

### 4.3.3.1 Conditioning

Mackiewicz *et al.* [35] propose in a similar problem to locate the sensors to minimize the ill-conditioning of the reconstruction process. This can be formalized here as the task of finding the sensor location $\pi$, which minimizes the following measure of ill-conditioning of the corresponding system matrix $\mathbf{D}_\pi^{\mathrm{f}}$:

$$q_1(\pi) = \log \frac{\sigma_{\max}(\mathbf{D}_\pi^{\mathrm{f}})}{\sigma_{\mathrm{median}}(\mathbf{D}_\pi^{\mathrm{f}})} \tag{60}$$

where $\sigma_{\max}(\mathbf{D}_\pi^{\mathrm{f}})$ and $\sigma_{\mathrm{median}}(\mathbf{D}_\pi^{\mathrm{f}})$ are the maximum and the median singular values of $\mathbf{D}_\pi^{\mathrm{f}}$. Note that the matrix condition number, which is the standard measure of conditioning of a matrix, involves the *minimum* singular value instead of the *median* used in Equation (60). However, with inaccurate floating-point computer arithmetic the median is more reliable, since for strongly ill-conditioned matrices the calculated minimum singular value usually lies at a predefined cut-off level, defined by the accuracy of the arithmetic used.

### 4.3.3.2 Accuracy

Besides ill-conditioning, in underdetermined systems along the nonuniqueness of solution the problem of accuracy of the reconstruction also arises. The reconstruction process may be very well-conditioned, but it is of no use if the reconstructed loading differs much from the actual loading. For a given loading, the accuracy of reconstruction can be quantified as a distance between the loading and its reconstructible component.

Therefore, the accuracy of the reconstruction process may be defined with respect to a given set of (expected or typical) unit loadings $\{\mathbf{f}^{(1)}, \mathbf{f}^{(2)}, \dots, \mathbf{f}^{(M)}\}$ as the mean square distance between the loadings and their reconstructible components (see Equation (51)):

$$q_2(\pi) = \frac{1}{M} \sum_{i=1}^{M} \left\| (\mathbf{I} - \mathbf{V}_{1\pi} \mathbf{V}_{1\pi}^{\mathrm{T}}) \mathbf{f}^{(i)} \right\|^2 = \frac{1}{M} \sum_{i=1}^{M} \left\| \mathbf{V}_{2\pi} \mathbf{V}_{2\pi}^{\mathrm{T}} \mathbf{f}^{(i)} \right\|^2 \tag{61}$$

where $\mathbf{V}_{1\pi}$ and $\mathbf{V}_{2\pi}$ denote the matrices $\mathbf{V}_1$ and $\mathbf{V}_2$ calculated for a given sensor location $\pi$. Note that the practical usefulness of $q_2$ depends heavily on the set of loadings $\{\mathbf{f}^{(1)}, \mathbf{f}^{(2)}, \dots, \mathbf{f}^{(M)}\}$, which should be well-suited to the loadings expected in the investigated system.

### 4.3.3.3 Compound Criterion

The conditioning ($q_1$) and the accuracy ($q_2$) criteria tend to be negatively correlated (see the numerical example later in Figure 4.30 (left)). Moreover, criterion $q_2$ assumes no measurement noise, which can significantly diminish the accuracy of reconstruction. Therefore, for practical reasons the ultimate criterion has to weight accuracy against conditioning, taking into account the relative measurement noise level $\delta$. This can be achieved by measuring the accuracy of the reconstruction process performed with the regularized system matrix $\mathbf{D}^{\mathrm{f}\delta}$ instead of the original matrix $\mathbf{D}^{\mathrm{f}}$:

$$q_2^\delta(\pi) = \frac{1}{M} \sum_{i=1}^{M} \left\| \mathbf{V}_{2\pi}^\delta \mathbf{V}_{2\pi}^{\delta\,\mathrm{T}} \mathbf{f}^{(i)} \right\|^2 \tag{62}$$

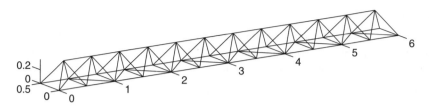

**Figure 4.29**  Elastic truss structure modeled in the numerical example

## 4.3.4 Numerical Example

This subsection illustrates the presented methodology for load reconstruction with a numerical example, which uses an elastic truss structure. A corresponding example for elastoplastic structures can be found in Reference [36].

### 4.3.4.1 Structure

Figure 4.29 shows the modeled elastic truss structure. It is 6 m long, the elements are 10 mm$^2$ in cross-section, 0.5 m or $0.5\sqrt{2}$ m long, and made of steel (7800 kg/m$^3$, 200 GPa). The two left-hand side corner nodes of the bottom plane were deprived of all degrees of freedom, while the two right-hand side corner nodes were deprived of the vertical degree of freedom and are free to move in the horizontal plane only.

Loading forces can occur only vertically in any/all of the 12 upper nodes of the structure. This assumption allows evolution of loading to be depicted in time-space in the convenient and illustrative form of a two-dimensional graph. The *measurement* time interval is $T = 10$ ms and has been discretized into 100 time steps of 0.1 ms each. The time shift is $\Delta T = 1$ ms, and hence the *reconstruction* time interval is 11 ms long (110 time steps). A total of $A$ strain sensors, $A \in \{1, 2, \ldots, 11\}$, can be located in any of the 11 upper elements, which join the 12 potentially loading-exposed nodes. The system transfer matrix $\mathbf{D}^f$ is hence $100A \times 1320$ and thus underdetermined.

### 4.3.4.2 Optimum Sensor Location

For each of 2047 possible sensor locations $\pi$ (nonempty subsets of the 11 strain sensors), both sensor location criteria $q_1$ and $q_2$ have been computed. Figure 4.30 (left) plots $q_2$ versus $q_1$ to illustrate the negative correlation; each dot corresponds to one sensor location $\pi$. A clear arrangement in groups corresponding to the number of sensors can be seen: the more sensors, the worse (larger) is the conditioning $q_1$ and the better (smaller) the accuracy $q_2$. Within the individual groups a slight positive correlation can be observed, although this is not always the case [19].

The criterion $q_2$ has been calculated with respect to a set of 276 simple impact unit loadings extending 9 time steps (0.9 ms) in time and 3 degrees of freedom in space, one half being distributed uniformly in time/space and the other half randomly. Figure 4.30 (right) shows the nonzero fragment of the force evolution pattern, which has been shifted in time and space to form all 276 test loadings.

A measurement noise level of 1 % rms (root mean square) has been assumed. Table 4.3 lists the computed best and worst locations of four, five and six sensors with respect to the compound

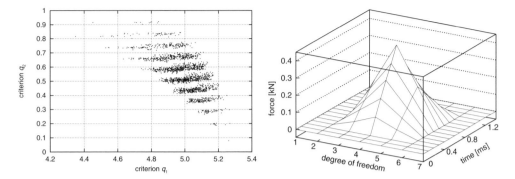

**Figure 4.30** Optimum sensor location: (left) correlation plot for the sensor location criteria $q_1$ and $q_2$; (right) loading pattern shifted in time and space to form all 276 test loadings used for calculating criterion $q_2$

criterion $q_2^{1\%}$ (Equation (62)). It can be seen that the worst locations group the sensors near the sliding supports on the right-hand side, while the best locations are more evenly distributed, although the left-hand side near the fixed supports is preferred. A comparable result was obtained for a cantilever beam in Reference [19].

In subsequent computations the $q_2^{1\%}$-best location of five sensors has been used, i.e. sensors in the upper horizontal elements 1, 2, 4, 6 and 10. Figure 4.31 plots the singular values of the corresponding system transfer matrix across the orders of magnitude. Only 136 loadings (corresponding to the columns of the matrix $\mathbf{V}_1^{1\%}$) fall above the 1 % noise limit.

### 4.3.4.3 Reconstructible Loadings

Figure 4.32 depicts examples of the columns of the matrix $\mathbf{V}_1^{1\%}$ computed for the assumed $q_2^{1\%}$-best sensor location $\pi = \{1, 2, 4, 6, 10\}$. Notice the increasing oscillations: the consecutive singular loadings introduce more and more high frequency components. As seen before, the loadings constitute an orthonormal basis for the space of reconstructible loadings.

**Table 4.3** Five best and worst locations of four, five and six sensors with respect to $q_2^{1\%}$ (each "o" denotes one sensor)

|  | Four sensors | Five sensors | Six sensors |
|---|---|---|---|
| $q_2^{1\%}$-best | oo-o-o------ | oo-o-o---o- | oo-o-o-o--o |
|  | oo-o---o--- | oo-o-o-o--- | oo-o-o--o-o |
|  | oo-o--o---- | oo-o-o----o | oo-o-o---oo |
|  | oo-o-----o- | oo-o---o-o- | oo-o-oo--o- |
|  | oo-o------o | oo-o-oo---- | oo-o-o-o-o- |
| $q_2^{1\%}$-worst | ------oooo | -----ooooo | ----oooooo |
|  | ------oooo- | ----ooooo- | ---oooooo- |
|  | -----ooo-o | ----o-oooo | ----o-ooooo |
|  | ----o-ooo- | ----oooo-o | ---o--oooooo |
|  | ------o-ooo | ---oo-ooo- | ----ooooo-o |

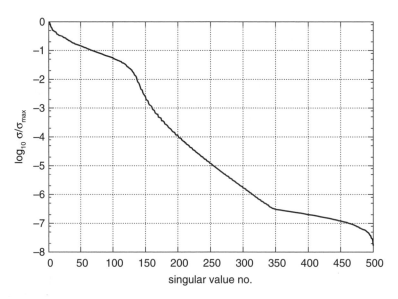

**Figure 4.31**  Singular values of the system transfer matrix plotted across the orders of magnitude for the $q_2^{1\%}$-best sensor location $\pi = \{1, 2, 4, 6, 10\}$

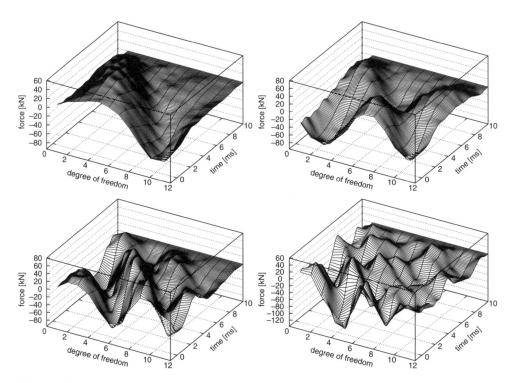

**Figure 4.32**  First (top left), second (top right), third (bottom left) and sixth (bottom right) singular loadings computed for sensor location $\pi = \{1, 2, 4, 6, 10\}$

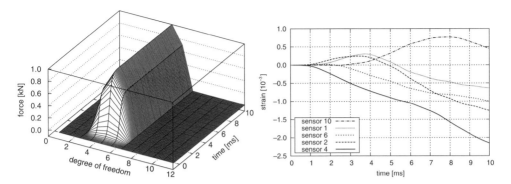

**Figure 4.33**  Numerical example of load reconstruction: (left) exact load to be reconstructed; (right) exact measurement data $\varepsilon^M$ for the $q_2^{1\%}$-best sensor location $\pi = \{1, 2, 4, 6, 10\}$

### 4.3.4.4  Loading and Identification

The load reconstruction procedure has been illustrated numerically. Figure 4.33 depicts the assumed exact loading and the computed exact responses of the five $q_2^{1\%}$-best sensors, $\pi = \{1, 2, 4, 6, 10\}$, which have been subsequently disturbed with numerically generated noise at the 1% rms level.

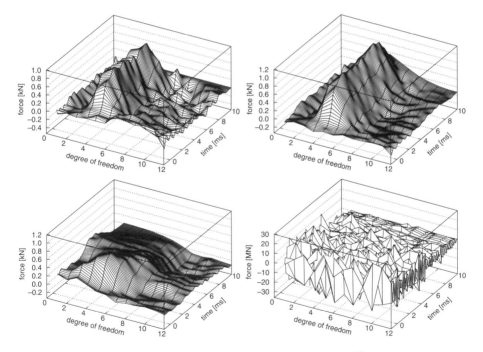

**Figure 4.34**  Numerical example of load reconstruction computed for the $q_2^{1\%}$-best sensor location $\pi = \{1, 2, 4, 6, 10\}$ at noise level 1% rms: (top left) reconstructible load component $\mathbf{f}_R^{1\%}$; (top right) heuristic reconstruction with Equations (63) and (57); (bottom left) heuristic reconstruction for the $q_2^{1\%}$-worst sensor location; (bottom right) reconstructible load component $\mathbf{f}_R$ (no numerical regularization, notice the difference in scale and units)

Using Equation (52), the disturbed sensor responses have been used to obtain the reconstructible component $\mathbf{f}_R^{1\%}$, shown in Figure 4.34 (top left). Full heuristic reconstruction (shown in the figure top right) was performed by Equation (57) with the heuristic assumption weighting smoothness w.r.t. (with respect to) time against smoothness w.r.t. space (degrees of freedom) by

$$\mathbf{B} = \begin{pmatrix} 1.0 & \mathbf{B}_{\text{time}} \\ 0.2 & \mathbf{B}_{\text{DOF}} \end{pmatrix} \tag{63}$$

The results of the corresponding full heuristic reconstruction for the $q_2^{1\%}$-worst location are shown for comparison in Figure 4.34 (bottom left). The figure bottom right confirms that the regularization is necessary by showing the dominated by noise reconstructible component $\mathbf{f}_R^{0\%}$, reconstructed with the full system transfer matrix $\mathbf{D}^f$ instead of its regularized form $\mathbf{D}^{f1\%}$ (Equation (48)). Note also the difference in scale and units.

# References

1. G. Mikułowski and J. Holnicki-Szulc, Fast controller and control algorithms for MR based adaptive impact absorbers – force based control, *Machine Dynamics Problems*, **30**(2), 2006, 113–122.
2. K. Sekuła and J. Holnicki-Szulc, On-line impact identification, *International Journal of Impact Engineering*, 2008 (submitted).
3. A. Tarantola, *Inverse Problem Theory and Methods for Model Parameter Estimation*, SIAM: Society for Industrial and Applied Mathematics, 2004.
4. R. Gryboś, *Teoria Uderzenia w Dyskretnych układach mechanicznych*, Państwowe Wydawnictwo Naukowe, 1969.
5. W. J. Stronge, *Impact Mechanics*, Cambridge University Press, Cambridge, 2000.
6. R. M. Brach, *Mechanical Impact Dynamics: Rigid Body Collisions*, John Wiley & Sons, Ltd, Chichester, 1991.
7. D. Luenberger, An introduction to observers, *IEEE Transactions on Automatic Control*, **AC-16**, December 1971, 596–602.
8. R. C. Dorf and R. H. Bishop, *Modern Control Systems*, 9th edition, Prentice-Hall, Englewood Cliffs, New Jersey, 2001.
9. G. F. Franklin, J. D. Powell and M. L. Workman, *Digital Control of Dynamic System*, Addison Wesley, Reading, Massachusetts, 1990.
10. G. C. Goodwin, S. F. Grabe and M. E. Salgado, *Control System Design*. Prentice-Hall, Englewood Cliffs, New Jersey, 2001.
11. J. Mendel, *Lessons in Estimation Theory for Signal Processing, Communications and Control*, Prentice-Hall, Englewood Cliffs, New Jersey, 1995.
12. M. Klinkov and C. P. Fritzen, Online force reconstruction using robust observers, in *Proceedings of 3rd European Workshop on Structural Health Monitoring*, July 2006, pp. 617–626.
13. M. Klinkov and C.-P. Fritzen, An updated comparison of the force reconstruction methods. *Key Engineering Materials*, **347**, 2007, 461–466.
14. Q. P. Ha and H. Trinh, State and input simultaneous estimation for a class of nonlinear systems. *Automatica*, **40**(10), October 2004, 1779–1785.
15. Y. Xiong and M. Saif, Unknown disturbance inputs estimation based on state functional observer design, *Automatica*, **39**, 2003, 1389–1398.
16. W. Gawronski, B. Bienkiewicz and R. E. Hill, Wind-induced dynamics of a deep space network antenna, *Journal of Sound and Vibration*, **178**(1), 1994, 67–77.
17. M. Hou and P. C. Müller, Design of observers for linear systems with unknown inputs, *IEEE Transactions on Automatic Control*, **37**(6), 1992, 871–875.

18. J. Holnicki-Szulc and J.T. Gierliński, *Structural Analysis, Design and Control by the Virtual Distortion Method*, John Wiley & Sons, Ltd, Chichester, 1995.
19. Ł. Jankowski, M. Wikło and J. Holnicki-Szulc, Robust post-accident reconstruction of loading forces, *Key Engineering Materials*, **347**, 2007, 659–664.
20. T. H. T. Chan, L. Yu and S. S. Law, Moving force identification studies: theory, *Journal of Sound and Vibration*, **247**(1), 2001, 59–76.
21. E. Jacquelin, A. Bennani and P. Hamelin, Force reconstruction: analysis and regularization of a deconvolution problem, *Journal of Sound and Vibration*, **265**, 2003, 81–107.
22. H. Inoue, J.J. Harrigan and S. R. Reid, Review of inverse analysis for indirect measurement of impact force, *Applied Mechanics Reviews*, **54**(6), 2001, 503–524.
23. E. Wu, C.-Z. Tsai and L.-H. Tseng, A deconvolution method for force reconstruction in rods under axial impact, *Journal of Acoustical Society of America*, **104**(3), 1998, 1418–1426.
24. H. Fukunaga and N. Hu, Experimental impact force identification of composite structures, in *Proceedings of 3rd European Workshop on Structural Health Monitoring*, July 2006, pp. 840–847.
25. S. S. Law, T. H. T. Chan and Q. H. Zeng, Moving force identification: a time domain method, *Journal of Sound and Vibration*, **201**(1), 1997, 1–22.
26. L. Gaul and S. Hurlebaus, Identification of the impact location on a plate using wavelets, *Mechanical Systems and System Processing*, **12**(6), 1997, 783–795.
27. M. T. Martin and J. F. Doyle, Impact force location in frame structures, *International Journal of Impact Engineering*, **18**(1), 1996, 79–97.
28. R. Adams and J. F. Doyle, Multiple force identification for complex structures, *Experimental Mechanics*, **42**(1), 2002, 25–36.
29. J. C. Simo and T. J. R. Hughes, *Computational Inelasticity*, Vol. 7, Interdisciplinary Applied Mathematics Series, Springer, New York, 1989.
30. R. Kress, *Linear Integral Equations*, Vol. 82, Applied Mathematical Sciences Series, Springer, New York, 1989.
31. P.Ch. Hansen, Deconvolution and regularization with Toeplitz matrices, *Numerical Algorithms*, **29**, 2002, 323–378.
32. A. N. Tikhonov and V.Y. Arsenin, *Solutions of Ill-Posed Problems*, John Wiley & Sons, Inc., New York, 1977.
33. J. Nocedal and S.J. Wright, *Numerical Optimization*, Springer Series in Operations Research, Springer, New York, 1999.
34. G. Dahlquist and A. Bjørck, *Numerical Methods in Scientific Computing*, SIAM, Philadelphia, 2006 (to be published).
35. A. Mackiewicz, J. Holnicki-Szulc and F. Lopez-Almansa, Optimal sensor location in active control of flexible structures, *American Institute of Aeronautical and Austronatical Journal*, **34**(4), 1996, 857–859.
36. Ł. Jankowski, Off-line identification of dynamic loads, *Structural and Multidisciplinary Optimization*, 2008 (in press).

# 5

# Adaptive Impact Absorption

Piotr K. Pawłowski, Grzegorz Mikułowski, Cezary Graczykowski,
Marian Ostrowski, Łukasz Jankowski and Jan Holnicki-Szulc

## 5.1 Introduction

The objective of this chapter is to present the concept of adaptive impact absorption (AIA) and some examples of its engineering applications. Assuming that the impact load can be identified (cf. *real-time dynamic load identification* discussed in Chapter 4, Section 4.1), the safety engineering based approach to structural design requires equipment of the structure with special devices allowing active control (in real time) of mechanical properties (e.g. local stiffness) in order to improve the dynamic response scenario (e.g. reduction of force or acceleration peaks). As a consequence, the desired AIA process is dissipative, with an optimal amount of impact energy absorption, while the applied control devices (actuators with small external energy consumption) modify only the local structural properties, without feeding the system with additional mechanical energy.

In general, the AIA system should be designed for random impact multiloads, which creates new research challenges due to optimal forming of structural geometry and the location of controllable devices. These problems will be discussed in Chapter 6. In this chapter it is assumed that the structural geometry has already been determined.

Another challenge is to invent innovative technologies applicable to the controllable devices mentioned above. One option discussed in this chapter deals with the concept of structural fuses with an elastoplastic type of overall performance and a controllable yield stress level, where the active device (controllable joints) itself can be based on various types of actuators, e.g. electromagnetic, piezoelectric, magnetostrictive or magneto-SMA. Another important option discussed below is based on the idea of so-called adaptive inflatable structures (AIS) with a controllable (in real-time) release of pressure. Shock absorbers based on MRFs or piezo-valves can be successfully used for AIA in repetitive exploitive impacts, which will be discussed in the case of adaptive landing gears. Finally, micro pyro-technique systems (MPS) can be used for detaching (in real time) selected structural joints in order to improve the structural response in emergency situations (e.g. in a crash of vehicles).

---

*Smart Technologies for Safety Engineering*   Edited by J. Holnicki-Szulc

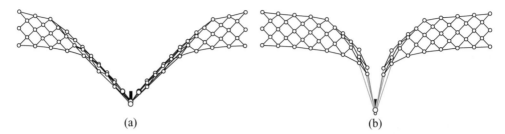

**Figure 5.1** Structural responses to two impacts with the same kinetic energy: (a) 'slow impact' case with a high mass/velocity ratio, (b) 'fast impact' case with a low mass–velocity ratio

Various strategies of adaptation to the identified impact can be proposed, depending on the particular problem, e.g. repetitive exploitive impacts versus critical emergency impacts. Minimization of an acceleration measure in the selected locations and time interval $< 0, T >$ for smoothing down the impact reception corresponds to the first case, when reduction of fatigue accumulation is an important issue. On the other hand, maximization of the impact energy dissipation in the selected time interval $< 0, T >$ for the most effective adaptation to the emergency situation corresponds to the second case. However, other desirable scenarios for AIA can also be proposed in particular situations. For example, the strategy of local structural degradation (e.g. due to provoked perforation in the impacted location) in order to minimize the damaged zone and preserve the structural integrity can also be an option in critical situations. The overall effect of such an action can be defined as improving structural critical load capacity through local debilitation.

Numerical simulations of the structural response to two different impact loads with the same impact energy are shown in Figure 5.1, where case (a) corresponds to slow dynamics (greater mass and smaller impact velocity), while case (b) corresponds to fast dynamics (smaller mass and greater impact velocity). A qualitatively different structural response in both cases confirms how important is the real-time impact load identification (its location and mass–velocity ratio), which decides the selection of proper control objective and the further control strategy.

The techniques applied to control the response of structural fuses allow various strategies of adaptation:

- The simplest one, passive, can be used, for example, if the expected impact load statistics are known. Then, the optimally distributed yield stress levels in structural fuses can be determined and applied as a fixed parameter.
- The semi-active strategy, taking advantage of the real-time impact load identification, can be used when tuning on-line the optimal but fixed during the impact process yield stress levels.
- The partially active strategy allows solutions for the semi-active one and with modifiable in time (but only decreasing) yield stress levels.
- Finally, the fully active strategy allows solutions for the semi-active one and with modifiable in time (both decreasing and increasing) yield stress levels. Going from passive to active strategies, the expected effectiveness of AIA is growing, together with exploitation problems due to timing of the hardware actions and the fact that the fully active strategy requires, for example, closing of the loaded valve. The feasibility of this last option depends on the energy sources available for the activation of structural fuses.

Finally, note that the impact load, even if adaptively received, causes some structural vibrations, which can be damped semi-actively with the use of structural fuses, controlled by the so-called 'pre-stress accumulation release' (PAR) strategy (cf. Chapter 7).

## 5.2 Multifolding Materials and Structures

### 5.2.1 Introduction

Motivation for the undertaken research comes from responding to increasing requirements for high-impact energy absorption in the structures exposed to the risk of extreme blast, tall and compliant offshore structures, etc. Requirements for optimal energy absorbing systems may be stated as follows:

- The system must dissipate the kinetic energy of an impact in a stable and controlled way.
- Displacements must not exceed the maximal allowable values.
- Extreme accelerations and forces of the impact should be reduced to the lowest possible level.

The majority of properly designed, passive energy-absorbing systems fulfill the first two of these requirements. The third one, because of the constant constitutive force–displacement relation, in this case can be realized only to some extent. Therefore, commonly applied passive protective systems are optimal for a limited range of loads.

In contrast to the standard solutions, the proposed approach focuses on active adaptation of energy-absorbing structures (equipped with a sensor system that can detect impact in advance and controllable semi-active dissipators – structural fuses) with a high ability of adaptation to extreme overloading.

The idea of multifolding microstructures (MFM) and numerical tools for analysis and optimization were discussed in detail in References [1] and [2]. The following section will focus on the most important features and results of experimental verification of the concept.

The proposed MFMs are honeycomb-like structures composed of truss elements equipped with special devices (so-called 'structural fuses') providing control over the force characteristics. A typical layout of the MFM is depicted in Figure 5.2. The size of the MFM can be reduced to one or two dimensions (the single-column structure, cf. Figure 5.4).

**Figure 5.2**   MFM structure

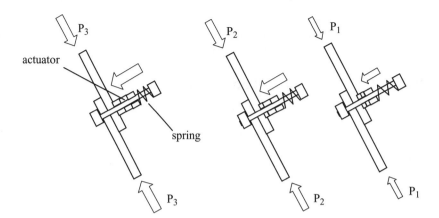

**Figure 5.3**   Controllable microfuses

As an example of a microfuse connection consider a friction joint presented in Figure 5.3. The two parts of an element of the MFM are connected by a specially designed microactuator based on smart materials, e.g. piezoelectric, giant magnetostrictive material (GMM) or shape-memory alloy. Activation of the actuator by an external field (e.g. electric, magnetic or thermal) changes the compressive force in the joint providing changes in the axial force in the element. The resultant constitutive behavior of the MFM element equipped with the structural fuse can be modeled in the first approximation as elastoplastic.

Assume that, without the activation, the compressive force able to start the yielding process in the microfuse is $P_1$ (cf. Figure 5.3). Two different levels of activation provide the forces $P_2 > P_1$ and $P_3 > P_2$ respectively, which are necessary to start yielding.

### 5.2.2 The Multifolding Effect

In order to present the synergy of the multifolding effect a simplified one-dimensional model will be analyzed of the truss structure depicted in Figure 5.4(a) composed of idealized elastoplastic members with the pre-selected yield stress levels (realized through properly activated fuses). When subjected to a load, the MFM will deform in a sequence of local collapses shown in Figures 5.4(b), (c) and (d), respectively.

The corresponding effect of energy dissipation (cf. Figure 5.4(f) related to the adopted initial distribution of yield stresses is over 300 % higher than for the same kind of microstructure, made of the same material volumes and with homogeneously distributed yield stress levels. As a consequence, when compressive forces in all members of an idealized truss-like structure shown in Figure 5.4(a) are the same, all members with the stress limit level $\sigma_2$ start to yield first, converting the structure into the configuration shown in Figure 5.4(b). Then the yield stress level for elements marked $\sigma_3$ is lower than for the tripled elements (with stress limit $2\sigma_1 - \delta$) and the subsequent structural configuration is shown in Figure 5.4(c). Following the same procedure, the next configurations, i.e. Figures 5.4(d) and (e), can be reached.

The crucial method used to obtain the additional value of energy dissipation (due to synergy of the repetitive use of dissipators) is to pre-design an optimal distribution of yield stress levels in all fuses, triggering the desired sequence of local collapses. The resultant force–displacement characteristics are strictly dependent on the distribution and can be optimally adjusted to the expected dynamic load.

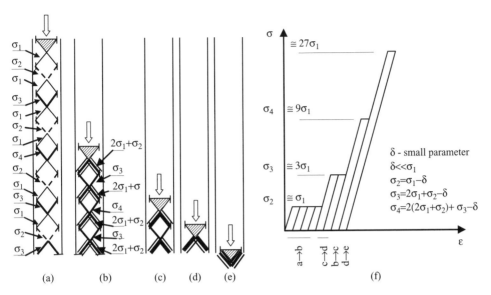

**Figure 5.4** Multifolding effect

## 5.2.3 Basic Model of the MFM

In order to understand the behavior of the multifolding structure better, the most basic model composed only of six elements will be analysed. The layout of the model is depicted in Figure 5.5. The yield stresses $\sigma_1$, $\sigma_2$ and $\sigma_3$ modeling the behaviour of structural fuses are assigned to elements in layers marked by 1, 2 and 3.

In the following discussion the focus will be only on the main deformation paths presented in Figures 5.5(a) to (c), caused by distribution of only three different values of yield stresses applied in each layer of elements. Deviation from symmetric folding can be achieved through application of control stresses individually to each element.

In the first folding pattern, the deformation starts from the top layer of elements and is followed by a quasi-snap-through of bottom elements. The distribution of yield stresses related to the pattern is the following:

$$\sigma_1 < \sigma_3 < \sigma_2 \tag{1}$$

From the point of view of energy dissipation, the first pattern is identical to the third one (with the inverted folding sequence, cf. Figure 5.5(c)), which is generated by the following relation of yield stresses:

$$\sigma_3 < \sigma_1 < \sigma_2 \tag{2}$$

The synergy of repetitive use of dissipators, which is the great advantage of the MFM, is clearly visible in the second folding pattern. Due to the final global snap-through of all elements, the second mode provides very high dissipation of energy related to plastic strains. Obviously this mode of deformation is caused by

$$\sigma_2 < \sigma_1 \leq \sigma_3 \tag{3}$$

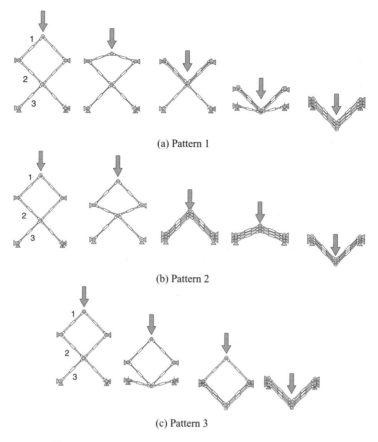

(a) Pattern 1

(b) Pattern 2

(c) Pattern 3

**Figure 5.5**   Folding patterns for the basic MFM structure

Even in the case of this basic structure, optimal control and adaptation to the identified impact load is possible by choosing the proper folding mode and adjusting the yield stresses according to particular requirements.

Figures 5.6(a) and (b) present the results of adaptation of the analysed structure to dynamic loads (impact at the top node) for two different scenarios (velocity of 15 m/s with masses of the impactor of 100 kg and 200 kg, respectively). Two different control strategies were applied for the simulation. In the case of the semi-active approach, optimal values of yield stresses are pre-selected before the impact and remain constant. Such formulation provides a very significant reduction of accelerations during the deformation, in comparison to the passive mode, in which the distribution of yield stresses is uniform and constant in time. Further improvement can be achieved by applying real-time changes in control parameters (active control). In the second scenario (Figure 5.6(b)), the MFM is very close to its limit load and so the effects of adaptation tend to be closer to the passive solution.

Numerical modeling and optimization of the system must include the elastoplastic constitutive model of material along with computation of the contact between nodes of the structure. Therefore simulation of the MFM performance requires geometrical and physical nonlinearities to be taken into account.

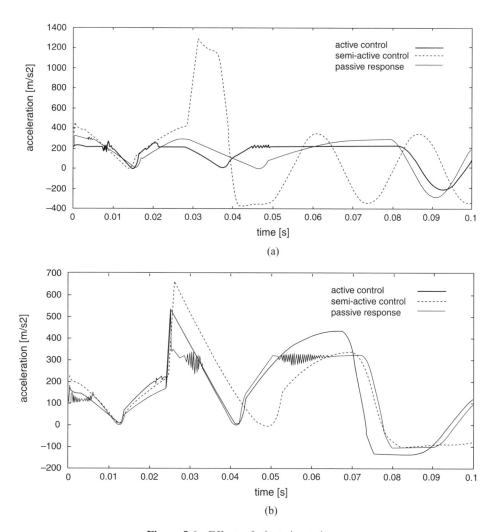

**Figure 5.6**  Effects of adaptation to impact

## 5.2.4 Experimental Results

The concept of multifolding was verified experimentally on a macroscale model. The layout of the experimental set-up representing the most basic, six-element structure is depicted in Figure 5.7. The control of the system is possible due to implementation of six magnetorheological fluid (MRF) dampers (RD-1005-3) manufactured by Lord Corporation.

Magnetorheological fluids belong to a class of smart fluids, which are sensitive to the applied magnetic field. They are a noncolloidal suspension of magnetically soft iron particles in a base fluid (e.g. mineral oil, water, etc.). Changes in the strength of magnetic field along a specially designed gap in the damper result in a fast change of the damping force. The rheology of MRF dampers is highly nonlinear, resulting in complicated mathematical models, but under special circumstances (e.g. velocity of the piston's movement close to constant) the response of the damper resembles the process of plastic yielding.

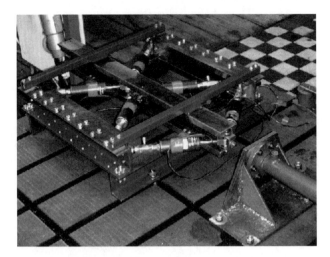

**Figure 5.7**   General view of the experimental set-up

Nodes of the structures are guided in specially designed bearings, which restrict horizontal displacements. The loading of the top node is applied by a hydraulic actuation system providing displacement control (with adjustable velocity up to 2 m/s). The maximal displacement is constrained to four levels of the structure. Variations of yield stress levels used in models of the MFM are simulated by changes of the magnetic field in magnetic heads of MRF dampers.

The results for three primary, symmetrical folding modes are depicted in Figures 5.8(a) to (d). The figures show the time evolution of normalized loading force and axial forces in elements obtained for three different deformation paths. The experimental loading force curve (cf. Figure 5.8(a)) is almost identical for the first and the third modes. In the second mode, a rapid increase of stiffness (cf. Figure 5.4(f)) is clearly visible.

The order of folding can be reconstructed from time–force profiles depicted in Figures 5.8(b) to (d). The profiles are adequate for the folding modes discussed earlier.

In the first pattern top elements deform first, followed by folding of the bottom layer. The middle elements remain rigid, only transferring the load. The second pattern is initiated by folding of middle layer with subsequent deformation of all elements. The last mode is similar to the first one with an inverted folding sequence.

## 5.3 Structural Fuses for Smooth Reception of Repetitive Impact Loads

In the previous section the concept of adaptive energy-absorbing microstructure (the MFM) was presented and discussed. A general conclusion can be drawn, that in order to minimize the consequences of heavy or even catastrophic dynamic loads, a process of structural adaptation to an impact should be carried out. Such an approach can be successfully applied also in the case of large-scale, compliant truss structures (e.g. offshore structures subjected to risk of collision with a ship or floating ice, cf. Section 5.5). The process mentioned above may consist of the following, subsequent stages:

- Impact detection. Impact detection is provided by a set of sensors, which respond in advance to a danger of collision (e.g. radar, ultrasonic devices) or are embedded into the structure within a small passive crush zone (e.g. piezo-sensors). Estimation of the impact energy is then based on an initial deformation of the passive zone.

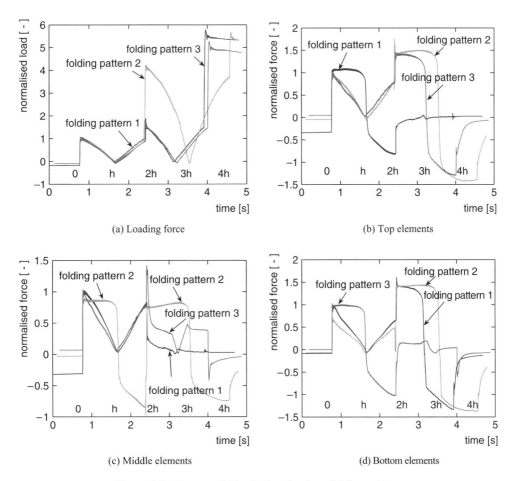

(a) Loading force                                        (b) Top elements

(c) Middle elements                                   (d) Bottom elements

**Figure 5.8**   Force evolution in time for three folding patterns

- Structural adaptation. The signal from the system of sensors must be directed to a controller unit, which selects an optimal distribution of yield forces in active zones containing elements equipped with structural fuses (cf. Figures 5.9(a) and 5.12(a)). The concept of the microfuse has already been discussed in Section 5.2 and can easily be scaled to the macroscale.
- Self-repair. Increasing requirements for structural durability and low-level operating costs create a need for new smart solutions. The results of an extreme dynamic load may very often be fatal for a structure. In the case of structures equipped with active elements, it is possible to remove residual distortions using low-level vibrations induced by an external load or embedded shaker.

## 5.3.1 Introductory Numerical Example

The 30 m high tower depicted in Figure 5.9, equipped with an active energy absorber, is subjected to an impact of a mass of 3000 kg with an initial velocity of 8 m/s. All structural members have a uniform cross-sectional area and elastic modulus. The yield stress level in active elements is adjusted according to the value of the kinetic energy of the impact, while in

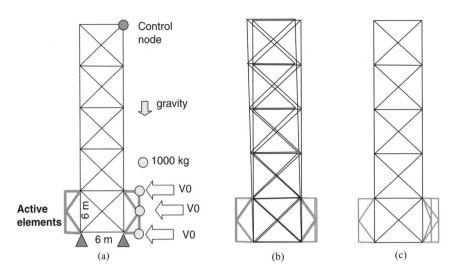

**Figure 5.9**   (a) Model of an adaptive structure, (b) displacements of the elastic solution (scaled × 1.5), (c) displacements of the optimal solution

all passive elements it is equal to $6 \times 10^8$ Pa. Maximal, allowable deformation in controlled elements is constrained to 50 % of their initial length.

The objective function in the optimization problem is to minimize the horizontal acceleration of the controlled node at the top of the tower. Deformed shapes of the structure calculated for two different thresholds of yield stresses in active elements are presented in Figures 5.9(b) to (c).

For the highest threshold of $6 \times 10^8$ Pa, the response of the structure is elastic, while thresholds of $1 \times 10^8$ Pa and $2.6 \times 10^7$ Pa (optimal solution) provide permanent plastic deformation. It is clearly visible that an adaptive strategy provides a very significant reduction of the acceleration level (cf. Figure 5.10) and that control parameters in active elements should be adjusted according to the severity of the impact.

## 5.3.2 Optimal Control

Two strategies of semi-active and active control might be considered. In the first strategy, yield stresses in structural members located in active zones remain unchanged during an impact. In the second one, a possibility of real-time changes in control parameters is assumed. The following discussion will be focused only on the semi-active approach, which provides a good balance between the expected results and complexity of the control strategy.

A position of active elements follows from an assumption that only selected parts of the considered structure would be exposed to a danger of an extreme dynamic load. A number of active elements and their structural properties should be chosen as a result of the separate optimization task.

The problem of the control may be formulated as follows: for a given structure $A$ and a given set of active elements $E_i \in A$, find the optimal distribution of control yield stresses $\sigma_i^p$ that minimize the objective function $f$ defined by the impact norm $I_2$ (providing information

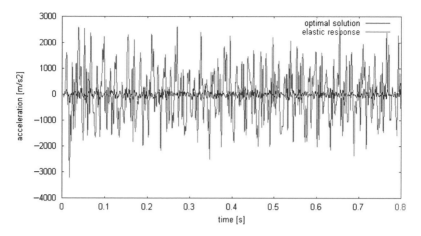

**Figure 5.10** Horizontal acceleration of the monitored node, for the passive and the optimal plastic solutions

about the average acceleration level at monitored degrees of freedom):

$$\min f(\sigma_i): \quad f \to I_2 \tag{4}$$

$$I_2 = \frac{1}{T} \sum_{t=1}^{T} \sum_{i=1}^{N_{\text{cntr}}} \left| \ddot{q}_i^{\text{cntr}}(t) \right| \tag{5}$$

with the following constraints, imposed on control stresses and displacements in active elements:

$$\sigma_i^P \in \langle \sigma^{\min}, \sigma^{\max} \rangle \tag{6}$$

$$\max\{q_i(t) \in A\} \le q_i^{\max} \tag{7}$$

Two additional measures of structural dynamic response, describing the overall acceleration level in the structure and maximal acceleration values at specified degrees of freedom (DOFs), may be introduced:

$$I_1 = \frac{1}{T N} \sum_{t=1}^{T} \sum_{i=1}^{N} \left| \ddot{q}_i(t) \right| \tag{8}$$

$$I_3 = \max_{t,i} \{ \left| \ddot{q}_i^{\text{cntr}}(t) \right| \} \tag{9}$$

where $T$ is the time of the analysis, $N$ the number of DOFs and $N_{\text{cntr}}$ the number of monitored DOFs.

## 5.3.3 Structural Recovery

After optimal impact absorption, permanent deformation should be localized in selected active elements. Assume that residual strain in an active element is equal to $\varepsilon_i^R$. Low-level vibrations

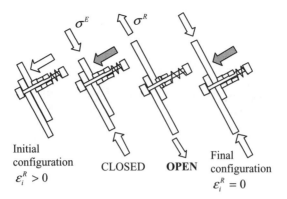

Initial configuration $\varepsilon_i^R > 0$     CLOSED     OPEN     Final configuration $\varepsilon_i^R = 0$

**Figure 5.11**   Structural fuse in the self-repair mode

induced by a shaker (external or embedded into the structure) generate strains $\varepsilon_i(t)$, which can be used to recover the initial length of distorted elements. When residual and actual strains have opposite signs the structural fuse opens and releases accumulated distortions. In the case of equal signs, the fuse remains closed:

$$\sigma_i(t) = \begin{cases} \sigma^O & \text{if} \quad \varepsilon_i(t)\varepsilon^R < 0, \quad \sigma^O \quad \text{open} \\ \sigma^C & \text{if} \quad \varepsilon_i(t)\varepsilon^R > 0, \quad \sigma^C \quad \text{closed: } \sigma^C \gg \sigma^O \end{cases} \tag{10}$$

Figure 5.11 presents the behavior of a structural fuse in the self-recovery mode. In order to ensure stability, only one active zone should be recovered at a time.

### 5.3.4 Numerical Example of Adaptation and Recovery

The following subsection presents an example of self-adaptation of a tall, compliant tower to a dynamic load ($M = 1500\,\text{kg}$, $v_0 = 8\,\text{m/s}$, $F = 20\,\text{kN}$), followed by the process of structural recovery. The 48 m high truss structure is depicted in Figure 5.12(a). It is assumed that an impact can take place only at specified nodes. Therefore, four active zones are located in the lower part of the structure. Values of the control yield stresses in structural fuses belong to the range of $< 1 \times 10^5\,\text{Pa}, 6 \times 10^8\,\text{Pa}>$.

All the passive elements have a uniform yield stress value of $6 \times 10^8\,\text{Pa}$. The cross-sectional area of the active elements is 5 times higher than in case of the passive ones. Stability of the structure is ensured by a constraint imposed on the maximal displacement of the loaded node: $q_{max} < 0.8\,\text{m}$. The passive and optimal deformations (heuristic solution) with the impact norms are presented in Figures 5.12(b) and (c).

Structural recovery was performed according to the procedure described in the previous subsection. A sine-shaped, horizontal dynamic load (amplitude $8 \times 10^4\,\text{N}$, frequency 20 Hz) was applied at the support level of the structure. The values $\sigma^C$ and $\sigma^O$ in structural fuses were chosen as $1 \times 10^6\,\text{Pa}$ and $1 \times 10^3\,\text{Pa}$, respectively. The residual distortions were subsequently removed from active zones 1 to 4. The results of the self-repair are depicted in Figures 5.12(d) to (g). Evolution in time of acceleration and displacement at the controlled node for the optimal and passive solutions is presented in Figure 5.13.

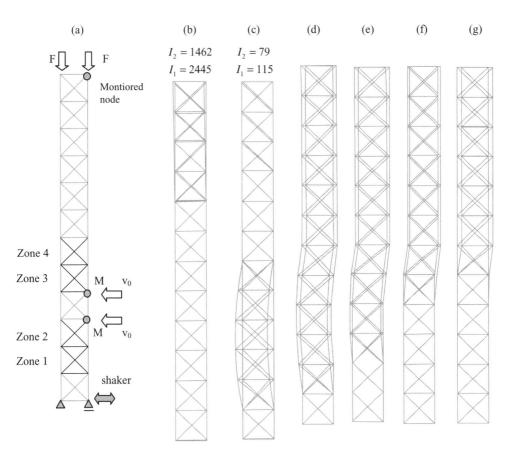

**Figure 5.12** (a) Adaptive structure, (b) passive response, (c) optimal solution, (d) to (g) structural recovery of active zones 1 to 4, respectively

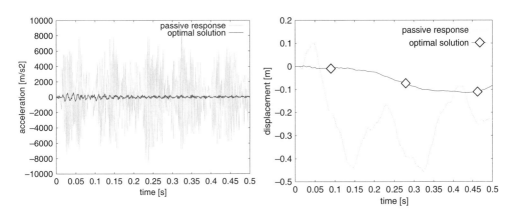

**Figure 5.13** Horizontal accelerations and displacements of the monitored node for passive and optimal solutions

## 5.4 Absorption of Repetitive, Exploitative Impact Loads in Adaptive Landing Gears

### 5.4.1 The Concept of Adaptive Landing Gear

An adaptive landing gear is a landing gear (LG) capable of active adaptation to particular landing conditions by means of a controlled hydraulic force. The objective of the adaptive control is to mitigate the peak force transferred to the aircraft structure during touchdown, and thus to limit the structural fatigue factor. The objective of shock absorbers is to mitigate undesirable dynamic effects caused by accidental impact forces acting on the protected structure. The impacts can be considered in catastrophic and noncatastrophic scales, but in both cases the proper energy absorption capabilities of the landing gears and structural elements are significant from the point of view of safety [3–5].In noncatastrophic cases, one of the force origins is kinetic excitations coming from the shock of the base, where it is fixed to the structure (e.g. road vehicles, rail vehicles, aircraft during landing and taxiing) [6,7]. Impacts result in the impulsive generation of acceleration levels within the suspension. The objective of the shock absorbers is then to minimize the acceleration peak levels being transferred to the structure.

The classical methodology of shock absorber conceptual design allows tuning of the damping and stiffness characteristics to one particular magnitude of the impact energy. However, in the case of landing gear shock absorbers, the impacts vary significantly between landings. According to industry regulations, all landing gears are optimized for the case of an impact with the maximum aircraft weight and sink speed (i.e. max. impact energy). This is purely for reasons of safety. This situation results in nonoptimal landing gear behavior for more common landings with lower impact energies. More specifically, this nonoptimal behavior results in the generation of very high damping forces and an unwanted reduction in the effective stroke. Consequently, acceleration of the protected structures is increased. These unnecessary overloads of the structure significantly influence the fatigue processes. A solution for the mentioned problem is the introduction of an adaptive landing gear, which has the possibility of fitting its characteristics to particular landing sink speeds and weights of the aircraft.

This adaptation of the shock absorber would allow optimal performance to be achieved for a wide range of impact velocities and weights of the structure. The introduction of such a system would reduce the fatigue of the protected structure, as a result of reduced accelerations that the structure is subjected to. The statistical benefit from the introduction of an active LG (ALG) was estimated at 16 % [8]. The force generated by the LG depends on the difference between the fluid pressures in the lower and upper chamber of the strut [9, 10] (Figure 5.14). This force can be controlled in three ways. The pressure drop can be modified via control of the

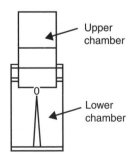

**Figure 5.14**   Classical shock strut scheme

gas pressure in the upper chamber or by regulating the fluid flow resistance across the orifice (using a fast actuated valve or by changing the rheological properties of the fluid in the gap). An interesting ALG concept for improving the energy dissipation of a landing impact was considered by NASA researchers for many decades. The researchers at the Langley Research Centre developed a system that was designed to change the pressure drop between the upper and lower chambers of the shock absorber. The system contained two gas accumulators, which were connected to the chambers by servo valves. The damping force of the strut was influenced by increasing or decreasing the pressure difference between the chambers. However, the valve operation was too slow when it faced the landing impact phenomenon [11–15].

In recent years conceptual research was performed on adaptive landing gears actuated via magnetorheological fluids, which solved the problem of time delays (cf. the ADLAND project [16, 17]). Implementation of the MRF into the shock absorber gives a unique opportunity to control the pressure drop between the upper and lower chambers. The fluid has a feature of changing its apparent viscosity when it is subjected to an external magnetic field [18]. By having the magnetic force generator incorporated around the hydraulic orifice, a controllable valve effect can be reached by means of control of the local apparent viscosity of the fluid. The time delays of the MRF valve's operation depend on the rate of generation of the magnetic excitation [19].

## 5.4.2 Control System Issues

The control system for the adaptive landing gear is a challenging task to design. The designer must take into account a series of aspects that cannot be neglected, which are the result of the aircraft's ground operations. The control system design must consider the following three important problems.

The first problem is related to the duration of the phenomenon. In general, the landing impact lasts between 50 and 200 ms, depending on the size of the landing gear and the landing conditions. This short time period makes it difficult to implement the control strategies effectively as the present actuators are not able to respond fast enough. High response valves currently available on the market offer the best time performance on the level of 10–12 ms operational delay in the case of hydraulic valves, directly operated, with electrical position feedback [20]. When it comes to the pneumatic solutions, the fastest designs for valves give the possibility of operation with the time delay equal to 12 ms in the version with an additional pressure accumulator [21]. An important disadvantage of the mentioned solutions is the fact that the standard valve weights are around 2.5 kg. These were strong limitations in the field of application of the active solutions into the adaptive impact absorbers. The proposed actuating systems collaborating with the designed hardware controllers are able to execute one control loop with a delay of approximately 4 ms. This means that the system is able to update the control signal around 13 times in the case of a 50 ms impact duration. If it is assumed that four or five loops would be consumed for recognition of the process conditions, then the remaining period would be insufficient for execution of an efficient control process. Another important factor is that the impact is random in nature. In contrast to a harmonic process, it is not possible to characterize an impact on the basis of one period and to implement the proper control law for the following periods of the process duration. In these circumstances it is necessary to apply a control system in which the feedback control strategy is assisted by an impact energy prediction unit. The energy prediction unit would process the sink speed, position and the actual weight of the aircraft, in order to provide an estimate of the coming impact. Estimation of the aircraft's sinking speed is presently achieved using the pressure-based altimeters.

However, a much higher accuracy is required for the above purpose. For this vertical velocity measurement (sink speed), the following instruments are considered to be feasible: photolaser, low-power radars, ultrasound sensors. The measurements of the actual mass of the aircraft can be conducted in a passive or an active routine. The passive routine consists of storing data about the aircraft's takeoff mass and its center of gravity. This requires an estimation of the fuel consumption before landing. The active method of mass estimation can be realized via introduction of the real-time mass identification system [22], which enables identification of the actual weight loading of each landing gear strut. When real-time mass identification is used with an integrated sink speed measurement, it is possible to assess precisely the energy of the coming impact for each wheel. This configuration would make it possible to establish the optimal strategy for active energy dissipation of the whole structure.

The second problem to be considered for the design of the active landing gear is calculation of the exact position of the aircraft during landing in relation to the runway. The position is important since the impact energy dissipation process must be significantly different, depending on whether the plane lands on one or both main landing gears. The position of the aeroplane is continuously monitored during flight by gyroscopic sensors, but the measurements give the absolute outcome and it is not possible to calculate the exact position of the aeroplane in relation to the surface of the runway. One method of conducting these measurements is to integrate the height sensors with sink speed sensors on each landing gear. This would enable monitoring of the 6 DOF position of the aeroplane, allowing the landing gears to adapt more effectively to the coming impact. The third problem that must be considered in the design of active landing gears is the springback force that occurs during touchdown. Springback forces come from the acceleration of the wheels after contact with the runway surface. The circumferential velocity of the wheels must be equalized with the horizontal velocity of the aircraft. The horizontal component of the load vector acting on the landing gear causes bending of the strut. The deflected strut springs back rapidly and increases seal friction within telescopic oleopneumatic landing gears. This phenomenon introduces significant friction damping, which acts parallel to the oleodamping generated by the orifice. The influence of friction damping is very difficult to predict since it varies with each landing and is dependent on the horizontal speed of the aeroplane, the sink speed of the aeroplane, runway adhesion, temperature and the exact 6 DOF position of the aeroplane. Prediction of the exact value of friction is a very complex task and the result can be estimated with a significant error. In the case of a control system for which the damping force would be treated as an input, the safest and most convenient solution is to use a sensor that measures the total force generated by the landing strut and to modify it with the adaptive component. Control systems used in such a routine were analysed and tested in the laboratory [22], but the measurement of the total axial force in the strut is a challenging problem due to technological limitations in real applications.

According to the presented discussion, the preliminary requirements for the active landing gear control system are as follows. In connection with the fact that the impact process duration does not exceed 50 ms in the most severe cases, the control system must have the capability of recognizing the impact energy before touchdown in order to adapt the system before the process starts. The second requirement is that the system (actuator + sensors + control hardware) must have the capability to update its state within 4 ms in order to keep the control system performance efficient. The third established requirement for the control system refers to the feedback signal on which the control is based. The signal must describe the total reaction of the landing gear during the process. One of the possible signals can be the total interface force between the strut and the aircraft structure [23], but the force sensor is difficult to assemble from the technological point of view. In the case of landing gear shock absorbers, it is possible to mount a pressure sensor of the hydraulic fluid inside the chambers. However, the pressure signal can give only information about the hydraulic shock-absorbing force acting on the structure. The signal does not give any information about the frictional forces, tire forces or the springback phenomenon. The parameter that gives absolute information on the reactions of the fuselage and landing gear is the acceleration measured at the top of each landing gear. This signal can be chosen as the input signal for the control system.

### 5.4.3 Modeling of ALG

As a reference, a standard oleopneumatic one-stage cantilever-type [14–16] passive nose LG of an I23 light aircraft [4] is used, with no structural modifications assumed besides the possibility of an instantaneous active control.

For the purposes of this analysis, the (A)LG is represented by a 2 DOF system, shown schematically in Figure 5.15, and modeled by the following set of equations (a modified version of the equations derived by Milwitzky and Cook in Reference [10]):

$$
\begin{aligned}
m_1\ddot{z}_1 &= m_1 g - F_S - mgL \\
m_2\ddot{z}_2 &= m_2 g + F_S - F_G \\
z_1(0) &= z_2(0) = 0 \\
\dot{z}_1(0) &= \dot{z}_2(0) = v_0
\end{aligned}
\tag{11}
$$

where $m$, $m_1$ and $m_2$ are respectively total, upper (aircraft) and lower (wheel) masses ($m = m_1 + m_2$, $m_2 = 8.71$ kg); $z_1$ and $z_2$ denote their vertical displacements from initial contact; $L$ denotes the lift factor ($L = 0.667$ [24]); $F_G$ is the vertical force acting on the tire at the ground; $F_S$ is the total axial strut force; and $v_0$ denotes the initial landing sinking velocity.

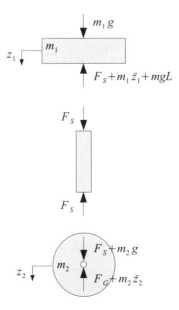

**Figure 5.15** Schematic diagram of forces acting in a landing gear

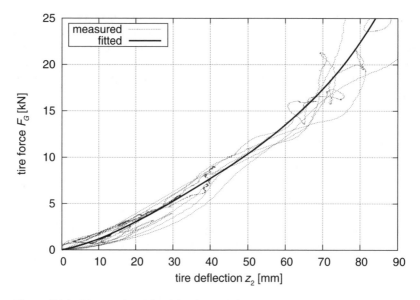

**Figure 5.16**   Measured and fitted tire force–deflection characteristics (I23 nose LG)

The total mass $m$ and the sinking velocity $v_0$ are limited [25] by

$$288 \text{ kg} = m_{\min} \leq m \leq m_{\max} = 422 \text{ kg}$$
$$0 \text{ m/s} = v_{0(\max)} \leq v_0 \leq v_{0(\max)} = 2.93 \text{ m/s} \tag{12}$$

For the assumed, detailed statistics of the landing mass $m$ and sinking velocity $v_0$, see Section 5.4.5, where $m$ is the *reduced* mass (mass per landing gear) and is less than the *total* landing mass of the aircraft, 422 kg corresponds to the maximum I23 design landing mass of 1117 kg and 282 kg was chosen to be proportional to the mass of an empty aircraft with a pilot [25].

The dynamic tire force–deflection characteristics of the I23 nose LG, denoted in Equation (11) by $F_G$, has been obtained by fitting experimental data measured in three dynamic tests (see Figure 5.16). The least squares fit is a fourth-order polynomial:

$$F_G(z_2) \approx (7.3 \times 10^4 + 5.4 \times 10^6 z_2 - 8.6 \times 10^7 z_2^2 + 6.4 \times 10^8 z_2^3) \max(z_2, 0)$$

where the last multiplier denotes symbolically that $F_G$ vanishes when the tire hovers above the ground.

The total axial strut force $F_S$ is modeled as a sum of four forces:

$$F_S = F_a + F_h + F_f + F_d \tag{13}$$

which are respectively strut pneumatic, hydraulic and friction forces, denoted by $F_a$, $F_h$ and $F_f$, and the delimiting force $F_d$, which prevents excessive strut elongation. Notice that introduction of the actually occurring delimiting force $F_d$ greatly simplifies modeling of the landing process, as compared to the approach of Milwitzky and Cook [10], since the LG can be uniformly modeled as a 2 DOF system during the whole landing, including its initial stage and – if necessary – rebounds. The total strut force and the component forces depend directly on the strut axial stroke $s$,

$$s = z_1 - z_2$$

The pneumatic force $F_a$ in Equation (13) is modeled in accordance with the polytropic law for compression of gases,

$$F_a(s) = p_0 A_a \left( \frac{V_0}{V_0 - s A_a} \right)^n \tag{14}$$

while the hydraulic force $F_h$ is modeled in the standard way [10] as

$$F_h(\dot{s}) = \text{sign}(\dot{s}) \frac{1}{2} \frac{\rho A_h^3}{C_d^2 A_o^2} \dot{s}^2 \tag{15}$$

The friction occurring in the strut is assumed to be dry friction only [10] and is modeled by

$$F_f(\dot{s}) = C_f \frac{2}{\pi} \arctan(10^4 \dot{s}) \tag{16}$$

where the inverse tangent function was used to ensure smooth variations of the friction force at the turning points and to enable numerical integrations of the equations of motion. The delimiting force $F_d$ prevents excessive elongation of the strut and attempts to model the actual force occurring on the strut delimiter. It acts within the last $l_d$ of the fully elongated strut and is modeled by a simple spring force as

$$F_d(s) = p_0 A_a \min \left( \frac{s - l_d}{l_d}, 0 \right) \tag{17}$$

where the coefficient $p_0 A_a$ has been chosen to obtain equilibrium at full elongation: $F_d(0) + F_a(0) = 0$. Possible oscillations of a fully elongated strut are damped directly by the hydraulic force $F_h$.

The symbols used in Equations (14) to (17) are explained in Table 5.1. The numerical values of $p_0$, $V_0$ and of the dry friction coefficient $C_f$ have been obtained by numerical fitting of LG quasi-static compression data. The trimming bounds on $A_o$, which is the controlling parameter, are chosen arbitrarily to model real technological constraints. Notice that the following simplifying assumptions concerning the friction have been made:

• The dynamic friction equals the quasi-static friction.
• The strut friction is not considerably affected by the normal loading occurring due to tire friction in the first milliseconds of the landing process at the wheel axle. This is an oversimplification in the case of a cantilever-type LG but can be legitimate in the case of levered trailing arm gears.

**Table 5.1**   Symbols used in Equations (14) to Equation (17) [10, 25, 26]

| Symbol | Numerical value | Explanation |
| --- | --- | --- |
| $A_a$ | $1.385 \times 10^{-3}$ m$^2$ | Pneumatic area |
| $p_0$ | 1.028 MPa | Initial air pressure in the upper chamber |
| $V_0$ | $171 \times 10^{-6}$ m$^3$ | Initial air volume in the upper chamber |
| $n$ | 1.1 | Polytropic exponent for the air compression process |
| $\rho$ | 872.6 kg/m$^3$ | Density of hydraulic fluid (Aeroshell 41) |
| $A_h$ | $1.018 \times 10^{-3}$ m$^2$ | Hydraulic area |
| $A_o$ | $A_{o(min)} \leq A_o$ $A_o \leq A_{o(max)}$ | Cross-sectional area of the discharge orifice |
| $A_{o(min)}$ | 5 mm$^2$ | Technological lower bound on $A_o$ |
| $A_{o(max)}$ | 40 mm$^2$ | Technological upper bound on $A_o$ |
| $C_d$ | 0.6 | Orifice discharge coefficient |
| $C_f$ | 559 N | Dry friction coefficient |
| $l_d$ | $500 \times 10^{-6}$ m | Delimiting force acting interval |

Accuracy of the model can be partly verified by comparison of calculated forces and displacements with the forces and displacements measured in tests of a real I23 nose LG, passive version. The Institute of Aviation (Warsaw, Poland) has made available two sets of measurement data suitable for the comparison:

(1) mass $m = 422$ kg, sinking velocity $v_0 = 2.93$ m/s, lift factor $L = 0.667$;
(2) mass $m = 422$ kg, sinking velocity $v_0 = 3.52$ m/s, lift factor $L = 1$.

The first case corresponds to the highest-energy design landing conditions, (see Equation (12)) while the second is even more demanding. Figure 5.17 compares the calculated and measured tire forces $F_G$, while Figures 5.18 and 5.19 compare the calculated and measured aircraft and tire displacements ($z_1$ and $z_2$). The discharge orifice area $A_o$ was assumed to equal 17.43 mm$^2$,

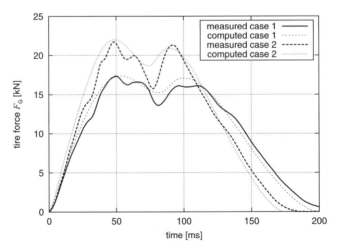

**Figure 5.17**   Measured and computed tire force $F_G$, passive LG: test case 1 ($m = 422$ kg, $v_0 = 2.93$ m/s, $L = 0.667$); test case 2 ($m = 422$ kg, $v_0 = 3.52$ m/s, $L = 1$)

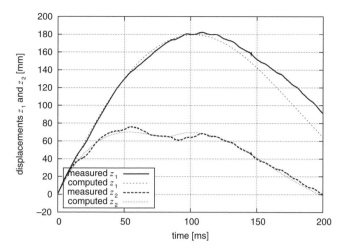

**Figure 5.18** Measured and computed displacements of aircraft $z_1$ and tire $z_2$, passive LG: test case 1 ($m = 422$ kg, $v_0 = 2.93$ m/s, $L = 0.667$)

which is the optimum value in the case of a passive LG (see Section 5.4.4). Simulations and measurements agree well in the case of the tire force $F_G$, the first 150 ms of tire displacement $z_2$ and the first 100 ms of aircraft displacement $z_1$, which corresponds to the strut compression phase. However, there is an increasing discrepancy between the displacements calculated and measured in the strut decompression phase, which starts approximately 100 ms after the impact. The discrepancy suggests additional factors coming into play during the strut decompression phase (possibly recoil orifices and hydraulic oil foaming), which cannot thus be modeled using a constant recoil orifice area. Nevertheless, the strut compression phase seems to be modeled reliably; hence all considerations of this paper concerning the peak strut force $F_{peak}$ and the proposed control strategies are valid.

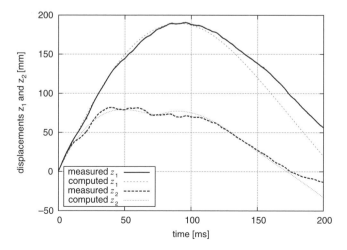

**Figure 5.19** Measured and computed displacements of aircraft $z_1$ and tire $z_2$, passive LG: test case 2 ($m = 422$ kg, $v_0 = 3.52$ m/s, $L = 1$)

## 5.4.4 Control Strategies

It is assumed that the landing scenario is fully defined by two basic parameters:

(1) total landing mass per strut $m$;
(2) initial strut sinking velocity $v_0$.

Their ranges are given in Equation (12), their distributions in Section 5.4.5. The common objective of all the investigated control strategies is to minimize the peak strut force $F_{peak}$ occurring during the landing. The evolution of the total strut force $F_S$ is fully determined by the control parameter, which is area $A_o$ of the discharge orifice.

Essentially, there are three control strategies possible [27]:

- *Passive LG* (PLG), i.e. no control. The orifice area $A_o$ is constant and cannot be adjusted to particular landing conditions. Nevertheless, its pre-set constant value is optimized to mitigate the peak strut force occurring at the highest-energy landing conditions (maximum design landing mass and sinking velocity).
- *Semi-active LG* (SLG). The orifice area $A_o$ is optimally set directly before each landing, based on the actual sinking velocity $v_0$ and/or mass $m$, which have to be measured or known in advance. $A_o$ remains constant during the landing process, which makes the strategy relatively easy to implement, since no quick closed control loops are necessary.
- *Active LG* (ALG). The orifice area $A_o$ changes continuously during the landing process, according to a strategy defined by actual values of the initial sinking velocity $v_0$ and/or mass $m$, which have to be measured or known in advance. This strategy potentially yields the highest improvement, but requires quick closed control loops and accurate real-time measurement data, which may result in instabilities.

The strategies apply to the strut compression phase only. During the decompression phase, the recoil orifices appear instead of the main discharge orifice, the numerical model has thus to be modified and the optimization goal redefined. However, as the (to-be-minimized) peak strut force $F_{peak}$ occurs within the strut compression phase, modeling of the decompression phase is outside the scope of this paper.

To apply the active or the semi-active control strategy, as defined above, the actual landing scenario has to be at least partially known in advance. In real conditions the sinking velocity $v_0$ can be relatively easily measured just before the touchdown by a dedicated sensor (e.g. ultrasonic, one per LG). However, it may not be possible to know the exact actual value of the total landing mass per strut $m$. Therefore, altogether five strategies are analysed and compared, differing in the control scheme and availability of $m$:

(1) passive LG (PLG);
(2) semi-active LG (SLG): both $v_0$ and $m$ are known before landing;
(3) velocity-driven semi-active LG (VD-SLG): only $v_0$ is known before landing;
(4) active LG (ALG): both $v_0$ and $m$ are known before landing;
(5) velocity-driven active LG (VD-ALG): only $v_0$ is known before landing.

To investigate and assess the strategies, the equations of motion (11) had to be solved numerically, which was done with explicit methods and the time step 5 μs or 50 μs (active strategies).

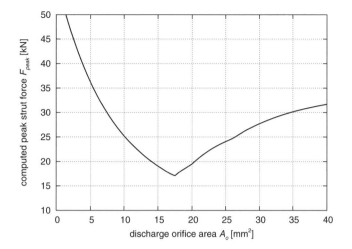

**Figure 5.20** Passive LG: computed dependence of the peak strut force $F_{peak}$ on the discharge orifice area $A_o$ at the highest-energy design landing scenario ($m = 422$ kg and $v_0 = 2.93$ m/s)

The peak strut forces computed at 5 μs and 50 μs differed by approximately 0.1–0.2 %, which is acceptable to compare the strategies reliably.

### 5.4.4.1 Passive LG

The pre-set constant discharge orifice area $A_o^{PLG}$ has to be chosen to minimize the peak force occurring during the highest-energy design landing scenario. Figure 5.20 shows the dependence of the peak strut force $F_{peak}^{PLG}(m_{max}, v_{0(max)}, A_o)$ on the discharge orifice area $A_o$. The left slope corresponds to the decreasing peak of the hydraulic force, while the right slope corresponds to the increasing peak of the pneumatic force. The minimum value of 17 021 N has been found at $A_o^{PLG} = 17.43$ mm², where both peaks are equal. The corresponding computed tire peak force equals 17 374 N, which is relatively very close to the measured value of 17 400 N (see Figure 5.17).

Thus, the maximum design strut force was assumed to be $F_{max} = 17 021$ N. The optimum discharge orifice area for a passive LG is $A_o^{PLG} = 17.43$ mm²:

$$F_{max} = 17 021 \text{ N}$$
$$A_o^{PLG} = 17.43 \text{ mm}^2$$

(18)

### 5.4.4.2 Semi-active LG

A graph of $F_{peak}$ in dependence on the orifice area $A_o$ (similar to that shown in Figure 5.20) can in fact be drawn for each combination of landing mass $m$ and vertical velocity $v_0$. Therefore, if both $m$ and $v_0$ are known or measured just before the touchdown, the discharge orifice area $A_o$ can be set to the optimum value $A_o^{SLG}(m, v_0)$, within the technological bounds $A_{o(min)}$ and $A_{o(max)}$ (Table 5.1), which yields the *technologically attainable* minimum peak force

$$F_{peak}^{SLG}(m, v_0) := F_{peak}\left(m, v_0, A_o^{SLG}(m, v_0)\right)$$

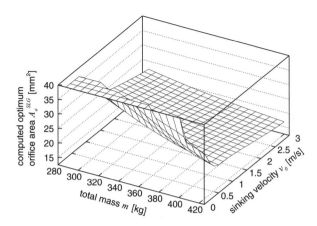

**Figure 5.21** Semi-active LG: computed dependence of the optimum discharge orifice area on the total mass $m$ and sinking velocity $v_0$

This is substantially advantageous to the passive LG, which is optimized only for the highest-energy design landing scenario.

Figure 5.21 shows the dependence of the computed optimum discharge orifice area $A_o^{SLG}(m, v_0)$ on the landing conditions. At standard landing conditions (low sinking velocity), the optimum orifice area $A_o^{SLG}$ considerably exceeds the constant value $A_o^{PLG}$ used in the passive LG (Equation (18)). Hence, at the same landing conditions, the peak strut force in the semi-active LG can be expected to be considerably lower than in the passive LG. The relative improvement is shown in Figure 5.22. There is obviously no improvement in the highest-energy landing scenario, since the SLG amounts then to the PLG. The effect of the semi-active control becomes apparent as the mass or sinking velocity decreases. However, at low sinking velocities, the advantage of the SLG over the PLG tends to diminish, which is due to the effect

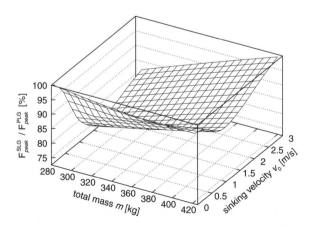

**Figure 5.22** SLG compared to PLG, relative improvement: computed ratio of the peak strut force in the optimally controlled SLG to the peak strut force in PLG, in dependence on the total mass $m$ and sinking velocity $v_0$

of the initial stiffness of the air spring: the landing energy is too low to compress considerably the strut and trigger the hydraulic force. Thus most of the vertical displacement is the tire deflection, and it is not possible to take advantage of the semi-active control scheme.

### 5.4.4.3 Velocity-Driven (VD) Semi-active LG

The semi-active control strategy requires both the landing mass $m$ and sinking velocity $v_0$ to be known before landing. However, in practice only the velocity $v_0$ can be relatively easily measured, and thus the discharge orifice area $A_o$ has to be chosen to minimize the peak strut force in the corresponding highest-energy landing scenario, i.e. at the highest mass $m_{max}$,

$$A_o^{\text{VD-SLG}}(v_0) := A_o^{\text{SLG}}(m_{max}, v_0)$$

The performance of the VD-SLG must thus suffer, compared to the SLG. However, as Figure 5.21 shows, the decisive parameter influencing $A_o^{\text{SLG}}$ is not the mass $m$ but the landing velocity $v_0$; hence the difference does not need to be very pronounced. A comparison of the performance of the VD-SLG and PLG is shown in Figure 5.23, which may be compared with Figure 5.22. At low sinking velocities the same effect of the initial stiffness of the air spring occurs.

### 5.4.4.4 Active LG

In an actively controlled LG the discharge orifice area $A_o$ is actively modified during the strut compression phase. The equations in Section 5.4.3 directly relate the total strut force $F_S$ to $A_o$ by

$$F_S = F_a + F_f + F_d + \frac{1}{2}\text{sign}(\dot{s})\frac{\rho A_h^3}{C_d^2 A_o^2}\dot{s}^2$$

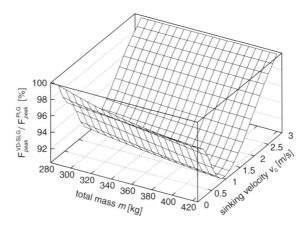

**Figure 5.23** VD-SLG compared to the PLG, relative improvement: computed ratio of the peak strut force in the optimally controlled VD-SLG and PLG, in dependence on the total mass $m$ and sinking velocity $v_0$

Hence, if instantaneous control is assumed, an obvious method to keep the total strut force $F_S$ at a desired limit value $F_{limit}$ is actively to set $A_o$ during the compression phase according to

$$A_o^2 = \begin{cases} A_{o(max)}^2 & \text{if } F_{limit} \leq F_a + F_f + F_d \\ \max\left[ A_{o(min)}^2, \min\left( A_{o(max)}^2, \dfrac{\rho}{2} \dfrac{A_h^3}{C_d^2} \dfrac{\dot{s}^2 \, sign(\dot{s})}{F_{limit} - F_a - F_f - F_d} \right) \right] & \text{otherwise,} \end{cases} \qquad (19)$$

where the first value offers the instantaneous minimization of $F_S$ in case it inevitably exceeds $F_{limit}$.

According to Equation (19), at the very beginning of the strut motion, when the total strut force $F_S$ is still low, the discharge orifice area $A_o$ is set to $A_{o(min)}$ and stays so untill $F_S$ attains $F_{limit}$. Thereafter $A_o$ is actively controlled within the given limits until the decompression phase begins. Therefore, the active approach of Equation (19) requires optimization of the peak force with respect to only one parameter $F_{limit}$, which has to be performed for all landing conditions defined by $m$ and $v_0$.

However, Equation (19) is only an approximation to the optimum active control. A finer control strategy, applied especially in the beginning (before attaining $F_{limit}$) and possibly at the end of the compression phase, could further reduce the peak force. To come closer to the optimum strategy, the strut can be softened in the beginning of the impact by setting the initial discharge orifice area to a given value $A_{o(ini)}$. The active control of Equation (19) begins first when the force limit $F_{limit}$ has been attained. In a real LG it will also reduce the initial tire–runway friction, allow for a gradual wheel spinup and reduce the springback effects. This results in optimization with respect to two parameters:

(1) initial area of the discharge orifice $A_{o(ini)}$;
(2) desired strut force limit $F_{limit}$, for triggering and managing the phase of active control according to Equation (19).

Figure 5.24 shows the relative improvement in comparison to the SLG. This is an additional improvement, which should be multiplied by the improvement of the SLG (Figure 5.22) to obtain the total improvement to the PLG.

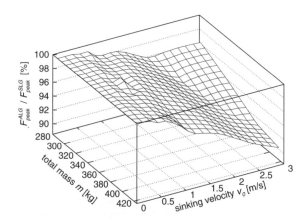

**Figure 5.24** ALG compared to the SLG, relative improvement: computed ratio of the peak strut forces in the optimally controlled ALG and SLG in dependence on the total mass $m$ and sinking velocity $v_0$

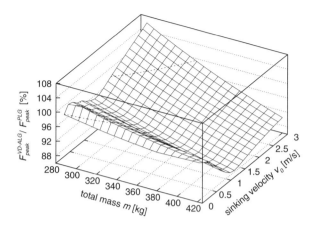

**Figure 5.25** VD-ALG compared to the PLG, relative change: computed ratio of the peak strut force in the optimally controlled VD-ALG and PLG in dependence on the total mass $m$ and sinking velocity $v_0$; the contour line marks the 100 % level

The advantage of the ALG over the SLG at standard landing conditions (low sinking speed) amounts to not more than 3 % and is rather insignificant. However, at the highest-energy landing conditions it attains the maximum of approximately 9 %, which is significantly better than the semi-active strategy (which in this case shows no improvement).

### 5.4.4.5 Velocity-Driven Active LG

Similarly to VD-SLG, a VD-ALG can be considered to study the practical scenarios when the mass $m$ is not known before the landing, and should be thus assumed to be the maximum $m_{max}$:

$$A_{o(ini)}^{VD\text{-}ALG}(v_0) := A_{o(ini)}^{ALG}(m_{max}, v_0)$$

$$F_{limit}^{VD\text{-}ALG}(v_0) := F_{limit}^{ALG}(m_{max}, v_0)$$

Figure 5.25 compares the performances of the VD-ALG and PLG. At low landing mass the peak force can be even *higher* than in the standard PLG, which renders the pure VD-ALG strategy useless. Therefore, it will be skipped in further parts of this paper.

### 5.4.4.6 Example

All sample simulations presented in this subsection are based on the landing conditions $m = 350$ kg and $v_0 = 1.5$ m/s, which are taken as examples and lie approximately in the middle of the design range of Equation (12).

Figure 5.26 compares the computed strut forces in the PLG and SLG during the first 200 ms of the landing process. Figure 5.27 shows the forces in the ALG along with the details of the applied active control. Two small temporary decreases of the total strut force at approximately 60 ms and 100 ms are results of attaining the lower limit $A_{o(min)}$ imposed on the orifice area (see the active control plot). The total strut force $F_S$ does not equal exactly the sum of the pneumatic $F_a$ and hydraulic forces $F_h$; the difference equals the friction and the delimiting force (see Equation (13)). Notice how the pneumatic and hydraulic force peaks, which are unequal in the PLG, are made equal in the SLG and additionally leveled in the ALG.

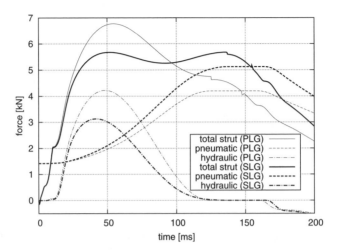

**Figure 5.26** Comparison of computed pneumatic, hydraulic and total strut forces in the PLG and SLG for $m = 350$ kg and $v_0 = 1.5$ m/s

Figure 5.28 compares the hysteresis (strut force $F_S$ versus strut deflection $s$) computed for the four considered strut types. The advantage of the SLG over the PLG is clear: the hydraulic and pneumatic force peaks could be made equal due to the optimally increased orifice area $A_o$. If the landing mass $m$ is not known (VD-SLG), the orifice area has to be tuned to the maximum mass and the peaks are not equal, although lower than in the PLG. The advantage of the ALG over the SLG is much less pronounced: both peaks could be additionally leveled a little. Further reduction of the peak strut force is possible only by increasing the force growth rate in the first 25 ms by decreasing the initial orifice area $A_{o(ini)}$. However, in a real LG this would considerably increase the tire–runway friction and the springback effect. It would also require a substantial increase of the orifice area upper limit $A_{o(max)}$ to maintain the constant force level in the subsequent 25 ms (see the first peak of the active control in Figure 5.27).

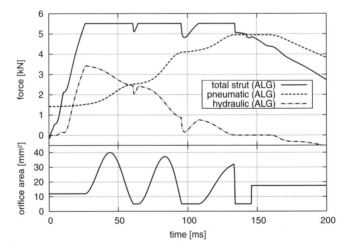

**Figure 5.27** Computed pneumatic, hydraulic and total strut forces in the ALG, and active control for $m = 350$ kg and $v_0 = 1.5$ m/s

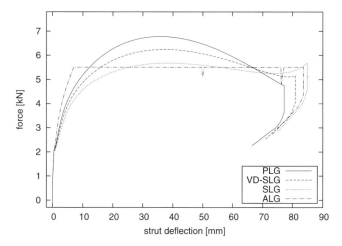

**Figure 5.28**   Strut force hystereses in the PLG, VD-SLG, SLG and ALG for $m = 350$ kg and $v_0 = 1.5$ m/s

## 5.4.5   Potential for Improvement

### 5.4.5.1   Air Spring Influence

A typical landing scenario involves a low sinking velocity (see Figure 5.29). However, the investigated control strategies reveal improvements only for medium-to-high sinking velocities (see Figures 5.22, 5.23 and 5.24). This is due to the effect of the pre-stressed air spring: to compress the strut further than the delimiting force acting range $l_d = 0.5$ mm, the strut force has to overcome the joint effect of the pneumatic and friction forces, which amounts to 1983 N (see Table 5.1, Equations (14) and (16) and is comparable to the peak force at $v_0 = 0$ m/s, which was computed to be 1993–2179 N, depending on the mass. Therefore, at low sinking velocities the strut is being barely compressed and there is practically no hydraulic force to be controlled. As a result no strategy based on hydraulic force control can yield any improvement at low sinking velocities.

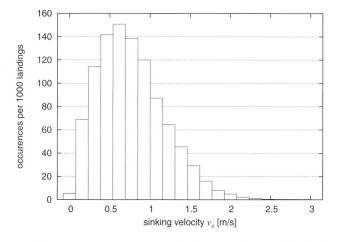

**Figure 5.29**   Assumed occurrences of sinking velocities per 1000 landings

An imminent idea is to alter the construction of landing gear and soften the air spring at the beginning of the strut stroke, e.g. by double-acting shock absorbers with a double air chamber [9]. A closer examination, however, reveals that the static runway loading is considerably higher than the landing peak force in the PLG at low sinking velocities. This is due to the lift factor $L = 0.667$, which occurs in Equation (11) and disappears in static or taxiing conditions. Therefore, the (semi)-active control strategies can in fact be considered necessary only in the cases when the peak force in the PLG is *significant*, i.e. exceeds the static loading, $F_{\text{peak}}^{\text{PLG}}(m, v_0) > mg$. This occurs at higher sinking velocities only as the limiting velocity ranges from 0.54 m/s to 0.85 m/s, depending on the landing mass $m$.

### 5.4.5.2 Mean and Median Peak Strut Force

Figures 5.22 to 5.25 compare the performance of the discussed LG types for each design landing condition separately. An overall comparison is possible by statistical means if the probability distributions of landing conditions are defined. The initial sinking velocity $v_0$ and the total mass $m$ are assumed to be independent. To ease the statistical computations, their ranges in Equation (12) have been discretized into 20 equally spaced values. The distribution of the landing mass has been assumed to be uniform in the whole range of 282–422 kg, which leads to 50 occurrences per 1000 landings for each of the 20 discretized values. The assumed discretized distribution of the initial sinking velocity $v_0$ is listed in Table 5.2 (cumulative occurrences) and illustrated in Figure 5.29 (occurrences).

Table 5.3 compares statistically the performances of four LG types in terms of the expected and median peak strut forces. Two cases have been considered, unconditional and conditional:

1. All landing conditions have been taken into account; *unconditional* $E[F_{\text{peak}}]$ and median$[F_{\text{peak}}]$ have been computed for the four control strategies considered.
2. Only landings with the PLG peak strut force exceeding the static load have been taken into account, which results in *conditional* probability distributions, expected values and medians

$$E\left[F_{\text{peak}} \mid F_{\text{peak}}^{\text{PLG}} > mg\right], \qquad \text{median}\left[F_{\text{peak}} \mid F_{\text{peak}}^{\text{PLG}} > mg\right]$$

being computed. This case reports on the statistical reduction of *significant* peaks.

**Table 5.2**  Assumed cumulative occurrences of sinking velocities per 1000 landings

| Sinking velocity $v_0$ (m/s) | Cumulative occurrences | Sinking velocity $v_0$ (m/s) | Cumulative occurrences |
|---|---|---|---|
| 0.00 | 1000.0 | 1.54 | 63.0 |
| 0.15 | 994.6 | 1.70 | 33.8 |
| 0.31 | 925.6 | 1.85 | 17.7 |
| 0.46 | 811.3 | 2.00 | 9.7 |
| 0.62 | 669.6 | 2.16 | 4.7 |
| 0.77 | 518.8 | 2.31 | 2.5 |
| 0.93 | 380.2 | 2.47 | 1.6 |
| 1.08 | 260.3 | 2.62 | 1.0 |
| 1.23 | 172.9 | 2.78 | 0.6 |
| 1.39 | 108.4 | 2.93 | 0.3 |

**Table 5.3** Performance of four LG control strategies, a statistical comparison: case 1 unconditional values (all landing scenarios); case 2, conditional values (landings with the PLG peak strut force exceeding the static load)

| | Peak strut force | | Relative improvement | |
|---|---|---|---|---|
| LG type | Expected value (kN) | Median value (kN) | Expected value (%) | Median value (%) |
| Case 1: unconditional (all landing scenarios) | | | | |
| PLG | 3890 | 3527 | — | — |
| VD-SLG | 3618 | 3284 | 7.0 | 6.9 |
| SLG | 3386 | 2992 | 12.9 | 15.2 |
| ALG | 3331 | 2959 | 14.4 | 16.1 |
| Case 2: conditional ($F_{\mathrm{peak}}^{\mathrm{PLG}} > mg$) | | | | |
| PLG | 4962 | 4613 | — | — |
| VD-SLG | 4565 | 4232 | 8.0 | 8.3 |
| SLG | 4202 | 3878 | 15.3 | 15.9 |
| ALG | 4106 | 3790 | 17.2 | 17.8 |

### 5.4.5.3 Safe Sinking Velocity Range

Table 5.3 evaluates the four discussed control strategies in statistical terms by comparing their performances in design landing conditions (Equation (12)). The optimum strategy for strut force control also allows the sinking velocity range to be extended beyond $v_{0(\mathrm{max})} = 2.93$ m/s without exceeding the peak strut force limit $F_{\mathrm{max}}$. Figure 5.30 compares, in terms of the landing mass $m$ at three control strategies (PLG, SLG, ALG), the maximum safe sinking velocities $v$, which are defined by

$$F_{\mathrm{peak}}(v, m) = F_{\mathrm{max}}$$

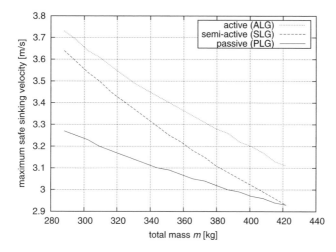

**Figure 5.30** Maximum safe sinking velocity at three control strategies in terms of the total landing mass

The velocity-driven semi-active strategy (VD-SLG) has been skipped, since it assumes no information about the landing mass $m$ and is hence bound by $v_{0(max)} = 2.93$ m/s, which occurs at $m_{max}$.

### 5.4.6 Fast Control of an MRF-Based Shock Absorber

Active adaptation systems that are intended to operate in the conditions of impact must respond faster than the duration of the controlled phenomenon. The actuation time delay should be at least 10 times faster than the total duration of the physical phenomenon. One of the potential actuation systems considered for the case of adaptive impact absorbers is magnetorheological fluid (MRF). The medium gives a unique opportunity of designing compact magnetorheological dampers (MRDs) with the possibility of smooth changing of its damping force. The fluid changes its properties when it is affected by an external magnetic field. However, the devices require very careful design in order to withstand all the demands. The main aspects, which have an important influence on the MRD, are: (1) the design of a mechanical structure, (2) the design of a magnetic circuit that should provide magnetic flux on the required level, (3) selection of the adequate MRF which should ensure the necessary yield stress, (4) selection of the proper size and shape of the housing in order to find the proper thermal balance and (5) the total response time of the magnetorheological (MR) device should satisfy the dynamic requirements of the system under control. The total response time of MR devices depends on the operation of the magnetic actuator. The actuating element in such a system is a compact electromagnet. Intensity of the magnetic field is the factor that influences the behavior of the MRF. At the same time the response time of the electromagnet is the parameter that determines the time delay of operation of the complete actuation system. Figure 5.31 presents the simplest circuit that describes the electromagnet. Two parameters are crucial for the element: resistance of coil R and inductance of coil L. The magnetic field generated by the coil is proportional to the value of current in the circuit. In the case of the step response of the RL circuit, the process of current rise has the character depicted in Figure 5.32. The time of response of this system is defined as reaching by the circuit 95 % of the demanded current value. The governing equation for the considered RL circuit can be written as follows:

$$L\frac{di}{dt} + Ri = U \tag{20}$$

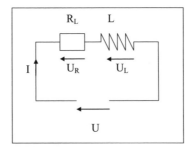

**Figure 5.31**   RL circuit of the electromagnet

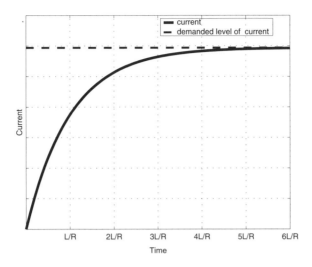

**Figure 5.32**   Step response of the RL circuit

and the solution can be found as

$$i\,(t) = \frac{U}{R}\left(1 - e^{(R/L)t}\right) \tag{21}$$

The ratio $L/R$ is called the time constant of the circuit and is proportional to the time required by the system to reach the steady-state value of the current:

$$\tau = \frac{L}{R} \tag{22}$$

As presented in Figure 5.32, the response time of RL circuits can be determined by the time constant. After a period equal to three time constants, the current reaches the level of 95 % of the demanded value. The time response of the circuit can be influenced by the designer in the following ways: by decreasing the inductance $L$ by limiting the turns on the coil in the electromagnet or by increasing the resistance of the coil. However, the inductance has two orders of magnitude lower values in comparison to the resistance values, which shows that adjusting the resistance gives more flexibility in optimization of the circuit. The resistance of the coil can be increased via reduction of its wire diameter, but the resistance of the whole circuit can also be increased by adding an external resistance in series to the coil. In this way the response time of the coil can be reduced. The consequence of higher resistance of the circuit is higher voltage, which must be provided to keep the demanded level of current and induction. In the case of fast systems, which require time delays no longer than 1 ms, the increase of the resistance would lead to voltages of the order of hundreds of volts. Such high voltages eliminate this solution from many practical applications. Another possible way of solving the problem of fast operation of the electromagnets is but using a closed-loop controller of the current level. The reduction of the time response can be obtained via a method of adapting a temporal overvoltage in the circuit. The RL circuits driven by higher voltage approach the adequately higher steady-state level of current with the same time delay as at the level of lower

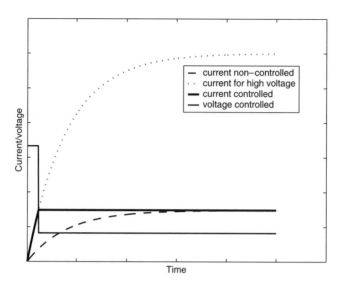

**Figure 5.33**   RL circuit performance

voltage. It is a consequence of the fact that the time constant does not depend on the voltage. As shown in Figure 5.33, the current rate is higher in the case of higher voltage. This feature can be utilized by adaptation of the current level controller. The basic idea of the controller's logical operation is to perform the algorithm as follows:

$$U(t) = \begin{cases} U_{\text{high}} & \text{if } I < I_{\text{demand}} \\ U_{\text{demand}} & \text{if } I \geq I_{\text{demand}} \end{cases} \tag{23}$$

An example of the controller operation is presented in Figure 5.33. An increase in voltage of 4 times in the preliminary period gave a reduction in the time response of the order of 15.

A practical example of the concept presented above was realized in laboratory conditions as a part of the full control system for the adaptive impact absorber. The conceptual adaptive impact absorbing system was considered to consist of three essential components: a set of sensors for recognition of the initial impact energy, a microcontroller for executing the established control sequence in real time and the adaptive actuator based on an MRF [28]. The set of photosensors determined the velocity of the structure and specified its mass before the instant of the impact. On the basis of the readings from the sensors, the microcontroller was able to recognize the magnitude of the impact energy. The objective of the control algorithm was then to adapt the actuator to the recognized impact energy. The adaptation of the absorption system was realized by taking advantage of the features of the MRF. In the case of systems devoted to absorption of the impact energy, an important fact was that the full period of the phenomenon was not longer than 50 ms (in severe cases), which specified that the update rate of the integrated system (sensors, control unit and actuators) should not exceed 4 ms. For this reason, the closed-loop controller for the current circuit was introduced. The controller was developed on the basis of the field programmable gate array (FPGA), a programmable hardware processing unit configured with a voltage type of power source. Using the FPGA processing unit allowed the period of execution of the logic operations to decrease to 5 $\mu$s per operation, which allowed a feedback regulation of the current source to be introduced. Figures 5.34(a) and (b) depict two

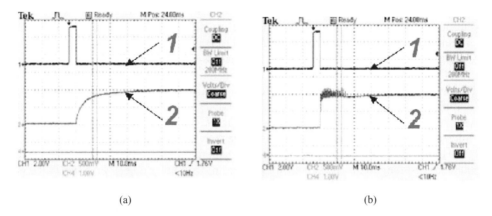

(a)                                                                    (b)

**Figure 5.34**    Performance of (a) a standard controller in the time domain and (b) the improved controller in the time domain

examples of operation of the developed controller with and without the regulation unit. The graphs present exemplary time histories of the velocity sensor input and the generated control signals. The tasks for the controller were identification of the vertical velocity on the basis of the signal from a photosensor, determination of the proper control signal sequence and application of the signal after a minimal time delay. The cases on both graphs are characterized by an identical initial input of velocity photosensor, which are represented by channel 1 in Figures 5.34(a) and (b). A dedicated numerical algorithm recalculated the signal readings in order to determine the impact velocity in the time between the end of the velocity signal and the beginning of the control signal, depicted as channel 2. The controller performed the required logic operations and generated the output signal after 40 µs from receiving the velocity input signal (Figures 5.34(a) and (b)). Channel 2 in both graphs in Figures 5.34(a) and (b) depicts the time history of the control current generation by the developed system. The response time of the current generation circuit determines the dominant time delays of the actuation system. In the case presented in the graphs (a) and (b), the desired current magnitude to generate was 0.5 A. The plot (a) represents the current generation process (channel 2) which is not regulated by the additional fast current generation unit, with the effect of the time delay equal to ca. 15 ms. In the graph (b) the regulated generation process is presented, which allowed the current generation time delay to be reduced to ca. 0.5 ms. Minimization of the response time made it possible to implement the MR device for the impact application, as the original response time of 15 ms gave no practical possibilities to control the process that lasts 50 ms in total.

## 5.5  Adaptive Inflatable Structures with Controlled Release of Pressure

### 5.5.1  The Concept of Adaptive Inflatable Structures (AIS), Mathematical Modeling and Numerical Tools

Adaptive inflatable structures are one of the special technologies for adaptive impact absorption. AIS are structures filled with compressed gas, the pressure of which is actively adjusted during the impact process. Pressure adjustment relies on appropriate initial inflation and controlled release of gas during the event. Such active control of internal pressure allows to change the dynamic characteristics of the inflatable structure and enables adaptation to various impact forces and scenarios. The form and shape of the pneumatic structure depends on its particular application. The inflated structure may be rigid (as a cylinder enclosed by the piston), it may be a thin-walled steel structure or completely deformable cushion made of fabric. Impact absorbing pneumatic structures, analysed in a further part of this chapter, are open-sea docking facilities, adaptive road barriers and external airbags for helicopter emergency landing.

Particular types of pneumatic structures (for instance a road barrier) are permanently inflated to a relatively low pressure which provides mitigation of weaker impacts. In other types of pressurized structures (such as an emergency airbag), the inflation is executed only when the collision occurs. Fast-reacting pyrotechnical gas generators based on deflagration of chemical mixture (similarly as in a car airbag) are used for immediate gas pumping. Initial pressure is adjusted according to the mass, velocity and area of the hitting object. Pressure of the gas increases the stiffness of the inflatable structure and prevents its excessive deformation during the impact. Compressed air also has a beneficial influence on the buckling behavior of the structure since it usually reduces the compressive forces, which may cause loss of the structure stability. Further improvement of the AIS can be achieved by dividing the structure into several pressurized packages separated by flexible walls. This method enables independent adjustment of initial pressure in different parts of the AIS without a significant increase in the structure weight.

During the collision with an external object, the release of pressure is executed by opening controllable piezo-valves. Such valves are mounted in external walls of inflatable structures and in diaphragms between the pressurized packages. Therefore, the gas can flow between internal chambers and outside the structure. Release of pressure allows the stiffness of the pneumatic structure in the subsequent stages of impact to be controlled. In this way the impacting object can be stopped using the whole admissible stroke and its acceleration can be significantly reduced. Another purpose of applying the release of compressed air is to dissipate the impact energy and to avoid the hitting object rebound.

Numerical analysis of the pneumatic structure subjected to an impact load requires consideration of the interaction between its walls and the fluid enclosed inside the chambers. An applied external loading causes deformation of the structure and change of the capacity and pressure of the fluid. The pressure exerted by the fluid affects, in turn, the deformation of the structure and its internal forces. The most precise method for solving above fluid–structure interaction problem is the arbitrary Lagrangian Eulerian (ALE) approach, where Navier–Stokes equations for the fluid are solved in the Euler reference frame and structural mechanics equations are solved in the Lagrange reference frame. Such an approach is applied to model extremely fast processes such as airbag deployment (cf. Reference [29]) or when recognition of the flow pattern is the main problem addressed. In the problem considered, the fluid part of the analysis is used to compute global forces acting on the walls of an inflatable structure and the exact distribution of the fluid pressure and velocity is not of interest. Moreover, the impacting object velocity is much lower than the speed of impulse propagation in the gas, so that the pressure becomes constant across the chambers relatively fast. Therefore, the so-called uniform pressure method (UPM) will be used, which assumes that the gas is uniformly distributed inside the chambers and the chamber walls are subjected to uniform pressure.

The dynamics of the inflatable structure is described by the nonlinear equation of motion, whose general form reads

$$\mathbf{M\ddot{q}} + \mathbf{C\dot{q}} + \mathbf{K(q)q} = \mathbf{F(p, q)} + \mathbf{F}_I$$
$$\mathbf{q}(0) = \mathbf{q}_0, \quad \mathbf{\dot{q}}(0) = \mathbf{V}_0$$

(24)

Vector $\mathbf{p} = \{p_1(t), p_2(t), \ldots, p_n(t)\}$ indicates gauge pressures in the cavities. The impact can be modeled by right-hand side force vector $\mathbf{F}_I$, by initial conditions or by the contact defined between the inflatable structure and other objects. In any case, the $\mathbf{F}(\mathbf{p}, \mathbf{q})$ vector is present, since it provides coupling with the fluid. The interaction of the structure and the fluid during a large deformation can be correctly taken into account only by assembling the equilibrium equations in the actual configuration so the equation of motion has to be considered in a nonlinear form. It is assumed that each cavity of the inflatable structure is filled with a compressible (pneumatic) fluid, which is described analytically by the equation of state for the ideal gas:

$$\overline{p}(t) = \rho(t)R\overline{T}(t)$$

or                                                                                                    (25)

$$\overline{p}(t)V(t) = m(t)R(T(t) - T_Z)$$

Absolute pressure $\overline{p}$ is defined as $\overline{p} = p + p_A$, where $p$ is the gauge pressure and $p_A$ is an ambient pressure. The absolute temperature is defined as $\overline{T} = T - T_Z$, where $T$ is the current temperature in the Celsius scale and $T_Z$ is the absolute zero temperature. The gas constant $R$ is related to the universal gas constant $\overline{R}$ and molecular weight MW by the formula: $R = \overline{R}/\text{MW}$. Moreover, the variables $\rho$, $V$ and $m$ indicate gas density, volume and mass, respectively. The initial conditions for the fluid are given by $\overline{p}(0) = p_0$, $T(0) = T_0$. In the case when fluid flow occurs, the change in fluid mass in the cavity is described by the equation

$$m(t) = m_0 + \int_0^t q(\overline{t})\mathrm{d}\overline{t} = m_0 + \int_0^t q_{\text{in}}(\overline{t}) - q_{\text{out}}(\overline{t})\,\mathrm{d}\overline{t}$$                                                                                    (26)

where $q_{\text{in}}$ is the mass flow rate into the cavity and $q_{\text{out}}$ is the mass flow rate outside the cavity. The direction of flow depends on the sign of pressure difference between the cavities. In the simplest model of the flow considered, the mass flow rate is related to the pressure difference according to the formula

$$\Delta p(t) = C_V q(t) + C_H q(t)|q(t)|$$                                                                                    (27)

where $\Delta p(t) = p_{\text{out}} - p(t)$, $C_V$ is the viscous resistance coefficient and $C_H$ is the hydrodynamic resistance coefficient. Both of these coefficients depend on the area of the orifice and can be found experimentally for a given type of valve. Under the assumption of an isothermal process (or arbitrarily given change of temperature), a set of equations (24 to 27) fully describes the coupling between the fluid and the surrounding structure.

In a more general case, the temperature of the gas is treated as a subsequent unknown of the problem. The balance of the heat transferred to the system $\mathrm{d}Q$, enthalpy of the gas added (removed) $\mathrm{d}m_{\text{in}}\overline{H}_{\text{in}}$ ($\mathrm{d}m_{\text{out}}\overline{H}_{\text{out}}$), change of gas internal energy $\mathrm{d}(m\overline{U})$ and the work done by gas $\mathrm{d}W$, is given by the first law of thermodynamics for an open system:

$$\mathrm{d}Q + \mathrm{d}m_{\text{in}}\overline{H}_{\text{in}} - \mathrm{d}m_{\text{out}}\overline{H}_{\text{out}} = \mathrm{d}(m\overline{U}) + \mathrm{d}W$$                                                                                    (28)

The specific gas energy and specific gas enthalpy are defined as

$$\overline{U}(t) = c_V\overline{T}(t), \qquad \overline{H}_{\text{in}}(t) = c_p\overline{T}_{\text{in}}(t), \qquad \overline{H}_{\text{out}}(t) = c_p\overline{T}(t)$$                                                                                    (29)

The flow of the heat across the cavity walls is described by

$$\frac{dQ(t)}{dt} = \lambda A(t)(T_{\text{ext}} - T(t)) \tag{30}$$

where $\lambda$ is the heat conductivity coefficient of the considered chamber wall and $A(t)$ is the area of that wall. Finally, the work done by the gas equals

$$dW(t) = \overline{p}(t)dV(t) \tag{31}$$

If the wall of the cavity is a perfect insulator ($\lambda = 0$) or when the process is relatively fast, conditions of adiabatic process are fulfilled and no heat transfer through the chamber walls occurs: $dQ = 0$. In the case when the gas flows into the considered chamber Equation (28) reads

$$\frac{\dot{T}(t)}{T(t)} = \left[ \chi \frac{\overline{T}_{\text{in}}(t)}{T(t)} - 1 \right] \frac{\dot{m}(t)}{m(t)} - (\chi - 1)\frac{\dot{V}(t)}{V(t)}, \qquad \chi = \frac{c_p}{c_V} \tag{32}$$

where $\chi$ is an adiabatic exponent defined as the ratio of constant pressure heat capacity $c_p$ and constant volume heat capacity $c_V$, and for the air $\chi = 1.41$. In the case when the gas flows out of the chamber, Equation (32) can be solved analytically:

$$\frac{\overline{T}(t)}{\overline{T}_0} = \left[ \frac{m_0}{V_0} \frac{V(t)}{m(t)} \right]^{1-\chi} \quad \text{or} \quad \frac{\overline{p}(t)}{\overline{p}_0} = \left[ \frac{m_0}{V_0} \frac{V(t)}{m(t)} \right]^{-\chi} \tag{33}$$

Finally, when no flow of the gas occurs, i.e. the mass of the gas in the cavity remains constant, the well-known equation is obtained:

$$\overline{p}(t)V(t)^\chi = \overline{p}_0 V_0^\chi = \text{constant} \quad \text{or} \quad T(t)V(t)^{\chi-1} = T_0 V_0^{\chi-1} \tag{34}$$

Adiabatic flow of the gas through the orifice is described by the model that assumes that the flow is blocked at the critical velocity. In such a case, the mass flow rate is defined by the Saint-Venant formula, (cf. Reference [30]):

$$|q(t)| = CA\frac{\overline{p}_e}{\sqrt{R(T - T^Z)}}\sqrt{\frac{2\chi}{\chi - 1}\left[ \left(\frac{\overline{p}_0}{\overline{p}_e}\right)^{2/\chi} - \left(\frac{\overline{p}_0}{\overline{p}_e}\right)^{(\chi+1)/\chi} \right]} \tag{35}$$

where $C$ is the discharge coefficient, $A$ is the orifice area, $\overline{p}_e$ is the absolute pressure in the upstream fluid cavity, $\overline{p}_0$ is the absolute pressure in the orifice given by

$$\begin{aligned} \overline{p}_0 &= \overline{p}_a && \text{if } \overline{p}_a \geq \overline{p}_c \\ \overline{p}_0 &= \overline{p}_c && \text{if } \overline{p}_a < \overline{p}_c \end{aligned} \tag{36}$$

where $\bar{p}_a$ is equal to the ambient pressure or the downstream cavity pressure. The critical pressure $\bar{p}_c$ at which chocked or sonic flow occurs is defined as

$$\bar{p}_c = \bar{p}_e \left( \frac{2}{\chi + 1} \right)^{\chi/(\chi-1)} \tag{37}$$

The main numerical tool used for the analysis of inflatable structures was the commercial finite element code ABAQUS (cf. Reference [31]), which allows the compressible fluids described by Equations (25) to (37) to be modeled. A controlled release of pressure was executed by changing the flow resistance coefficients from Equation (27) or the orifice area from Equation (35), according to the assumed control strategy. For this purpose FORTRAN subroutines cooperating with ABAQUS/Standard were implemented. When ABAQUS/Explicit (cf. References [31] and [32], was used, the connection to MATLAB software was established and its optimization procedures were utilized.

The basic example of an adaptive inflatable structure is a cylinder enclosed by a piston and equipped with a controllable valve. In the simplest case the gas is released from the cylinder directly to the environment (Figure 5.35(a)). Alternatively, the gas can flow into the chamber adjacent to the cylinder, which serves as an accumulator (Figure 5.35(b)). In another option, gas is released to the chamber located above the piston (Figure 5.35(c)) and its pressure influences the global force acting on the piston.

The device presented in Figure 5.35(a) will now be analyzed more precisely. The system is described by a simple equation of piston motion and moreover Equations (25) to (33) hold. It is assumed that at the beginning of impact, the mass $M$ and weightless piston have a common initial velocity $V_0$ and the initial pressure in the cylinder is set to $\bar{p}_0$. The objective of the applied control is to protect both the hitting object (minimize its acceleration and rebound) and the impacted structure (minimize pressure arising during impact). According to the equation of piston equilibrium, the problems of acceleration minimization and pressure minimization are for this simple model equivalent. The optimal adaptation strategy is to keep the valve closed until the gas pressure achieves a certain level and then to control the valve opening to maintain the pressure constant. The value of the pressure at which the adaptation begins is chosen in such a way that the mass is stopped using the whole stroke of inflatable structure. Equations

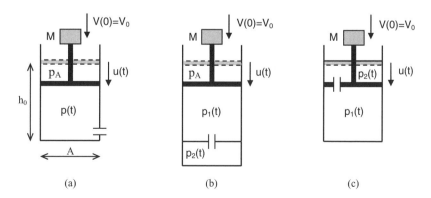

(a)                                  (b)                                  (c)

**Figure 5.35**   Various options for the design of an adaptive pneumatic cylinder

describing the optimal system read

$$M\frac{d^2u(t)}{dt^2} + \left\{\frac{\overline{p}_0 h_0^\chi}{[h_0 - u(t)]^\chi} - p_A\right\} A = 0 \quad \text{for } u(t) \in (0, u_x) \tag{38}$$

$$M\frac{d^2u(t)}{dt^2} + \left[\frac{\overline{p}_0 h_0^\chi}{(h_0 - u_x)^\chi} - p_A\right] A = 0 \quad \text{for } u(t) \in (u_x, h_0) \tag{39}$$

The variable $u_x$ indicates the piston displacement at the time instant when the valve is opened and can be calculated by integrating Equations (38) and (39) over displacement in the range from 0 to $h_0$:

$$u_x = h_0 - h_0\left[\frac{(\chi - 1)}{\chi}\left(\frac{MV_0^2}{2\overline{p}_0 h_0 A} + \frac{p_A}{\overline{p}_0}\right) + \frac{1}{\chi}\right]^{1/(1-\chi)} \tag{40}$$

The result for the isothermal process can be obtained by calculating the limit of Equation (40) when $\chi$ approaches 1. The corresponding optimal pressure and optimal acceleration of the hitting mass can be computed using Equation (39). In the case when the initial pressure is also adjusted, the optimal pressure and acceleration remain constant during the whole process:

$$\overline{p}^{\text{opt}} = \frac{MV_0^2}{2Ah_0} + p_A, \qquad \ddot{u}^{\text{opt}} = -\frac{V_0^2}{2h_0} \tag{41}$$

These formulas hold until the mass displacement equals $h_0$ and its velocity is reduced to zero. At this moment, no gas remains in the chamber, there is no force pushing the mass away and its rebound is completely mitigated. The optimal mass flow rate $q$ during the active stage of impact can be calculated using geometrical relations, the ideal gas law and, in the case of the adiabatic process, the energy equation. The model of flow (27) with coefficient $C_H = 0$ is applied so the viscous resistance coefficient $C_V$ is the main control parameter. Finally,

$$q(t) = -\frac{\overline{p}^{\text{opt}} A\dot{u}^{\text{opt}}(t)h_0^{1-\chi}}{R\overline{T}_0(h_0 - u_x)^{1-\chi}}, \qquad C_V(t) = \frac{(\overline{p}^{\text{opt}} - p_A)R\overline{T}_0(h_0 - u_x)^{1-\chi}}{\overline{p}^{\text{opt}} A\dot{u}^{\text{opt}}(t)h_0^{1-\chi}} \tag{42}$$

Formulas (42) clearly define the dependence of control variables in terms of the impact energy and impact velocity. The above solution holds when there are no restrictions imposed on the valve opening.

### 5.5.2 Protection against Exploitative Impact Loads for Waterborne Transport

One of the most relevant challenges in waterborne transport are docking operations in the open sea. The problem concerns small service ships that dock to offshore wind turbine towers for the purpose of maintenance and monitoring. Collisions occur especially often during rough sea conditions and can lead to serious damage to both the wind generator tower and the ship. Therefore, use of an inflatable structure attached to the wind turbine tower to provide safety of the docking operation is proposed (cf. Reference [33]).

The considered pneumatic structure is torus-shaped and surrounds the tower at water level (cf. Figure 5.36). The dimensions of the inflatable structure are restricted to 2 m in height and 0.7 m

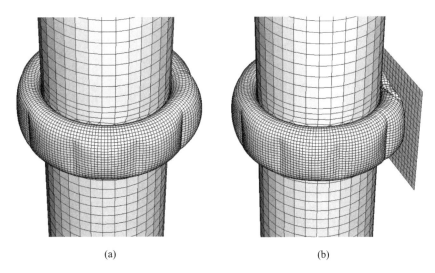

**Figure 5.36**  Inflatable structure surrounding the tower: (a) initial inflation of the chambers, (b) deformation during the collision

in width, due to the necessity of fast inflation and pressure release. The walls of the pneumatic structure are made of rubber reinforced by steel rods, which provide high durability and allow large deformations during ship impact. The inflatable structure is divided into several separate air chambers distributed around the tower. Inflation is executed for each chamber separately by a compressor located inside the tower or, alternatively, by a fast-reacting pyrotechnic system. The value of the initial pressure is restricted to 3 atm because of high stresses arising in the rubber walls and excessive deformation of the whole pneumatic structure after inflation. Adaptation to a particular impact is obtained by adjusting the initial internal pressure according to the velocity and mass of the hitting object and by varying its level between the chambers. During the collision, piezoelectric valves mounted in external and internal walls of the inflatable structure are activated and controlled release of pressure is executed.

The numerical model contains only the lower part of the wind turbine tower (cf. Figure 5.36); the upper part is modeled by additional masses and forces applied at the top edge. The tower consists of shell elements with the thickness increased on the water level and the torus-shaped inflatable structure consists of membrane elements. Gas inflating the chambers is modeled by using 'surface-based fluid cavities' available in ABAQUS software. Impact of the ship is defined as a contact problem with the ship modeled as a rigid surface approaching the tower with an initial velocity. The two-dimensional model presented in Figure 5.37 was implemented to reduce the time of analysis and to examine various options for the inflatable structure design. The additional mass obtained from reduction of the full model (cf. References [34] and [35]) was located in the middle of the structure. The stiffness of the tower was modeled by an additional element connected to its middle point.

The purpose of applying a pneumatic structure is to mitigate the response of both the ship and the wind turbine tower. In particular, the inflatable structure helps to minimize the ship deceleration, dissipate the impact energy, decrease stresses arising at the location of the collision and reduce the tower vibrations. Adaptation of an inflatable structure to each of the mentioned objectives will be considered by using semi-active and active control strategies for an exemplary impact of a 60 ton ship moving with a velocity of 6 m/s. The distribution of the initial pressures

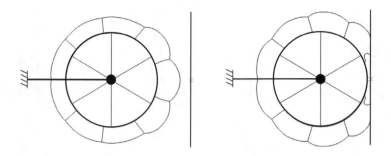

**Figure 5.37**   Two-dimensional model of collision: initial state and resulting deformation

in the cavities will be fixed and equal to 100 % in the front cavity, 50 % in adjacent cavities
and 10 % in others. In all considered examples, the initial inflation of the chambers will be
executed during 200 ms and the ship will impact the structure directly afterwards.

### 5.5.2.1 Minimization of Ship Acceleration

The first objective of the pressure adjustment is to decrease ship accelerations. In the case
of an inflatable torus, the equivalence of acceleration and pressure minimization problems
is disrupted due to the influence of an initial change in the chamber volume, forces arising
during rubber deformation and a change in the contact area between the ship and the inflatable
structure during collision. In the simplest semi-active case, only the initial pressure is
adjusted so after the inflation phase, the mass of the gas in the chambers remains constant. A
minimal ship acceleration of 58.84 m/s$^2$ (cf. Figure 5.38, dark line) is obtained for the highest
admissible initial pressure (3 atm), which is the consequence of the significant expansion of
the front chamber after inflation.

   A more sophisticated type of semi-active system includes the exhaust valves, the opening of
which can be adjusted for a particular impact scheme, but it remains constant during the impact

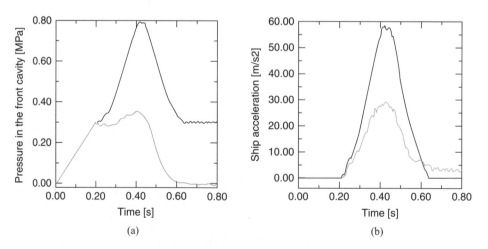

**Figure 5.38**   Semi-active adaptation without pressure release (dark line) and with pressure release
(bright line): (a) pressure in the front chamber, (b) corresponding ship acceleration

time. In the considered case, the air is released from three front chambers intrinsic for impact absorption. By letting the gas out of the cavities the global force acting on the ship is reduced and simultaneously the depth of inflatable structure compression is increased. Minimal ship acceleration equals $29.14\,\text{m/s}^2$ (cf. Figure 5.38, bright line) and is obtained for the case when maximal ship displacement achieves its limit, i.e. the ship is stopped just before the tower wall. The computed value of the corresponding flow resistance coefficient $C_V = 275\,\text{kPa s/kg}$ and maximal mass flow rate $q = 1.27\,\text{kg/s}$ allows a proper type of valve to be chosen.

A further decrease in ship acceleration can be obtained by active control of the orifice diameter during impact. In an active control strategy the valve remains closed until the acceleration achieves the level required to stop the ship using the distance remaining to the tower. Therefore the time of valve opening is based on the actual ship velocity, acceleration and distance to the tower. The pressure value will be used as a main control parameter since it directly influences ship acceleration. The optimal change of pressure can be calculated by comparing actual and optimal states of the system for each time step of the finite element analysis. Pressure modification that has to be applied at the following time step equals

$$\Delta p(t) = \frac{M(a^* - a(t))}{A(p,t)} \tag{43}$$

where $a^*$ and $a(t)$ indicate optimal and actual ship acceleration, respectively. The area of contact between the ship and inflatable structure $A(p,t)$ is estimated in terms of ship distance to the tower wall and pressure. The required mass flow rate can be calculated afterwards by using the optimal pressure and chamber volume obtained from the finite element analysis.

An alternative numerical approach utilizes the flow resistance coefficient, instead of pressure, as a control variable. In this strategy, the flow coefficient is proportionally adjusted in several time intervals (cf. Figure 5.39) to achieve ship acceleration possibly close to the desired level. The advantage of this method is the possibility of imposing constraints on the maximal orifice area. In the numerical example, the adaptation procedure is started 110 ms after the contact between the ship and inflatable structure occurs (cf. Figure 5.39, dark line). The maximal acceleration is reduced to $21.56\,\text{m/s}^2$, but a high mass flow rate $q = 3.9\,\text{kg/s}$ is required at the initial stage of impact. A disadvantageous slow increase in acceleration in the passive stage of impact (200–310 ms) can be avoided by additional inflation of the main chamber after the ship approaches the pneumatic structure. The pressure has to be increased to 6.6 atm and as a result a sudden growth of acceleration is obtained (cf. Figure 5.39, bright line). Due to the fact that the way of stopping the ship is longer than in the previous case, the required level of acceleration is decreased to $18.76\,\text{m/s}^2$, which constitutes 32 % of ship acceleration in the basic semi-active case.

### 5.5.2.2 Minimization of Ship Rebound

The considered torus-shaped adaptive inflatable structure serves as a docking facility and ship rebound is not a desired phenomenon. Thus, the next control objective is to reduce the ship rebound, i.e. to minimize the ship velocity after collision, possibly to zero. Under adiabatic conditions, the equation of energy balance for the process of a ship collision with an inflatable torus reads

$$-\Delta E^{\text{ship}} + \Delta W^{\text{ext}} = \Delta E^{\text{tower}} + \Delta U + \Delta E^{\text{AIS}} - \Delta H \tag{44}$$

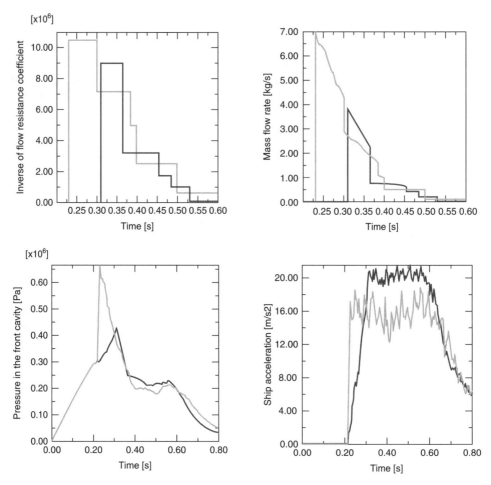

**Figure 5.39** Two strategies of active control: with pressure release (dark line), additional inflation at the beginning of the impact (bright line)

In a passive system with no pressure release, the main part of the ship kinetic energy ($E^{\text{ship}}$) is converted into internal gas energy ($\Delta U$) and the small part is converted to AIS walls strain energy ($\Delta E^{\text{AIS}}$). The kinetic and potential energy of the tower ($\Delta E^{\text{tower}}$) can be neglected due to small tower displacement and velocity. During the rebound stage of impact the gas expands, its internal energy decreases and it is changed back into kinetic energy of the ship. Thus, in a passive system, the absolute value of the final ship velocity is close to the initial one. In contrast to a passive system, in a semi-active or active system with pressure release, internal gas energy is not accumulated but dissipated by letting the air out of the cavities ($\Delta H$). Hence, the value of gas exergy (its ability to perform useful work) at the moment when the ship is stopped is decreased and the ship rebound is mitigated.

Semi-active adjustment of the initial pressure does not affect the final ship velocity since it does not cause energy dissipation. However, the ship rebound is significantly decreased by using inflatable structure comprising a valve with a constant opening. The minimal final ship velocity of 1.25 m/s is obtained in the case when the whole stroke of an inflatable structure is

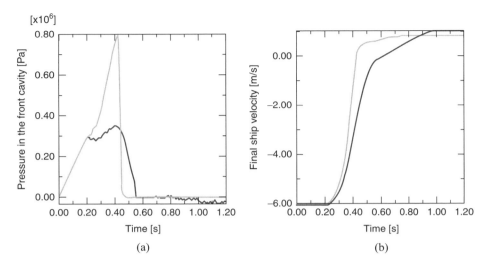

**Figure 5.40** Active mitigation of the ship rebound: (a) pressure of the gas in the cavity, (b) corresponding final velocity of the ship

used. The optimal flow resistance coefficient and pressure change during impact are the same as in the case of semi-active acceleration mitigation (cf. Figure 5.38(a)). In the considered example, the pressure at the moment when the ship is stopped is not completely reduced, but the chamber volume is minimal and work that can be done by gas is relatively small. Therefore the semi-active system is very efficient and allows almost 96 % of the initial ship energy to be dissipated.

In the active strategy, the valve opening remains constant while the ship is approaching the tower, as in a semi-active approach. When the ship velocity decreases to zero (at time 0.54 s), the valves are fully opened to release a surplus of pressure (cf. Figure 5.40(a), dark line). Then the valve is closed again and backward deformation of the chambers is reduced due to arising underpressure. As a result of this strategy, the final ship velocity is diminished to 1.05 m/s (cf. Figure 5.40(b), dark line). Nonzero final rebound velocity is the consequence of ship interaction with strongly deformed AIS walls, which repel the ship.

The influence of inflatable structure walls can be further reduced by minimizing their deformation and thereby the amount of strain energy accumulated. In the present control strategy the valves remain closed during the whole compression stage of impact in order to achieve the highest possible stiffness of the front chambers of the inflatable torus. The ship is stopped after crushing only a part of the pneumatic structure at time 0.42 s and then an immediate pressure release is executed (cf. Figure 5.40(a), bright line). Finally, the rebound velocity is reduced to 0.82 m/s, which means that more than 98 % of the initial ship energy is dissipated.

### 5.5.2.3 Minimization of Stresses in the Tower Wall

Another purpose of applying the inflatable structure is mitigation of tower response to impact. In particular, the pneumatic structure reduces local stresses in the front tower wall by preventing direct contact of the ship and the tower. The front tower wall is subjected to bending caused by pressure loading and the level of stress depends approximately on the actual value of pressure. Therefore, minimization of stresses is equivalent to minimization of front chamber pressure.

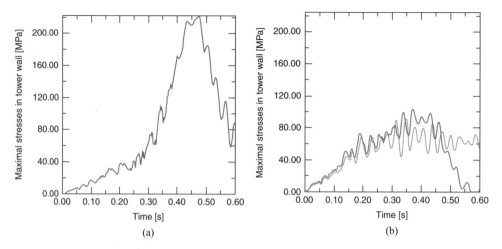

**Figure 5.41** Stress in the tower wall: (a) semi-active control without pressure release, (b) semi-active (dark line) and active control with pressure release

In a semi-active system, without gas release tensile stresses can be reduced to 221.9 MPa (cf. Figure 5.41(a)) by applying optimal initial pressure equal to 1.7 atm. The semi-active system with pressure release is most efficient when maximal initial pressure is applied and the whole stroke of the inflatable structure is utilized. In such a system maximal pressure is decreased to 3.49 atm and stresses in the tower wall are reduced to 104.5 MPa (cf. Figure 5.41(b), dark line).

In a fully active system, pressure is adjusted to a constant, possibly lowest level. For impact energies lower than 1.25 MJ, the level of such a pressure does not exceed the maximal allowable initial value and thus pressure may remain constant during the whole process. For impact of energy higher than 1.25 MJ, the optimal strategy assumes an initial stage of pressure increase and activation of control when the pressure achieves the level that enables the ship to be stopped in the vicinity of the tower wall. For the considered impact of a ship of 60 tons with velocity of 6 m/s, an optimal constant pressure of 2.68 atm was found by performing multiple finite element analysis. The flow resistance coefficient that provides appropriate mass exchange under the given conditions of pressure difference was calculated using Equations (25) to (27). In the considered active system, pressure is significantly reduced in comparison to the semi-active case, but high-frequency vibrations of the tower wall arise and hinder significant minimization of stresses. Finally, the stresses are reduced to 91.9 MPa and the active system is more effective by 12 % than the semi-active one.

### 5.5.2.4 Mitigation of Tower Vibrations

The last goal of pressure adjustment is to minimize the amplitude of tower vibrations after impact. This objective can be alternatively understood as a minimization of the energy transmitted to the tower during collision. Due to the fact that the impact time is relatively short in comparison to the period of tower vibration, the maximal tower displacement depends on the ship impulse (cf. Reference [36]). The impulse transmitted to the tower is proportional to the mass of the ship and the difference between its initial and final velocities. Thus, minimization of tower vibrations is approximately equivalent to minimization of the ship rebound and vibration amplitude can be reduced maximally by 50 %.

In a semi-active system without pressure release, the maximal tower displacement is almost independent of the initial pressure and equals 4.58–4.60 mm. By using the semi-active system with pressure release, the amplitude of tower vibration is gradually reduced along with the decrease in the flow resistance coefficient. The best result (2.67 mm) is obtained for the lowest flow resistance coefficient for which the ship does not hit the tower. The most effective active adaptation procedure assumes an additional release of pressure at the moment when the ship velocity approaches zero. By using such a strategy, the maximal tower displacement is reduced to 2.51 mm. Keeping the valve closed and performing pressure release only when the ship is stopped (which is most effective for rebound mitigation) gives a slightly larger tower displacement, probably due to the shorter impact time.

### 5.5.3 Protective Barriers against an Emergency Crash for Road Transport

Impact absorbers for road and railway transport are usually designed as thin-walled steel structures due to their huge durability and small weight. They dissipate the energy of front impact very efficiently by a folding mechanism (cf. Reference [37]). However, in the case of lateral impact, thin-walled structures easily undergo large deformation and local plastic yielding and usually only a small part of the impact energy is dissipated. A significant increase in durability to lateral loading can be obtained by filling a thin-walled structure with compressed air and by controlling its release during the impact (cf. Reference [38]). The examples of structures that can be effectively improved by using compressed gas are a door of the passenger car (cf. Figure 5.42(a)) and a protective barrier against an emergency crash for road transport (cf. Figure 5.42(b)).

This section is aimed at developing a strategy for optimal initial distribution and release of gas within a simple two-dimensional frame (cf. Figure 5.43), which could serve as a basis for a protective barrier design. The structure is divided into several pressurized packages equipped with piezoelectric exhaust valves. The elastoplastic material model with hardening is assumed

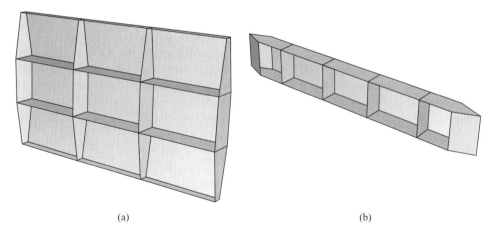

(a)                                                                                  (b)

**Figure 5.42**  Examples of inflatable thin-walled structures: (a) door of the car divided into pressurized chambers, (b) adaptive inflatable road barrier

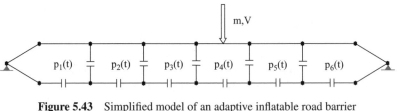

**Figure 5.43** Simplified model of an adaptive inflatable road barrier

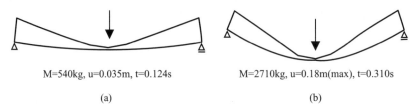

M=540kg, u=0.035m, t=0.124s                    M=2710kg, u=0.18m(max), t=0.310s

(a)                                            (b)

**Figure 5.44** Limiting deformation and load capacity of the empty and inflated barrier

and large deformations are taken into account. The frame is subjected to lateral impact modeled by a mass with initial velocity.

The load capacity of the barrier is defined as the maximal impacting mass, which can be applied to the structure with established velocity and does not violate two kinematic conditions imposed on its deformation. The first condition states that no collision between the mass and the lower span occurs and the second one confines maximal displacements of the lower span $u$ to a limiting value $u_{max}$. The influence of pressure on the load capacity will be estimated in the case of a barrier with a single chamber and one sliding support. The maximal mass that can be applied to such a structure equals 540 kg since it causes a collision between the spans (cf. Figure 5.44(a)). Sealing of the structure and filling it with gas under optimal constant pressure increases the load capacity to 2710 kg. By using the inflated structure, the impact time is elongated 2.5 times and the distance of stopping the mass is extended to the maximal allowable value (cf. Figure 5.44(b)).

In the case of a multichamber barrier, maximization of the load capacity can be formulated as an optimization problem in which the mass of the hitting object is maximized with respect to the values of pressure inside the cavities $\mathbf{p} = \{p_1(t), p_2(t), ..., p_n(t)\}$. This problem was solved for a three-chamber inflatable structure fixed with no sliding (cf. Figure 5.45). In such a case vector $\mathbf{p}$ has two components, $p_1$ and $p_2$, which indicate pressures in the lateral and middle cell, respectively. Initially, a constant value of pressure during the impact was considered. The highest increase in load capacity (6.1 times) was obtained for maximal allowable pressure in lateral cells and significantly smaller pressure in the middle cell (cf. Table 5.4). In the second

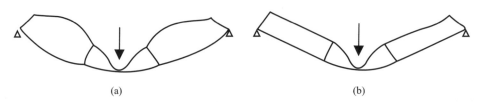

(a)                                            (b)

**Figure 5.45** Deformation of the optimally inflated structure loaded by the maximal mass: (a) constant pressure; (b) linear decrease of pressure

**Table 5.4** Comparison of the structure load capacity for various schemes of pressure adjustment

| $p_1(0)$ (kN/m) | $p_2(0)$ (kN/m) | $p_1(t_{stop})$ (kN/m) | $p_2(t_{stop})$ (kN/m) | $u(t_{stop})$ (m) | $t_{stop}$ (s) | $M$ (kg) |
|---|---|---|---|---|---|---|
| 0 | 0 | 0 | 0 | 0.04 | 0.152 | **7596** |
| 1600 | 1147 | 1600 | 1147 | 0.18 | 0.213 | **46374** |
| 400 | 3925 | 0 | 0 | 0.18 | 0.265 | **68489** |

example, a linear decrease of pressure was assumed. Gas was completely released at time $t_{stop}$ when the velocity of hitting the object was decreased to zero. In the optimal solution, the middle chamber is almost tenfold more inflated than the lateral chambers. The load capacity is increased 9.02 times in comparison to the generic barrier. Detailed results are presented in Table 5.4 and the corresponding deformation of the structure is depicted in Figure 5.45.

Another strong advantage of using inflatable road barriers instead of passive ones is the possibility of adaptation to a particular impact for its optimal mitigation. An active pressure adjustment allows the kinematics of the hitting object to be controlled. In particular, it helps to reduce the impacting object acceleration to the desired, possibly constant level or to confine its maximal displacement. A corresponding problem of optimal pressure distribution can be solved for several packages in the structure, but inflating the chamber to which the impact is applied is usually the most efficient.

The structure considered in the numerical example contains only one pressurized package and has one sliding support. The objective is to achieve a constant, formerly assumed level of acceleration during the whole process. The numerical methods for pressure adjustment are similar to those described for the inflatable torus (cf. Section 5.5.2.1). Adequate precision of the solution (cf. Figure 5.46(b)) is obtained by adjusting the pressure in ten uniformly distributed time instants. A high value of pressure is necessary at the beginning of the impact and then the curve of pressure gradually declines (cf. Figure 5.46(a)).

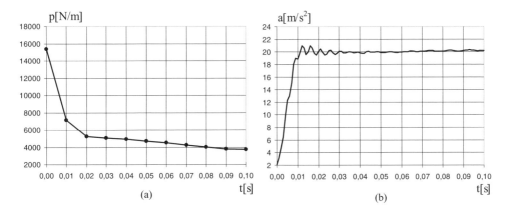

**Figure 5.46** Minimization of accelerations: (a) optimal change of pressure, (b) resulting acceleration of the hitting object

### 5.5.4 Adaptive Airbag for Emergency Landing in Aeronautic Applications

The last application considered is an adaptive external airbag for a helicopter (cf. Reference [39]). The system is designed to mitigate extremely severe emergency landing. It consists of a multichamber air-filled cushions attached to the helicopter undercarriage (cf. Figure 5.47(a)). Deployment of these cushions is executed just before touchdown by means of pyrotechnic inflators. The value of the initial pressure and its distribution among airbags is adjusted to the landing direction and velocity, which are identified by using ultrasonic velocity sensors. Additionally, the actual value of pressure during the emergency touchdown is measured by piezoelectric sensors located inside the airbags. Pressure is released by fabric leakage as in the classical airbag and by additional controllable high-speed and stroke valves. Such an emergency system can be linked with (also controllable) landing gear of the helicopter by using a common hardware controller.

In a simplified numerical example, the helicopter is modeled by deformable beams and point mass (cf. Figure 5.47(b)). The total weight of the model is comparable to the weight of a typical civil helicopter. The velocity at the beginning of the emergency landing equals 10 m/s, which is equivalent to a free fall from 5.1 m. Moreover, it is arbitrarily assumed that maximal average acceleration during an emergency landing does not exceed 10 g. That condition can be satisfied by applying airbags of a total projection area 4 m$^2$ and width 0.5 m. A simplified equation of energy balance indicates that constant pressure not higher than 2.25 atm (1.25 atm overpressure) is sufficient to avoid direct collision with the ground.

Control strategy is oriented towards minimization of total forces and accelerations acting on the helicopter, its stabilization during landing and mitigation of stresses arising in the helicopter undercarriage. These objectives are fulfilled when the total force acting on the helicopter is kept at a constant level and the airbags are completely compressed during touchdown. The control strategy assumes initial inflation of the airbags to 0.9 atm overpressure (cf. Figure 5.48(a)). Briefly after the beginning of the collision, the valves are opened to control the release of gas (Figure 5.48(b)) and maintain the pressure at a constant level. The release of gas is continued when the helicopter velocity drops to zero, which results in a sudden decrease of pressure. The subsequent peaks on the pressure plot are related to the rebounds of the helicopter after the main impact. Corresponding helicopter accelerations are reduced several times in comparison to the case when no airbag is applied. Conducted simulations prove that the proposed adaptive airbags efficiently decrease forces and accelerations and significantly increase the safety of an emergency landing.

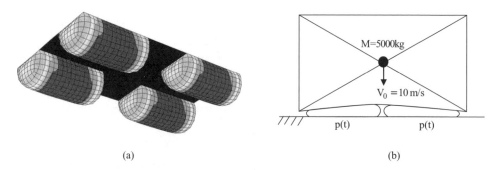

(a)  (b)

**Figure 5.47**  (a) Deployment of emergency airbags; (b) simplified simulation of emergency landing

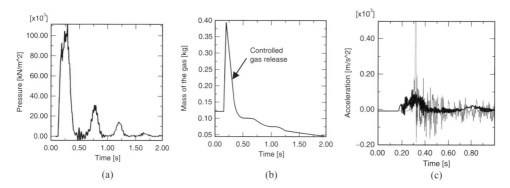

**Figure 5.48** Applied control strategy: (a) change of pressure inside the airbag, (b) mass of the gas inside the airbag, (c) obtained reduction of accelerations

## 5.6 Adaptive Crash Energy Absorber

### 5.6.1 Low-Velocity Impacts

Most of the real low-velocity impacts, involving elastic–plastic structures, can be considered with several simplifications. Since the influence of the wave propagation phenomenon on the impact process is small, all complex elastic and plastic wave propagation effects can be neglected; therefore a simple quasi-static approach can lead to accurate results. Loads acting on the impacting objects depend on their impact interface (the absorber), described by the force–displacement $P(\delta)$ characteristics, produced during crushing of the protective structure. For the sake of simplification, all velocity-sensitive hardening effects will be disregarded in the following pages. For further considerations, an introduction of the equivalent average crushing force $P_m$ will be useful. This handy parameter is defined by the integral mean from the crashing interface characteristics on the length of the total crushing distance $\delta_{\max}$:

$$P_{\mathrm{m}} = \frac{1}{\delta_{\max}} \int_0^{\delta_{\max}} P(\delta)\,\mathrm{d}\delta \qquad (45)$$

The idea of the adaptive impact absorption system will be demonstrated on a simple, one-dimensional central impact problem. A rigid object, modeled by the lumped mass and equipped with elastic–plastic protective structure, becoming an isolated system, is assumed. Force–displacement characteristics of the protective structure is approximately of a constant type, where the force is oscillating around the mean value, which can be modeled by the constant average force $P_m$ and maximal force value $P_{\max}$. The object is hitting into a rigid wall, within the possible domain of masses and initial velocities, where mass is equal to $m \in \Re^+$ and the initial velocity of the impacting objects is $V \in \Re^+$. The maximal acceptable deceleration $a_{\max}$ that the structure can survive is assumed to be constant for all impact cases, imposing a limit to acting deceleration $a < a_{\max}$. The maximal crush length $L$ of the protective structure is also limited and constant. Hence, an upper boundary of allowable impact velocity, meeting the deceleration limit, is given by the basic kinematic dependence:

$$V_{\max} = \sqrt{2a_{\max}L} \qquad (46)$$

being also the most effective solution for the maximal energy absorption on the supposed length of crushing. Another limit, driven by the maximal force produced by the impact interface $P_{max}$, imposes minimal mass condition

$$m_{min} = \frac{P_{max}}{a_{max}} \tag{47}$$

In the considered case of a one-dimensional central impact, the initial kinetic energy of the impacting object must be dissipated during the impact interface crush, which equals the force integrated over the displacement $\delta$ up to the maximal displacement $\delta_{max}$:

$$E_d = \int_0^{\delta_{max}} P(\delta)\, d\delta = P_m \delta_{max} \tag{48}$$

The boundary of safe impact conditions $\Omega$, within which none of the limits is exceeded, is successively defined:

$$\Omega = \begin{cases} m = \left\langle m_{min}, \dfrac{2E_d}{V^2} \right) \\[2ex] V = \left\langle 0, \left\{ V_{max} \vee \sqrt{\dfrac{2E_d}{m}} \right\} \right) \end{cases} \tag{49}$$

Introducing a structure equipped with a special crash sensor (i.e. for a rigid body impact – the mass and approach velocity sensor), a fast control system and a controlled energy-absorbing device (the controllable impact interface), an improved crash behavior with adaptation to the impact loads can be obtained. Assume the impact interface, with two selectable crushing characteristics, the average crushing force of which can be readjusted to the impact conditions. The defined structure can extend the initial impact conditions of the coverage boundary, forming an extended region of allowable parameters $\Omega_a = \Omega_1 \cup \Omega_2$ (Figure 5.49).

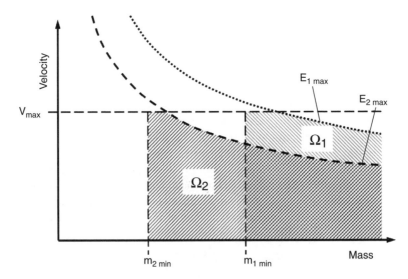

**Figure 5.49**   Example of initial impact conditions coverage by the double-stiffness adaptive absorber

## 5.6.2  Energy Absorption by the Prismatic Thin-Walled Structure

One of the most important issues of the energy-absorbing structure's properties are their technical and economical efficiencies. Technical efficiency of the impact energy absorber, in the case of lightweight means of transport applications, can be measured as a ratio of its maximal energy-absorption capability to the gross mass of the absorber. The defined indicator is known as the SEA (J/kg), the specific energy absorption parameter. The significant economical efficiency index can be formulated as the quotient of the protective structure cost to its SEA. One of the best known SEA performance energy-absorbing processes is the axial crushing of the thin-walled tube. Its huge advantage is the fact that all internal forces generated during the process of crushing are self-balanced; hence tubes do not need any external supports to provide stable and progressive deformation. Such profiles can be used in many engineering objects as structural members, where the most popular examples are the automotive crash zones.

The first scientific studies devoted to the problem of crash behavior focused on a simple process of crushing of the thin-walled circular tube, being loaded along the direction of its axis of symmetry. During experimental studies, two characteristic deformation patterns were observed: axisymmetric, called the 'concertina' mode and nonaxis-symmetric mode known as the 'diamond' pattern. An approximate theoretical formula, describing the concertina folding mode, was derived and published by Alexander in 1960 [40] and modified later by several scientists for better accuracy. The mentioned family of analytical solutions was based on the rigid–perfectly plastic material model. Following the experimental observations, Alexander proposed the kinematics of the axisymmetric deformation pattern with stationary plastic hinges, contributing to bending dissipation combined with the contribution of the circumferential stretching of a shell (cf. Figure (5.50). The described approach, with later modifications introducing a more complex description of the folding pattern, gave the answer to important parameters of the quasi-static crushing process: the average crushing force $P_m$ and the length of the plastic folding wave $2H$. Alexander's theory has given a base to the general macroelements method developed by Abramowicz and Wierzbicki [41] in the last two decades of the twentieth century.

The macroelement method is applicable to more complex structures and deformation forms, allowing axial, bending and torsional responses of assemblies of macroelements to be solved. Crushing of the profile, with the prismatic thin-walled cross-section, can be discretized into a set of super-folding elements (SFE). For a simple rectangular tube, three basic folding patterns were isolated: symmetric, asymmetric and inverted modes. The asymmetric mode has the

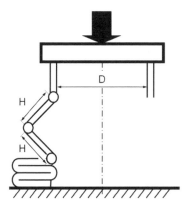

**Figure 5.50**  Alexander's model of concertina mode crushing of a round thin walled tube [40]

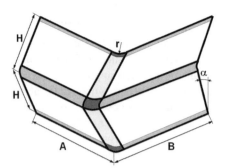

**Figure 5.51**    Abramowicz and Wierzbicki's single super-folding element [42]

lowest energy path, what makes it the most stable one before switching to the dangerous
inverted mode, causing global buckling instability. Such a pattern can be initiated by the
prefabricated triggering mechanism, which helps to form the first folding lobe and reduce the
initial limit force. Its total energy absorption is a sum of contributions of the following types of
energy dissipation mechanisms: rolling of the material through the traveling hinges, bending on
the stationary plastic hinges lines and the material flow through the toroidal surface in the area
of plastic fold lobes corners (cf. Figure 5.51). The general description of power of dissipation
of the internal energy (internal energy dissipation rate) in a deformed shell structure can be
formulated as follows [42]:

$$\dot{E}_{int} = \int_S \left( M_{\alpha\beta}\, \dot{\kappa}_{\alpha\beta} + N\, \dot{\varepsilon}_{\alpha\beta} \right) dS + \sum_{i=1}^n \int_{L_i} M_i(\dot{\theta}_i) dl_i \qquad (50)$$

The first integral describes the power of dissipation in a constant velocity field integrated over
the midshell surface $S$. The symbol $\dot{\kappa}$ is the rate of curvature and $\dot{\varepsilon}$ denotes the rate of extension
tensor, $M$ and $N$ are the conjugate generalized stresses (plate and membrane modes), $\theta_i$ is the
sum of current power dissipated in bending of the stationary hinges, integrated on the length
of hinge line $L_i$, $M_i$ is the bending moment and $\theta_i$ is the jump of the rate of rotation across
the hinge. Extension of the formulation (50) to the assumed dissipation mechanism leads to
the expression for the mean crushing force $P_\mathrm{m}$ of the single super folding element (SFE) [42],
describing the behavior of a single corner:

$$P_\mathrm{m} = \frac{1}{4}\left[ t_1^2 \sigma_0^\mathrm{N}(\bar{\varepsilon}_1) A_1 \frac{r}{t_1} + A_2 \left( \sigma_0^\mathrm{M}(\bar{\varepsilon}_2)\frac{at_a^2}{H} + \sigma_0^\mathrm{M}(\bar{\varepsilon}_2)\frac{bt_b^2}{H} \right) + t_3^2 \sigma_0^\mathrm{M}(\bar{\varepsilon}_3) A_3 \frac{H}{r} \right.$$
$$\left. + t_4^2 \sigma_0^\mathrm{M}(\bar{\varepsilon}_4) A_4 \frac{H}{t_4} + t_5^2 \sigma_0^\mathrm{N}(\bar{\varepsilon}_5) A_5 \right] \frac{2H}{\delta_\mathrm{eff}} \qquad (51)$$

where

$t$ is the thickness of the shell
$r$ is the average rolling radius
$\sigma_0$ is the material flow stress
$2H$ is the length of the folding wave
$C = A + B$ is the length of the initial lateral cross-section of the SE
$\delta_\mathrm{eff}$ is the effective crushing length
$A_1, A_2, A_3, A_4, A_5$ are the functionals depending of the central angle $\Theta$ and switching
    parameter $\alpha^*$

Unknown parameters, like the length of the folding wave, can be obtained using the methods of the variational calculus. A set of three nonlinear algebraic equations, ensuring the stability of the super-element, comes from the following minimum conditions [42]:

$$\frac{\partial P_{\mathrm{m}}}{\partial H} = 0, \qquad \frac{\partial P_{\mathrm{m}}}{\partial r} = 0, \qquad \frac{\partial P_{\mathrm{m}}}{\partial \alpha^*} = 0 \qquad (52)$$

The symbol $\alpha^*$ denotes the 'switching parameter' and its significance, due to its wider meaning, will be not discussed in this book [42]. To characterize the material modeled as rigid–perfectly plastic, the full plastic bending moment term is introduced. From the basics of the theory of plasticity, its physical meaning is the stationary plastic hinge moment of bending resistance per unit length:

$$M_{\mathrm{y}} = \frac{\sigma_{\mathrm{y}} t^2}{4} \qquad (53)$$

where $\sigma_{\mathrm{y}}$ denotes the yield stress (for the perfectly plastic material model). For the internal corner angle $\Theta$ equal to $90°$, Equation (51) yields to

$$\frac{P_{\mathrm{m}}}{M_{\mathrm{y}}} = 3\sqrt[3]{A_1 A_2 A_3} \sqrt[3]{\frac{C}{t}} \frac{2H}{\sigma_{\mathrm{y}}} \qquad (54)$$

Constants $A_1$, $A_2$ and $A_3$ represent elliptic integrals unable to be solved analytically. The nondimensional ratio $\eta$ expressed by the fraction of the effective crushing distance to the length of the plastic folding wave is equal to

$$\eta = \frac{\delta_{\mathrm{eff}}}{2H} \approx 0.73 \qquad (55)$$

Next, the length of the folding wave formula can be derived:

$$2H = 2\sqrt[3]{C^2 t} \qquad (56)$$

which leads to the final formula describing the average crushing force $P_{\mathrm{m}}$ to the plastic bending moment $M_{\mathrm{y}}$ ratio, for one corner SFE:

$$\frac{P_{\mathrm{m}}}{M_{\mathrm{y}}} = 13.052 \sqrt[3]{\frac{C}{t}} \qquad (57)$$

The average crushing force $P_{\mathrm{m}}$ of the square cross-section profile is the sum of the $P_{\mathrm{m}}$ of four super-elements treated independently and is finally given by the following formula [41]:

$$P_{\mathrm{m}} = M_{\mathrm{y}} \left( 52.22 \sqrt[3]{\frac{C}{t}} \right) \qquad (58)$$

### 5.6.3 Use of Pyrotechnic Technology for the Crash Stiffness Reduction

The idea of the absorber crushing resistance force control presented on the following pages uses the concept of the structural connections decoupled by pressure generated in deflagration of the pyrotechnic material. Technology for the controllable increase of the energy-absorbing capability is much more complex than its reduction from the initial value. The total average crushing force and impact energy dissipation capacity can be decreased, through controlled disconnection of the additional structural members from the main absorber's profile. In the particular example the rectangular cross-section for the absorbing structure was arbitrarily selected due to its technological simplicity. The additional members were designed as two C-shaped profiles connected to the structure by eight detachable pyroconnections.

Absorber energy dissipation capabilities can be assessed using the following arbitrary assumptions. The additional member negligibly disturbs the length of the folding wave. Hence, the total energy dissipated in the absorber consists of the sum of individual dissipation energies of the absorber's members. Each section is treated separately; thus any interaction between them during folding is not taken into account. Geometric differences between them were also neglected, so one cross-section $P_m$ (cf. Equation (58)) will be used for the main absorber as well as for the additional members. Hence, the maximal energy absorption capability depends only of the length of each cross-section profile. For the high-energy coupled members mode, it can be calculated from the following formula:

$$E_{high} = P_{ms}\eta(d_b + d_a) \tag{59}$$

where

$P_{ms}$ is the average crushing force of the single cross-section
$d_b$ is the length of the base member
$d_a$ is the length of the additional member
$\eta = 0.73$ is the effective crushing length factor

When the structural connections are decoupled, causing separation of the additional members from the main absorber's profile, they will not take part in the crushing process, reducing the total amount of possible energy dissipation to the lower $E_{low}$ level:

$$E_{low} = P_{ms}\eta d_b \tag{60}$$

Obviously, if $d_b = d_a$, the total average crushing force of the assembly in the high-energy mode is equal to $2P_m$, while in the low-energy mode it reduces to $P_m$.

To assess the concept feasibility, as well as to demonstrate the new approach to structural control, an explicit finite element simulation and laboratory demonstrative experiment were performed. The finite element simulations were performed in the explicit dynamics code LS-Dyna v971. Deformation pictures, depicting the state 40 ms after the impact beginning, are shown in Figure 5.52 (left) for the high-energy mode and Figure 5.52 (right) for the low-energy state. Comparison of deceleration of the head of the hammer for both modes is shown in Figure 5.53. On the internal energy plot (Figure 5.54), the influence of the disconnection of the additional members is clearly visible. The time and distance of dissipation for the same amount of initial energy are longer for the low-energy active mode. The calculated maximal energy difference between the modes was 35 %; however, it can be seen that the decoupled members were long only for 75 % of the base member height.

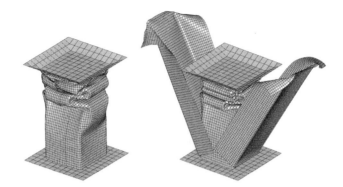

**Figure 5.52**  Results of simulation of the absorber crash in the high-(left) and low-(right) energy modes, 40 ms after the beginning of the impact

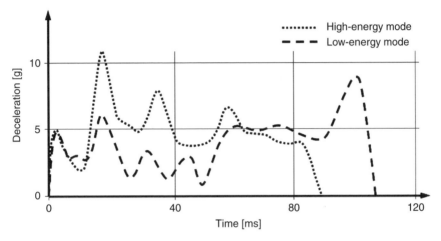

**Figure 5.53**  Absorber in the high- and low-energy modes in the FE simulations: impacting mass deceleration versus time [43]

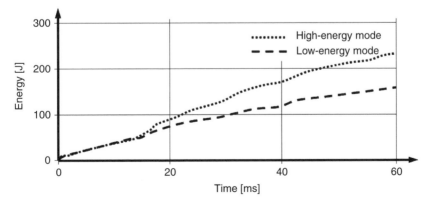

**Figure 5.54**  Internal energy versus time for high- and low-energy modes obtained in the FE simulations [43]

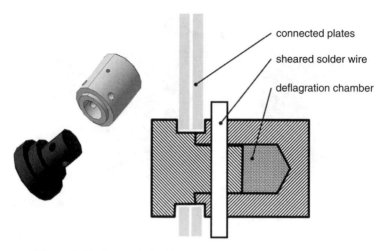

**Figure 5.55** Pyrotechnic-driven structural connection design [43]

In the small-scale tests low-impact energies were highly desirable; therefore to decrease the energy-absorbing capability and to reduce the forces acting on the laboratory pyroconnections, a lead alloy was used for the absorber's body. Measured quasi-static mechanical properties of the used material were: yield stress $\sigma_Y = 8$ MPa, ultimate tensile strength $\sigma_U = 21$ MPa, elongation at break $A_5 = 45\,\%$ and Young's modulus $E = 14$ GPa. The specimen was placed on the testing hammer's anvil and was crushed by the dropping head of the hammer. The measured velocity at the point of first contact with the specimen was 4.45 m/s. The total absorbed energy in the tested absorber was estimated by means of formulas (60) and (59) to 175 J for the base member and 271 J for the assembly. The initialization of the pyrotechnic material's deflagration was performed by a custom-designed, computer-controlled, high-voltage ignition system. The fuse was of the exploding bridge wire (EBW) type and ignited $\frac{3}{10}$ mg of black powder, filling the

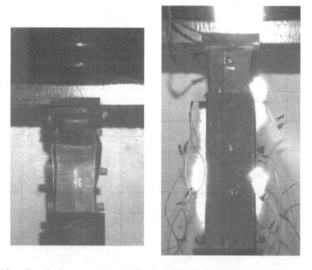

**Figure 5.56** Absorber during experiments in the high-(left) and low-(right) energy modes [43]

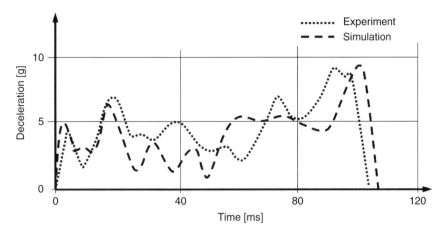

**Figure 5.57** Comparison of FE and test results: impacting mass deceleration versus time for the active mode [43]

deflagration chamber (Figure 5.55). Rapidly growing pressure acting on the pyroconnection's piston broke the sheared pin made of soldering alloy, causing unlocking of the absorber's members. A battery of capacitors, pre-charged to 311 V, on the triggering signal was rapidly discharged through the initiator wire. The initiator wire vaporized in a time shorter than 250 µs after receiving the signal coming from the real-time control system. The initiation process was controlled by the electrical control circuit, which was optically separated from the controller. The silicon-controlled rectifiers (SCR) were used for fast response switching of the initiating current. The acceleration sensor measured the deceleration of the impacting mass.

During the high-energy mode test, all pyroconnections remained locked through the crushing time (Figure 5.56, left). The test of the low-energy mode (Figure 5.56, right) was conducted with the same initial conditions as the passive one. A photocell activated by the dropping head of the hammer sent the signal to the control system. When the impulse was received by the control system, the initiation circuit was triggered with a pre-set time offset, causing deflagration of the powder, opening of the connections and disconnection of the additional members. The average delay time between the initiation and explosion, due to statistical dispersion, was around 3–4 ms. A comparison of FE and test results is given in Figure 5.57.

The demonstrated example shows another possible direction of the AIA system development, which may be applied in wide range of applications.

## References

1. J. Holnicki-Szulc, P. Pawłowski and M. Wikło, High-performance impact absorbing materials – the concept, design tools and applications, *Smart Materials and Structures*, **12**(3), 2003, 461–467.
2. J. Holnicki-Szulc and C. A. Mota Soares, *Advances in Smart Technologies in Structural Engineering*, Vol. 1, Computational Methods in Applied Sciences Series, Springer, New York, 2004.
3. N. Jones and T. Wierzbicki, *Structural Crashworthiness*. Butterworths, London, 1983.
4. N. Jones and T. Wierzbicki, *Structural Failure*, John Wiley & Sons, Inc., New York, 1989.
5. N. Jones and T. Wierzbicki, *Structural Crashworthiness and Failure*, Elsevier Applied Science, 1993.
6. C. M. Harris and A. G. Piersol, *Harris' Shock and Vibration Handbook*, McGraw-Hill, 2002.
7. R. Grybos, *Teoria uderzenia w dyskretnych układach mechanicznych* (in Polish), PWN, 1969.

8. G. Mikułowski and L. Jankowski, Adaptive landing gear: optimum control strategy and improvement potential, in *Proceedings of ISMA 2006* (ed. P. Sas M. De Munck), Leuven, 2006.

9. N. S. Currey, *Aircraft Landing Gear Design: Principles and Practices*, AIAA, Washington, DC, 1988.

10. B. Milwitzky and F. E. Cook, Analysis of landing gear behavior, Technical Report 1154, NACA, 1953.

11. R. Freymann, Actively damped landing gear system, in *AGARD CP-484 Ref. 20, Proceedings of the 71st Meeting of the AGARD Structures and Materials Panel*, October 1990.

12. J. R. McGehee and H. D. Carden, Active control landing gear for ground load alleviation, in *AGARD Conference Proceedings 384 FMP Symposium*, Toronto, 1984.

13. J. R. McGehee and H. D. Carden, Analytical investigation of the landing dynamics of a large airplane with a load-control system in the main landing gear, Technical Report 1555, NASA, 1979.

14. L. G. Horta, R. H. Daugherty and V. J. Martinson, Actively controlled landing gear for aircraft vibration reduction, Technical Report NASA-99-ceas-lgh, NASA, 1999.

15. L. G. Horta, R. H. Daugherty and V. J. Martinson, Modeling and validation of a Navy A6-Intruder actively controlled landing gear system, Technical Report TP-1999-209124, NASA, 1999.

16. Adaptive landing gears for improved impact absorption, ADLAND, EU FP6 Project IST-FP6-2002-Aero-1-502793-STREP, http://smart.ippt.gov.pl/adland.

17. D. Batterbee, N. D. Sims, Z. Wołejsza and A. Lafitte, Magnetorheological landing gear design: a feasibility study for small and large-scale aircraft, in *Proceedings of ISMA 2006* (ed. P. Sas M. De Munck), Leuven, 2006.

18. J. D. Carlson, Introduction to magnetorheological fluids, in *Proceedings of SMART"01 Workshop*, Warsaw, 2001.

19. H. Gavin, J. Hoagg and M. Dobossy, Optimal design of MR dampers, in *Optimal Design of MR Dampers, Proceedings of U.S.–Japan Workshop on Smart Structures for Improved Seismic Performance in Urban Regions* (ed. K. Kawashima, B.F. Spencer and Y. Suzuki), Seattle, Washington, 2001.

20. Rexroth Bosch Group, 2007, http://www.boschrexroth.com.

21. Moog Inc., 2007, http://www.moog.com.

22. K. Sekuła, G. Mikułowski and J. Holnicki-Szulc, Real time dynamic mass identification, in *Proceedings of the Third European Workshop on Structural Health Monitoring* (ed. A. Guemes Granada), 2006.

23. G. Mikułowski and J. Holnicki-Szulc, Fast controller and control algorithms for MR based adaptive impact absorbers – force based control, *Machine Dynamics Problems*, **30**(2), 2006, 113–122.

24. Federal Aviation Regulations (FAR), Part 23 – Airworthiness Standards: Normal, Utility, Acrobatic and Commuter Category Airplanes.

25. I23 Technical Specification, Institute of Aviation, Warsaw, Poland.

26. D. Batterbee, N. D. Sims and R. Stanway, ADLAND Report: Annex USFD-1(a): oleo-pneumatic shock absorber modeling and initial MR device sizing, Technical Report, University of Sheffield, 2004.

27. G. Mikułowski and L. Jankowski, Adaptive landing gear: optimum control strategy and potential for improvement, *Shock and Vibration* (submitted).

28. G. Mikułowski and J. Holnicki-Szulc, Adaptive landing gear concept – feedback control validation, *Smart Materials and Structures* (to be published).

29. I. Yeh, L. Chai and N. Saha, Application of ALE to airbag deployment simulation, *International Journal of Vehicle Safety*, **1**(4), 2006, 348–365.

30. G. J. Van Wylen and R. E. Sonntag, *Fundamentals of Classical Thermodynamics*, John Wiley & Sons, Ltd, Chichester, 1978.

31. ABAQUS, User's Manual, Version 6.5, Hibbitt, Karlsson and Sorensen, Providence, Rhode Island, 2005.

32. T. Belytschko, W. K. Liu and B. Moran, *Nonlinear Finite Elements for Continua and Structures*, John Wiley & Sons, Ltd, Chichester, 2000.

33. C. Graczykowski and J. Holnicki-Szulc, Protecting offshore wind turbines against ship impacts by means of adaptive inflatable structures shock and vibration, in revision.
34. A. Mróz and T. Kärnä, Mitigation of ice loading. Feasibility study of semi-active solution. Technical Report, VTT Working Papers 39, VTT, 2005.
35. J. F. Wilson, *Dynamics of Offshore Structures*, John Wiley & Sons, Inc., New York, 1984.
36. R. W. Clough and J. Penzien, *Dynamics of Structures*, McGraw-Hill, Maidenhead, 1975.
37. N. Jones, *Structural Impact*, Cambridge University Press, Cambridge, 1989.
38. C. Graczykowski and J. Holnicki-Szulc, Optimization of dynamic properties of adaptive pressurized structures subjected to impact loads, in *II ECCOMAS Thematic Conference on Smart Structures and Materials*, 18–21 July 2005.
39. C. Graczykowski and J. Holnicki-Szulc, Aeronautical applications of adaptive inflatable structures, in *ECCOMAS Thematic Conference on Textile Composites and Inflatable Structures*, 17–19 September 2007.
40. J.M. Alexander, An approximate analysis of the collapse of thin cylindrical shells under axial loading, *Quarterly Journal of Mechanical Applied Mathematics*, **4**(13), 1960, 10–15.
41. J. Ambrosio , Macro element method on crashworthiness of vehicles, in Crashworthiness and Energy Management and Occupant Protection. (ed. W. Abramowicz), Springer, Wien, New York, 2001.
42. W. Abramowicz, Thin-walled structures as impact energy absorbers, *Thin-Walled Structures*, **41**, 2001, 91–107.
43. P. Griskievicius, M. Ostrowski and J. Holnicki-Szulc, Adaptive crashworthiness of front-end structure of motor vehicles, in *SAE World Congress*, 16–19 April 2007.

# 6

# VDM-Based Remodeling of Adaptive Structures Exposed to Impact Loads

Marcin Wikło, Łukasz Jankowski, Małgorzata Mróz and Jan Holnicki-Szulc

The structural design for dynamic loads is still a challenge, particularly in the case of impact loads, which demand a very quick dynamic analysis to model the structural response. There are several commercially available numerical tools that are capable of providing a reliable simulation of the structural response to a determined impact scenario. The majority is devoted to the *crashworthiness* analysis for vehicle collisions. Nevertheless, these tools are not helpful in searching for the best design in the case of an unknown *a priori* impact scenario. The trial-and-error approach seems to be the unique option in this case. On the other hand, the AIA concept discussed in the previous chapter takes advantage of real-time impact load identification to trigger the pre-computed optimum *plastic-like adaptation* of active elements (*structural fuses*) to the actual impact load.

The main motivation for this chapter is thus to propose a methodology for the optimum design of *adaptive structures* for *adaptive impact absorption* (AIA). The first part deals with remodeling (material redistribution) of elastic structures exposed to impact loads, the second part extends the analysis to elastoplastic structures and the third part introduces the concept of adaptive structures controllable via structural fuses, which are special members able to mimic the plastic-like behavior. A methodology for the optimum design of such structures is proposed and numerically tested. Its effectiveness is demonstrated in both increasing the impact absorption capacity and smoothing the response of the considered structure, which are both nonstandard engineering problems. Finally, the fourth part generalizes the approach to take additionally the damping effect into account. Since damping is a strongly frequency-dependent phenomenon, the formulation in the frequency domain and harmonic loads are used.

From the time when the idea of optimum structural remodeling (including topological optimization) was presented in Reference [1] as a problem of material distribution, the subject has been continuously investigated for increasingly complex structures, taking nonlinear behavior into account and including the impact resistance design of vehicles. A formulation of the problem for many loading states was presented in Reference [2]. The minimization of the maximum compliance can be achieved with the application of the $\beta$ method [3, 4] or the KS (Kreisselmeier–Steinhauser) function method [5]. In the analysis of the impact resistance of vehicles, a long list of specific problems can be expected: nonlinear materials (plasticity, hardening), nonlinear geometry (large displacements, deformations, buckling), dynamics and contact. The first papers addressing the design for impact resistance were References [6] to [11].

The design of complex structures, such as the structures of automobiles, is an iterative process due to safety aspects. Each time the design is changed, the structure has to undergo the simulation. Duration of a single analysis of the whole structure can take many hours or even days [12]. The evaluation of influence of each modification requires a renewed analysis of the system. Manual changes of parameters and application of a precise model make the simulation process tedious, not effective, numerically expensive and time-consuming. Therefore, in the early stage of design and redesign, it is recommended that simplified models are used, which are numerically less expensive and allow the remodeling process to be automated. For linear systems the technique of simplified processes has been developed and applied for many years, while for nonlinear systems, especially those subjected to dynamic loads, there have been only a few papers published until now. Models with concentrated masses have been used in the automobile industry for safety improvements in crash tests since the early 1970s [13, 14]. In these models, the nonstructural elements are modeled by concentrated masses, whereas the essential structural elements, which undergo the deformation, are modeled as nonlinear elastic elements with the load–displacement characteristics of crushed tubes. These models have been also applied to simulations of passenger–automobile dynamics [15]. In Reference [16] three simulation methods (hybrid, analytical and mixed) of automobile structures exposed to impacts have been presented with examples, where the structural elements have been modeled by nonlinear spring elements with experimentally obtained load–displacement characteristics. Reference [17] describes in detail a time-effective method for simulation and design of automobile structure frames, called the V-CRASH method. More information about the development of simplified models can be found in References [18] and [19]. An indispensable stage of application of such models is the identification of the model itself. The subject has been developed in many researches for different structures (nuclear reactors, buildings exposed to earthquakes, etc. Several papers also consider the impact resistance of automobile vehicles [20–24].

The virtual distortion method (VDM) is a robust and versatile numerical tool for quick and exact structural modifications, which is able to simulate material redistribution (including vanishing of structural elements) and piecewise linear approximations to nonlinear constitutive laws. It is also capable of exact (analytical) sensitivity analysis and thus enables effective, gradient-based redesign algorithms, without the need for modifications of global stiffness and mass matrices. At the present state of development (Chapter 2), the method includes the dynamic problems mentioned above. However, the coupled problem of structural modifications (material redistribution in elastoplastic structures, which causes mutually interacting modifications of mass, stiffness and plastic zone distribution) and its sensitivity analysis have not yet been discussed. Therefore, it is considered in this chapter.

The necessity to restrict the considerations to small deformations and to approximate physical relations by piecewise linear characteristics results in (an increasing with time) error of the simulated dynamic response. However, the simplifications give a chance to create an effective tool to determine the optimum material distribution for the initial phase of the dynamic response, when the deformations are still below an arbitrarily assumed value. When the simplified analysis is accomplished, the precise numerical simulation of the dynamic structural response can be performed, including large deformations, with one of the commercially available codes. The simplest truss model is used in this chapter due to the illustrative simplicity of the derivations. However, the proposed approach is also applicable to other types of structures (see Reference [25] for frames and plates), provided additional distortion states are introduced to model additional potential modifications of the structure.

## 6.1 Material Redistribution in Elastic Structures

The structural modifications described in Chapter 2 refer to the changes of structural material in selected elements, which affect their stiffness or mass. The formulas in Equations (31) and (50) in Chapter 2 allow the virtual distortions (strains and forces) to be determined, which simulate the intended corresponding structural modifications. However, the cross-section of an element influences both its stiffness and mass; hence in the problem of material redistribution both virtual distortions are coupled with each other, which has to be accounted for.

### 6.1.1 VDM Formulation

The equation of motion for the system with *modified* cross-sectional areas of the elements can be stated using the Einstein summation convention as

$$\hat{M}_{NM} \ddot{u}_M(t) + G_{Ni}^{\mathrm{T}} l_i \hat{S}_{ii} G_{iM} u_M(t) = f_N(t)$$

where $\hat{M}_{NM}$ are the elements of the modified mass matrix, $G_{Ni}^{\mathrm{T}} l_i \hat{S}_{ii} G_{iM}$ are the elements of the modified stiffness matrix expressed in terms of the geometric matrix $\mathbf{G}$, which transforms global degrees of freedom to local strains, and of the diagonal matrix $\hat{\mathbf{S}}$ of element modified stiffnesses $E_i \hat{A}_i$, where the underlined indices are exempt from the Einstein summation convention. Young's modulus is denoted by $E_i$, $\hat{A}_i$ is the modified cross-sectional area and $l_i$ is the length of the $i$th element; lowercase indices denote the elements and capitals the degrees of freedom. On the other hand, the equation of motion formulated for the *modeled* structure can be stated using force distortions $f_N^0(t)$ and strain distortions $\varepsilon_i^0(t)$ to simulate respectively the modifications of mass and stiffness in the following manner:

$$M_{NM} \ddot{u}_M(t) + G_{Ni}^{\mathrm{T}} l_i S_{ii} \left[ G_{iM} u_M(t) - \varepsilon_i^0(t) \right] = f_N(t) + f_N^0(t) \tag{1}$$

where $\mathbf{S}$ is the diagonal matrix of the element original stiffnesses $E_i A_i$. Equation (1) can also be stated using the strain vector as

$$M_{NM} \ddot{u}_M(t) + G_{Ni}^{\mathrm{T}} l_i S_{ii} \left[ \varepsilon_i(t) - \varepsilon_i^0(t) \right] = f_N(t) + f_N^0(t) \tag{2}$$

where the strain $\varepsilon_j(t)$ is defined by Equation (2) in Chapter 2.

The virtual distortion method assumes the small deformation case and expresses the dynamic structural response as the superposition of the response $u_N^{\mathrm{L}}(t)$ of the original unmodified structure and the response due to the distortions $\varepsilon_i^0(t)$ and $f_N^0(t)$, which simulate the modifications (compare with Equation (27) in Chapter 2)

$$u_N(t) = u_N^{\mathrm{L}}(t) + \sum_{\tau \le t} B_{Nj}^{\varepsilon}(t - \tau) \varepsilon_j^0(\tau) + \sum_{\tau \le t} B_{NM}^{\mathrm{f}}(t - \tau) f_M^0(\tau) \tag{3}$$

where the index $N$ refers to the degrees of freedom, $j$ to the modified elements and $M$ to the degrees of freedom related to the modified elements. As in Equation (28) in Chapter 2, the strain $\varepsilon(t)$ can be obtained by premultiplying Equation (3) by $G_{iN}$:

$$\varepsilon_i(t) = \varepsilon_i^{\text{L}}(t) + \sum_{\tau \le t} D_{ij}^{\varepsilon}(t - \tau)\varepsilon_j^0(\tau) + \sum_{\tau \le t} D_{iM}^{\text{f}}(t - \tau)f_M^0(\tau) \qquad (4)$$

The member forces in the modified and modeled structures can be respectively expressed as

$$\begin{aligned} p_i(t) &= E_{\underline{i}}\hat{A}_{\underline{i}}\varepsilon_{\underline{i}}(t) \\ p_i(t) &= E_{\underline{i}}A_{\underline{i}}\left(\varepsilon_{\underline{i}}(t) - \varepsilon_{\underline{i}}^0(t)\right) \end{aligned} \qquad (5)$$

By postulating the identity of the member forces in both structures, Equations (5) yield the following formula for the strain distortion:

$$\varepsilon_i^0(t) = \left(1 - \mu_{\underline{i}}^{\text{A}}\right)\varepsilon_{\underline{i}}(t) \qquad (6)$$

where the cross-section modification parameter $\mu_i^{\text{A}}$ is defined in Equation (30) in Chapter 2 as $\mu_i^{\text{A}} = \hat{A}_{\underline{i}}/A_{\underline{i}}$. Substitution of Equation (6) into Equation (4) leads to

$$\left[\delta_{ij} - \left(1 - \mu_{\underline{i}}^{\text{A}}\right)D_{\underline{i}j}^{\varepsilon}(0)\right]\varepsilon_j^0(t) - \left(1 - \mu_{\underline{i}}^{\text{A}}\right)D_{\underline{i}M}^{\text{f}}(0)f_M^0(t) = \left(1 - \mu_{\underline{i}}^{\text{A}}\right)\varepsilon_{\underline{i}}^{\ne t}(t) \qquad (7)$$

where $\varepsilon_i^{\ne t}(t)$, similarly to Equation (32) in Chapter 2, describes the strains without the effect of the distortions in the current time step $t$:

$$\varepsilon_i^{\ne t}(t) = \varepsilon_i^{\text{L}}(t) + \sum_{\tau < t} D_{ij}^{\varepsilon}(t - \tau)\varepsilon_j^0(\tau) + \sum_{\tau < t} D_{iM}^{\text{f}}(t - \tau)f_M^0(\tau) \qquad (8)$$

Besides the equality of the member forces, the VDM also postulates the equality of the inertia forces in the modified and modeled structures. This leads to the relation Equation (45) in Chapter 2, which after substitution of the following analog of Equation (49) in Chapter 2:

$$\ddot{u}_N(t) = \ddot{u}_N^{\text{L}}(t) + \sum_{\tau \le t} \ddot{B}_{NM}^{\text{f}}(t - \tau)f_M^0(\tau) + \sum_{\tau \le t} \ddot{B}_{Nj}^{\varepsilon}(t - \tau)\varepsilon_j^0(\tau) \qquad (9)$$

yields

$$\Delta M_{NM}\ddot{B}_{Mj}^{\varepsilon}(0)\varepsilon_j^0(t) + \left[\delta_{NK} + \Delta M_{NM}\ddot{B}_{MK}^{\text{f}}(0)\right]f_K^0(t) = -\Delta M_{NM}\ddot{u}_M^{\ne t} \qquad (10)$$

where the influence matrices $\ddot{B}_{MK}^{\text{f}}(t)$ and $\ddot{B}_{MK}^{\varepsilon}(t)$ describe the history of accelerations resulting from a unitary impulse of respectively the force distortion and the strain distortion, $\Delta M_{NM}$ is related to the cross-section modification parameter $\mu_i^{\text{A}}$ as in Equation (46) in Chapter 2 and

$$\ddot{u}_N^{\ne t} = \ddot{u}_N^{\text{L}}(t) + \sum_{\tau < t} \ddot{B}_{Nj}^{\varepsilon}(t - \tau)\varepsilon_j^0(\tau) + \sum_{\tau < t} \ddot{B}_{NM}^{\text{f}}(t - \tau)f_M^0(\tau) \qquad (11)$$

Finally, Equations (7) and (10) yield together the following linear system of equations, which can serve in successive time steps to determine the virtual strain distortions and force distortions, which model the modifications $\mu_i^A$ of the element cross-sectional areas:

$$
\begin{bmatrix}
\delta_{ij} - \left(1 - \mu_{\underline{i}}^A\right) D_{\underline{i}j}^\varepsilon(0) & -\left(1 - \mu_{\underline{i}}^A\right) D_{\underline{i}K}^f(0) \\
\Delta M_{NM} \ddot{B}_{Mj}^\varepsilon(0) & \delta_{NK} + \Delta M_{NM} \ddot{B}_{MK}^f(0)
\end{bmatrix}
\begin{bmatrix}
\varepsilon_j^0(t) \\
f_K^0(t)
\end{bmatrix}
$$
$$
=
\begin{bmatrix}
\left(1 - \mu_{\underline{i}}^A\right) \varepsilon_{\underline{i}}^{\neq t}(t) \\
-\Delta M_{NM} \ddot{u}_M^{\neq t}
\end{bmatrix}
\tag{12a}
$$

which can be stated in the following shorter form:

$$
\mathbf{F}\mathbf{x}^0 = \mathbf{b}
\tag{12b}
$$

where the vector $\mathbf{x}^0$ collects the virtual distortions. The principal matrix $\mathbf{F}$ is time-independent and hence it is determined and decomposed only once. Moreover, it is also indispensable in the sensitivity analysis.

The algorithm for the material redistribution analysis is shown in Table 6.1.

**Table 6.1**   Algorithm for material redistribution in elastic structures

---

`Data and initial calculations`
Input data:

- Construction under external load
- Cross-section modification parameters $\mu_i^A$

Calculations:

- Linear response $\varepsilon_i^L(t)$ and $\ddot{u}_N^L$
- Dynamic influence matrices $D_{ij}^\varepsilon$, $D_{iN}^f$, $\ddot{B}_{Nj}^\varepsilon$ and $\ddot{B}_{NM}^f$
- Principal time-independent matrix $\mathbf{F}$ by Equation (12).

`Calculations in each time step` $t$
(a) Strains $\varepsilon_i^{\neq t}(t)$ and accelerations $\ddot{u}_N^{\neq t}$ by Equations (8) and (11)
(b) Virtual distortions $\varepsilon_i^0(t)$ and $f_N^0(t)$ by Equation (12)
(c) Actual strains and accelerations:

$$
\varepsilon_i(t) = \varepsilon_i^{\neq t}(t) + D_{ij}^\varepsilon(0)\varepsilon_j^0(t) + D_{iM}^f(0)f_M^0(t)
$$
$$
\ddot{u}_N(t) = \ddot{u}_N^{\neq t}(t) + \ddot{B}_{Nj}^\varepsilon \varepsilon_j^0(t) + \ddot{B}_{NM}^f f_M^0(t)
$$

(d) If necessary, by the corresponding influence matrices compute the response: the displacements $u_N(t)$, the velocities $\dot{u}_N(t)$, the stresses $\sigma_i(t)$, etc.
(e) If necessary, calculate the derivatives of the response (Section 6.1.2)
(f) $t := t + 1$

---

## 6.1.2 Sensitivity Analysis

The gradient of an analytically defined objective function can usually be expressed in terms of the derivatives of the response. Thus, an important advantage of the proposed reanalysis method is the availability of precise, analytical derivatives of the response with respect to the modified cross-sectional area $\hat{A}_i$, which plays a crucial role in optimization procedures. The derivatives of both strains (Equation (4)) and accelerations (Equation (9)) are expressible in terms of the derivatives of the virtual distortions $\partial \varepsilon_i^0(\tau)/\partial \hat{A}_l$ and $\partial f_N^0(\tau)/\partial \hat{A}_l$:

$$
\frac{\partial \varepsilon_i(t)}{\partial \hat{A}_l} = \sum_{\tau \leq t} D_{ij}^{\varepsilon}(t - \tau) \frac{\partial \varepsilon_j^0(\tau)}{\partial \hat{A}_l} + \sum_{\tau \leq t} D_{iM}^{f}(t - \tau) \frac{\partial f_M^0(\tau)}{\partial \hat{A}_l}
$$

$$
\frac{\partial \ddot{u}_N(t)}{\partial \hat{A}_l} = \sum_{\tau \leq t} \ddot{B}_{Nj}^{\varepsilon}(t - \tau) \frac{\partial \varepsilon_j^0(\tau)}{\partial \hat{A}_l} + \sum_{\tau \leq t} \ddot{B}_{NM}^{f}(t - \tau) \frac{\partial f_M^0(\tau)}{\partial \hat{A}_l}
$$

The components containing the current-step derivatives are separable by Equations (8) and (11):

$$
\frac{\partial \varepsilon_i(t)}{\partial \hat{A}_l} = \frac{\partial \varepsilon_i^{\neq t}(t)}{\partial \hat{A}_l} + D_{ij}^{\varepsilon}(0) \frac{\partial \varepsilon_j^0(t)}{\partial \hat{A}_l} + D_{iM}^{f}(0) \frac{\partial f_M^0(t)}{\partial \hat{A}_l}
$$

$$
\frac{\partial \ddot{u}_N(t)}{\partial \hat{A}_l} = \frac{\partial \ddot{u}_N^{\neq t}(t)}{\partial \hat{A}_l} + \ddot{B}_{Nj}^{\varepsilon}(0) \frac{\partial \varepsilon_j^0(t)}{\partial \hat{A}_l} + \ddot{B}_{NM}^{f}(0) \frac{\partial f_M^0(t)}{\partial \hat{A}_l}
\tag{13}
$$

Differentiation of the virtual distortion formulas of Equation (17) and Equation (45) in Chapter 2 yields

$$
\frac{\partial \varepsilon_i^0(t)}{\partial \hat{A}_l} = -\frac{\partial \mu_{\underline{i}}^{A}}{\partial \hat{A}_l} \varepsilon_{\underline{i}}(t) + \left(1 - \mu_{\underline{i}}^{A}\right) \frac{\partial \varepsilon_{\underline{i}}(t)}{\partial \hat{A}_l}
$$

$$
\frac{\partial f_N^0(t)}{\partial \hat{A}_l} = -\frac{\partial \Delta M_{NM}}{\partial \hat{A}_l} \ddot{u}_M(t) - \Delta M_{NM} \frac{\partial \ddot{u}_M(t)}{\partial \hat{A}_l}
\tag{14}
$$

where $\partial \Delta M_{NM}/\partial \hat{A}_l$ can be computed by direct differentiation of Equation (46) in Chapter 2 to be

$$
\frac{\partial \Delta M_{NM}}{\partial \hat{A}_l} = \frac{M_{NM}^l}{A_{\underline{l}}}
\tag{15}
$$

where $M_{NM}^l$ is the mass matrix of the $l$th element in the global coordinate system. Substitution of Equations (13) into Equations (14) yields two formulas similar to Equations (7) and (10), which combined together yield a formula similar to Equation (12a):

$$
\begin{bmatrix} \delta_{ij} - \left(1 - \mu_{\underline{i}}^{A}\right) D_{ij}^{\varepsilon}(0) & -\left(1 - \mu_{\underline{i}}^{A}\right) D_{\underline{i}K}^{f}(0) \\ \Delta M_{NM} \ddot{B}_{Mj}^{\varepsilon}(0) & \delta_{NK} + \Delta M_{NM} \ddot{B}_{MK}^{f}(0) \end{bmatrix} \begin{bmatrix} \dfrac{\partial \varepsilon_j^0(\tau)}{\partial \hat{A}_l} \\ \dfrac{\partial f_K^0(\tau)}{\partial \hat{A}_l} \end{bmatrix}
$$

$$
= \begin{bmatrix} (1 - \mu_{\underline{i}}) \dfrac{\partial \varepsilon_{\underline{i}}^{\neq t}(t)}{\partial \hat{A}_l} - \dfrac{\varepsilon_{\underline{i}}(t)}{A_{\underline{l}}} \delta_{\underline{i}l} \\ -\Delta M_{NM} \dfrac{\partial \ddot{u}_M^{\neq t}(t)}{\partial \hat{A}_l} - \dfrac{M_{NM}^l}{A_{\underline{l}}} \ddot{u}_M(t) \end{bmatrix}
\tag{16}
$$

**Table 6.2**  Material redistribution in elastic structures (numerical example): modifications of element cross-sections

| Element number | Modification parameter $\mu^A$ | New cross-section area (m²) |
|---|---|---|
| 1 | 1.2 | $1.2 \times 10^{-5}$ |
| 2 | 0.8 | $0.8 \times 10^{-5}$ |
| 3 | 0.1 | $0.1 \times 10^{-5}$ |
| 4 | 1.2 | $1.2 \times 10^{-5}$ |
| 5 | 0.2 | $0.2 \times 10^{-5}$ |

where an obvious relation $\partial \mu_i^A(t)/\partial \hat{A}_l = \delta_{il}/A_l$ has been used. The principal matrices in Equations (16) and (12a) are identical, which considerably simplifies the optimization process.

## 6.1.3 Numerical Testing Example

Consider the truss cantilever structure presented in Figure 2.4. The dynamic loading is realized by the initial vertical velocity $V_0 = -20$ m/s of the right bottom node. Material properties are uniformly distributed; Young's modulus and density are respectively $E = 2.1 \times 10''$ N/m² and $\rho = 7800$ kg/m³. The initial cross-section of all elements is $A = 1 \times 10^{-5}$ m², while the modified cross-sections are listed in Table 6.2.

The response of the modified structure has been computed using the described VDM approach and compared with the results of a standard FEM-based analysis. The discrepancies have been quantized by the following measure of the relative error:

$$\Delta e_i(t) = \frac{e_i^R(t) - e_i^{VDM}(t)}{e_i^R(t)} \tag{17}$$

where $e$ is the value being compared (displacement, velocity, acceleration etc.) and the superscripts VDM and R denote respectively the VDM result and the reference FEM result, which have been obtained using the Newmark integration scheme. Figure 6.1 plots the relative errors of displacements, velocities and accelerations of the second node in the vertical direction. The maximum errors do not exceed $4 \times 10^{-10}$, which is close to the numerical error level.

Figure 6.2 plots the discrepancies (Equation (17)) between the response gradients obtained analytically by Equation (16) and the gradients computed with the finite difference method (FDM). As an example, they have been computed for strains in all elements with respect to the cross-sectional area in the fourth element (left-hand side figure) and for accelerations in all degrees of freedom with respect to the cross-sectional area in the second element (right-hand side). The maximum values do not exceed 0.05 % for both diagrams, which confirms proper implementation of the VDM-based algorithm for sensitivity analysis. However, in more sensitive cases the analytical results obtained via the VDM-based techniques will be more accurate than those based on the FDM. This difference can be crucial for more challenging optimization problems.

Other examples for truss and beam structures have been presented in References [26] to [29] and deal with one of the most promising applications of the VDM in dynamic analysis – the inverse problem of parameter identification.

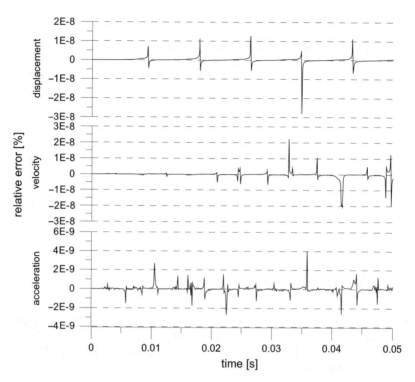

**Figure 6.1** Material redistribution in elastic structures (numerical example): error of VDM-based re-analysis relative to the standard FEM analysis of the structure in Figure 2.4 (the second node in the horizontal direction)

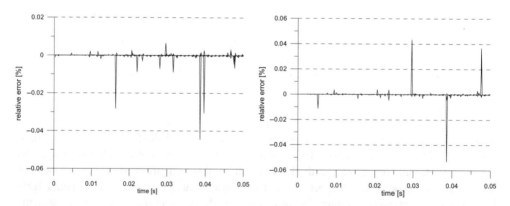

**Figure 6.2** Material redistribution in elastic structures (numerical example): errors of response gradients obtained analytically by Equation (16), relative to gradients obtained with the finite differences method: (left) strain gradients $\partial \varepsilon_i(t)/\partial \hat{A}_4$ for all elements $i$; (right) acceleration gradients $\partial \ddot{u}_N(t)/\partial \hat{A}_2$ for all degrees of freedom $N$

## 6.2  Remodeling of Elastoplastic Structures

This section considers the coupled problem of simulation of material redistribution for structures with elastoplastic material characteristics. Two cases are considered: modifications of the element cross-sectional areas $A_i$, which couple the stiffness and mass matrices, and separate independent modifications of element stiffnesses and masses.

### 6.2.1  VDM Formulation

The equations of motion for the modeled structures with virtual distortions simulating modifications of material distribution and physical nonlinearities can be expressed similarly to Equation (1) as

$$M_{NM}\ddot{u}_M(t) + G_{Ni}^{T}l_i S_{ii}\left[G_{iM}u_M(t) - \varepsilon_i^0(t) - \beta_i^0(t)\right] = f_N(t) + f_N^0(t)$$

which, as in Equation (2), can also be expressed via strains rather than displacements:

$$M_{NM}\ddot{u}_M(t) + G_{Ni}^{T}l_i S_{ii}\left[\varepsilon_i(t) - \varepsilon_i^0(t) - \beta_i^0(t)\right] = f_N(t) + f_N^0(t)$$

The virtual distortion method expresses the dynamic structural response as a superposition of the linear response $u_N^L$ and terms due to modifications, which are simulated by virtual distortions:

$$u_N(t) = u_N^L(t) + \sum_{\tau \leq t} B_{Nj}^{\varepsilon}(t - \tau)\varepsilon_j^0(\tau) + \sum_{\tau \leq t} B_{Nk}^{\varepsilon}(t - \tau)\beta_k^0(\tau) + \sum_{\tau \leq t} B_{NM}^{f}(t - \tau)f_M^0(\tau)$$

$$(18)$$

where the distortions $\varepsilon_i^0(t)$, $\beta_i^0(t)$ and $f_N^0(t)$ simulate respectively the modifications of element stiffness (strain distortions), physical nonlinearities (plastic distortions) and element mass (force distortions). The index $N$ denotes all degrees of freedom, $j$ denotes the elements with modified cross-sections, $k$ denotes the plastified elements and $M$ denotes the degrees of freedom related to the elements with modified cross-sections. The corresponding formula for strain is obtained by premultiplying Equation (18) by $G_{iN}$:

$$\varepsilon_i(t) = \varepsilon_i^L(t) + \sum_{\tau \leq t} D_{ij}^{\varepsilon}(t - \tau)\varepsilon_j^0(\tau) + \sum_{\tau \leq t} D_{ik}^{\varepsilon}(t - \tau)\beta_k^0(\tau) + \sum_{\tau \leq t} D_{iM}^{f}(t - \tau)f_M^0(\tau)$$

which, by directly separating the increments of plastic distortions $\Delta\beta_k^0$ in successive time steps, takes the form

$$\varepsilon_i(t) = \varepsilon_i^L(t) + \sum_{\tau \leq t} D_{ij}^{\varepsilon}(t - \tau)\varepsilon_j^0(\tau) + \sum_{\tau \leq t}\sum_{\kappa \leq \tau} D_{ik}^{\varepsilon}(t - \kappa)\Delta\beta_k^0(\kappa) + \sum_{\tau \leq t} D_{iM}^{f}(t - \tau)f_M^0(\tau)$$

$$(19)$$

Let $\varepsilon^{\neq t}(t)$ denote the strains without the effect of strain distortions $\varepsilon_i(t)$, distortion forces $f_N^0(t)$ and plastic distortion increments $\Delta\beta_i(t)$ in the current time step $t$:

$$\varepsilon_i(t) = \varepsilon_i^{\neq t}(t) + D_{ij}^{\varepsilon}(0)\varepsilon_j^0(t) + D_{ik}^{\varepsilon}(0)\Delta\beta_k^0(t) + D_{iM}^{f}(0)f_M^0(t) \qquad (20)$$

which can be compared with Equation (8). Formulas of Equations (5) for the member forces in the modified and modeled structures have to be modified in order to take into account the plastic distortions. In the considered elastoplastic case they take the form

$$p_i(t) = E_{\underline{i}} \hat{A}_{\underline{i}} \left[ \varepsilon_{\underline{i}}(t) - \beta_{\underline{i}}^0(t) \right] \tag{21a}$$

$$p_i(t) = E_{\underline{i}} A_{\underline{i}} \left[ \varepsilon_{\underline{i}}(t) - \varepsilon_{\underline{i}}^0(t) - \beta_{\underline{i}}^0(t) \right] \tag{21b}$$

and yield the following formula for the virtual distortion (see Equation (6)):

$$\varepsilon_{\underline{i}}^0(t) = \left( 1 - \mu_{\underline{i}}^A \right) \left[ \varepsilon_{\underline{i}}(t) - \beta_{\underline{i}}^0(t) \right]$$

where the modification parameter $\mu_i^A$ can represent the modification of either the element cross-section $\mu_i^A = \hat{A}_{\underline{i}} / A_{\underline{i}}$ or the element stiffness only. In the latter case it is denoted by $\mu_i^E = \hat{E}_{\underline{i}} / E_{\underline{i}}$. The increment of the plastic distortion $\Delta\beta^0(t)$ can be written explicitly:

$$\varepsilon_{\underline{i}}^0(t) = \left( 1 - \mu_{\underline{i}}^A \right) \left[ \varepsilon_{\underline{i}}(t) - \beta_{\underline{i}}^0(t-1) - \Delta\beta_{\underline{i}}^0(t) \right] \tag{22}$$

Substitution of Equation (20) into Equation (22) leads to the first of the three conditions necessary to determine the virtual distortions being sought:

$$\left( 1 - \mu_{\underline{i}}^A \right) \left[ \varepsilon_{\underline{i}}^{\neq t}(t) - \beta_{\underline{i}}^0(t-1) \right] = \left[ \delta_{ij} - \left( 1 - \mu_{\underline{i}}^A \right) D_{ij}^\varepsilon(0) \right] \varepsilon_j^0(t) - \left( 1 - \mu_{\underline{i}}^A \right) D_{iM}^f(0) f_M^0(t)$$
$$- \left( 1 - \mu_{\underline{i}}^A \right) \left[ D_{ik}^\varepsilon(0) - \delta_{ik} \right] \Delta\beta_k^0 \tag{23}$$

The postulate of identity of the inertia forces in the modified and modeled structures leads to the relation in Equation (45) in Chapter 2, which after the substitution of the following analog of Equation (20):

$$\ddot{u}_N(t) = \ddot{u}_N^{\neq t}(t) + \ddot{B}_{Ni}^\varepsilon(0)\varepsilon_i^0(t) + \ddot{B}_{Nk}^\varepsilon(0)\Delta\beta_k^0(t) + \ddot{B}_{NM}^f(0) f_M^0(t)$$

yields the second condition:

$$-\Delta M_{NM} \ddot{u}_M^{\neq t} = \Delta M_{NM} \ddot{B}_{Mj}^\varepsilon(0)\varepsilon_j^0(t) + \left[ \delta_{NM} + \Delta M_{NL} \ddot{B}_{LM}^f(0) \right] f_M^0(t)$$
$$+ \Delta M_{NM} \ddot{B}_{Mk}^\varepsilon(0)\Delta\beta_k^0 \tag{24}$$

where $\Delta M_{NM}$ is related to the cross-section (or density) modification parameter $\mu_i^A$ (or $\mu_i^\rho$) as in Equation (46) in Chapter 2.

The third condition necessary to determine the virtual distortions is based on the constitutive relation. As it makes use of element stresses, the two following cases should be distinguished:

1. For modifications of element cross-sections $A_i$, which couple element stiffnesses and masses, the stresses are computed based on Equation (21a):

$$\sigma_i(t) = \frac{p_{\underline{i}}(t)}{\hat{A}_{\underline{i}}} = E_{\underline{i}} \left[ \varepsilon_{\underline{i}}(t) - \beta_{\underline{i}}^0(t) \right] \tag{25a}$$

2. For independent modifications of both element Young's moduli and densities, the stresses have to be computed based on Equation (21b):

$$\sigma_i(t) = \frac{p_{\underline{i}}(t)}{A_{\underline{i}}} = E_{\underline{i}}\left[\varepsilon_{\underline{i}}(t) - \varepsilon_{\underline{i}}^0(t) - \beta_{\underline{i}}^0(t)\right] \tag{25b}$$

### 6.2.1.1 Modifications of Element Cross-Sections

The problem of material redistribution considered here assumes modifications of element cross-sections $A_i$, which couple both the element stiffness and mass. In this case the stresses (Equation (25a)) have to be used. The virtual distortions in the current time step $t$ can be explicitly separated by Equation (20):

$$\sigma_i(t) = E_{\underline{i}}\left[\varepsilon_{\underline{i}}^{\neq t}(t) + D_{\underline{i}j}^\varepsilon(0)\varepsilon_j^0(t) + D_{\underline{i}M}^f(0)f_M^0(t)\right.$$
$$\left. + D_{\underline{i}k}^\varepsilon(0)\Delta\beta_k^0(t) - \Delta\beta_{\underline{i}}^0(t) - \beta_{\underline{i}}^0(t-1)\right] \tag{26}$$

Substitution of the stresses (Equation (26)) to the constitutive relation (Equation (35) in Chapter 2) leads to the following formula, which is valid only for the currently yielding elements $i$:

$$\left(1 - \gamma_{\underline{i}}\right)\sigma_{\underline{i}}^{\mathrm{TR}}(t) - \mathrm{sign}\left(\sigma_{\underline{i}}^{\mathrm{TR}}(t)\right)\left[\left(1 - \gamma_{\underline{i}}\right)\sigma_{\underline{i}}^\star + \gamma_{\underline{i}}E_{\underline{i}}\Psi_{\underline{i}}(t)\right]$$
$$= -\left(1 - \gamma_{\underline{i}}\right)E_{\underline{i}}D_{\underline{i}j}^\varepsilon(0)\varepsilon_j^0(t) - \left(1 - \gamma_{\underline{i}}\right)E_{\underline{i}}D_{\underline{i}M}^f(0)f_M^0(t)$$
$$+ E_{\underline{i}}\left[\delta_{\underline{i}k} - \left(1 - \gamma_{\underline{i}}\right)D_{\underline{i}k}^\varepsilon(0)\right]\Delta\beta_k^0(t) \tag{27}$$

Finally, Equations (23), (24) and (27) combined together yield the following linear system, similar to Equations (12), which can be iteratively used in successive time steps to determine all virtual distortions:

$$\mathbf{F}^S\mathbf{x}^0 = \mathbf{b}^S \tag{28}$$

where the principal matrix $\mathbf{F}^S$ is given by

$$\begin{bmatrix} \delta_{ij} - \left(1 - \mu_{\underline{i}}^A\right)D_{\underline{i}j}^\varepsilon(0) & -\left(1 - \mu_{\underline{i}}^A\right)D_{\underline{i}M}^f(0) & -\left(1 - \mu_{\underline{i}}^A\right)[D_{\underline{i}k}^\varepsilon(0) - \delta_{\underline{i}k}] \\ \Delta M_{NM}\ddot{B}_{Mj}^\varepsilon(0) & \delta_{NM} + \Delta M_{NL}\ddot{B}_{LM}^f(0) & \Delta M_{NM}\ddot{B}_{Mk}^\varepsilon(0) \\ -\left(1 - \gamma_{\underline{i}}\right)E_{\underline{i}}D_{\underline{i}j}^\varepsilon(0) & -\left(1 - \gamma_{\underline{i}}\right)E_{\underline{i}}D_{\underline{i}M}^f(0) & E_{\underline{i}}[\delta_{\underline{i}k} - \left(1 - \gamma_{\underline{i}}\right)D_{\underline{i}k}^\varepsilon(0)] \end{bmatrix}$$

the vector $\mathbf{x}^0$ collects the virtual distortions,

$$\mathbf{x}^0 = \begin{bmatrix} \varepsilon_j^0(t) & f_M^0(t) & \Delta\beta_k^0(t) \end{bmatrix}^\mathrm{T}$$

and

$$\mathbf{b}^S = \begin{bmatrix} \left(1 - \mu_{\underline{i}}^A\right)\left(\varepsilon_{\underline{i}}^{\neq t}(t) - \beta_{\underline{i}}^0(t-1)\right) \\ -\Delta M_{NM}\ddot{u}_M^{\neq t} \\ \left(1 - \gamma_{\underline{i}}\right)\sigma_{\underline{i}}^{\mathrm{TR}}(t) - \mathrm{sign}\left(\sigma_{\underline{i}}^{\mathrm{TR}}(t)\right)[(1 - \gamma_{\underline{i}})\sigma_{\underline{i}}^\star + \gamma_{\underline{i}}E_{\underline{i}}\Psi_{\underline{i}}(t)] \end{bmatrix}$$

Note, however, that although the elements of the full principal matrix $\mathbf{F}^S$ do not depend on time and can thus be computed only once, its size is time-dependent, since the third equation of (28) is valid only for the currently yielding elements and the extent of the plastic zone changes in time. Thus, the principal matrix $\mathbf{F}^S$ and the system of Equation (28) have to be updated each time the plastic zone changes. The algorithm for simulation of material redistribution in elastoplastic structures is shown in Table 6.3.

**Table 6.3** Algorithm for material redistribution in elastoplastic structures

---

`Data and initial calculations`
Input data

- Construction under external load
- Initially empty plastic zone $\mathfrak{B} = \emptyset$
- Yield stress limits $\sigma_i^*$, hardening parameters $\gamma_i$, cross-section modification parameters $\mu_i^A$

Calculations

- Linear response $\varepsilon_i^L(t)$ and $\ddot{u}_N^L(t)$
- Dynamic influence matrices $D_{ij}^\varepsilon(t)$, $D_{iN}^f(t)$, $\ddot{B}_{Nj}^\varepsilon(t)$ and $\ddot{B}_{NM}^f(t)$
- Principal matrix $\mathbf{F}^S$ by Equation (28)

`Calculations in each time step` $t$

Strains $\varepsilon_i^{\neq t}(t)$ and accelerations $\ddot{u}_N^{\neq t}(t)$

Estimate the trial stresses $\sigma_i^{\mathrm{TR}}(t) = E_i\left[\varepsilon_i^{\neq t}(t) - \beta_i^0(t-1)\right]$

Estimate the extent of the plastic zone $\mathfrak{B}$ by the condition $|\sigma_i^{\mathrm{TR}}(t)| > \sigma_i^* + \frac{\gamma_i E_i}{1-\gamma_i}\Psi_i(t)$

NONEMPTY PLASTIC ZONE

(a) Update the principal matrix $\mathbf{F}^S$ according to the plastic zone $\mathfrak{B}$ and $\mathrm{sign}(\sigma_i^{\mathrm{TR}})$
(b) Virtual distortions $\Delta\beta_k^0(t)$, $\varepsilon_i^0(t)$ and $f_M^0(t)$ by Equation (28)
(c) Plastic distortions: $\beta_k^0(t) := \beta_k^0(t-1) + \Delta\beta_k^0(t)$
(d) Total plastic strain for isotropic hardening: $\Psi_k(t) := \Psi_k(t-1) + |\Delta\beta_k^0(t)|$
(e) Actual strains by Equation (20) and accelerations
(f) Stresses by Equation (26)
(g) Verify the sign of $\Delta\beta_k^0(t)\sigma_k(t)$. If negative, reestimate the extent of the plastic zone $\mathfrak{B}$ using the computed stresses $\sigma_i(t)$ instead of the trial stresses $\sigma_i^{\mathrm{TR}}(t)$ and go back to (a)

EMPTY PLASTIC ZONE

(a) Update the principal matrix $\mathbf{F}^S$ according to the empty plastic zone $\mathfrak{B} = \emptyset$
(b) Virtual distortions $\varepsilon_j^0(t)$ and $f_M^0(t)$
(c) Plastic distortions $\beta_k^0(t) := \beta_k^0(t-1)$
(d) Actual strains by Equation (20) and accelerations
(e) Stresses by Equation (26)

If necessary, by the corresponding influence matrices compute the response: the displacements $u_N(t)$, the velocities $\dot{u}_N(t)$, etc.

If necessary, calculate the derivatives of the response (Section 6.2.2)

$t := t + 1$

---

**Table 6.4** Material redistribution in elastoplastic structures (numerical example): modifications of element cross-sections

| Element number | $\sigma^*$ (MPa) | $\gamma$ | $A$ (m²) | $\mu^A$ | $\hat{A}$ (m²) |
|---|---|---|---|---|---|
| 1 | 50 | 0.01 | $1 \times 10^{-5}$ | 0.8 | $0.8 \times 10^{-5}$ |
| 2 | 50 | 0.01 | $1 \times 10^{-5}$ | 1.1 | $1.1 \times 10^{-5}$ |
| 3 | 50 | 0.01 | $1 \times 10^{-5}$ | 0.6 | $0.6 \times 10^{-5}$ |
| 4 | 50 | 0.01 | $1 \times 10^{-5}$ | 0.2 | $0.2 \times 10^{-5}$ |
| 5 | 50 | 0.01 | $1 \times 10^{-5}$ | 0.7 | $0.7 \times 10^{-5}$ |

### 6.2.1.2 Numerical Testing Example

The testing example (Figure 2.4 in Chapter 2) has been used to test the presented reanalysis procedure based on virtual distortions and to verify it against a reanalysis performed with the commercially available package ANSYS (the two-dimensional truss element LINK1 and bilinear isotropic hardening).

The uniformly distributed material properties, the original and the modified element cross-sections are listed in Table 6.4. The dynamic excitation is realized by the initial vertical velocity $V_0 = -20$ m/s of the right bottom node.

Figure 6.3 compares the computed plastic distortions $\beta_k^0(t)$ with the reference ANSYS results and shows their relative discrepancies (Equation (17)), which stay below 1.2 %. Figure 6.4 shows the distortion forces $f_N^0(t)$, which model the modifications of mass. Figure 6.5 plots the virtual strain distortions $\varepsilon_j^0(t)$, which model the modifications of stiffness and compares the computed energy balance with the reference ANSYS results.

### 6.2.1.3 Independent Modifications of Element Mass and Stiffness

If structural remodelling proceeds via direct modifications of the element material (density $\rho$ and Young's modulus $E$) rather than modifications of cross-sections, then element stresses are expressed by Equation (25b), which should be used with the constitutive relation Equation (35)

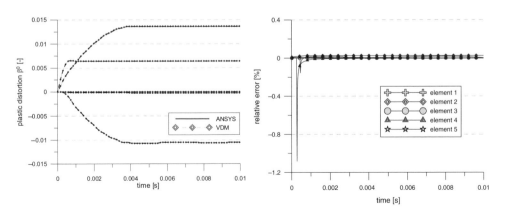

**Figure 6.3** Material redistribution in elastoplastic structures, modifications of element cross-sections (numerical example): (left) computed plastic distortions compared to reference ANSYS results; (right) errors of computed plastic distortions relative to reference ANSYS results

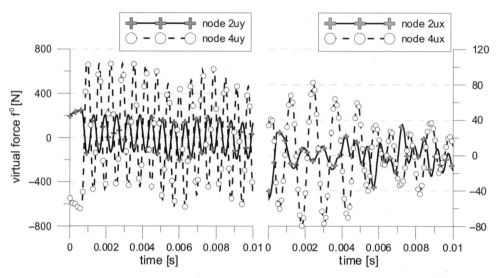

**Figure 6.4**  Material redistribution in elastoplastic structures, modifications of element cross-sections (numerical example): virtual distortion forces simulating modifications of element mass

in Chapter 2. In consequence, the third condition, which builds the linear system Equation (28), takes, instead of Equation (27), the following form:

$$
\left(1 - \gamma_{\underline{i}}\right) E_{\underline{i}} \left[\varepsilon_{\underline{i}}^{\neq t}(t) - \beta_{\underline{i}}^{0}(t-1)\right] - \text{sign}(\sigma_{\underline{i}}^{\text{TR}}(t)) \left[\left(1 - \gamma_{\underline{i}}\right) \sigma_{\underline{i}}^{\star} + \gamma_{\underline{i}} E_{\underline{i}} \Psi_{\underline{i}}(t)\right]
$$

$$
= \left(1 - \gamma_{\underline{i}}\right) E_{\underline{i}} \left[\delta_{\underline{i}j} - D_{\underline{i}j}^{\varepsilon}(0)\right] \varepsilon_{j}^{0}(t) - \left(1 - \gamma_{\underline{i}}\right) E_{\underline{i}} D_{\underline{i}M}^{f}(0) f_{M}^{0}(t)
$$

$$
+ E_{\underline{i}} \left[\delta_{\underline{i}k} - \left(1 - \gamma_{\underline{i}}\right) D_{\underline{i}k}^{\varepsilon}(0)\right] \Delta \beta_{k}^{0}(t)
$$

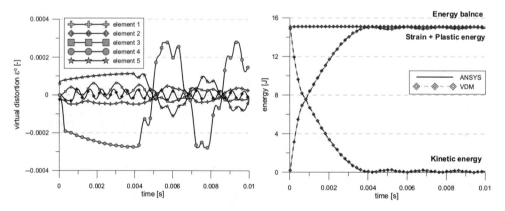

**Figure 6.5**  Material redistribution in elastoplastic structures, modifications of element cross-sections (numerical example): (left) virtual strain distortions simulating modifications of element stiffness; (right) computed energy balance compared to reference ANSYS results

The algorithm in Table 6.3 has also to be modified: in the considered case the trial stresses, similarly to the actual stresses of Equation (25b), depend on the virtual strain distortions $\varepsilon_i^0(t)$ and are expressed by

$$\sigma_i^{TR}(t) = E_{\underline{i}}\left[\varepsilon_{\underline{i}}^{\neq t}(t) - \varepsilon_{\underline{i}}^0(t) - \beta_{\underline{i}}^0(t-1)\right]$$

However, in order to find the extent of the plastic zone $\mathfrak{B}$, the trial stresses have to be estimated *before* the strain distortion (and other virtual distortions) are computed. Nevertheless, they can be initially estimated by assuming

$$\varepsilon_i^0(t) \approx \left(1 - \mu_{\underline{i}}^E\right)\left[\varepsilon_{\underline{i}}^{\neq t}(t) - \beta_{\underline{i}}^0(t-1)\right]$$

### 6.2.1.4 Numerical Testing Example

Consider the same five-element truss structure as in the previous examples (Figure 2.4) with the same dynamic excitation in the form of the initial vertical velocity $V_0 = -20$ m/s of the bottom right node. A piecewise-linear constitutive law was included and the reanalysis was performed for the case of direct modifications of element material properties, i.e. independent modifications of stiffness and mass, which makes two independent modification parameters $\mu_i^E$ and $\mu_i^\rho$ necessary. The material properties of the elements are listed in Table 6.5; the changes correspond to the replacement of steel in the third element with magnesium and with copper in the fourth and fifth elements.

Figure 6.6 plots the distortion forces $f_i^0$, which act in the second and fourth node and model the assumed mass modifications. The five plastic distortions $\beta_i^0(t)$ are illustrated in Figure 6.7 and compared to the reference ANSYS results. For better visualization of the discrepancies, the relative errors are depicted on the right-hand side, showing 0.6 % for the third element and less than 0.2 % for the other elements, which confirms the validity of the presented VDM-based reanalysis method. Figure 6.8 (left) shows the virtual strain distortions $\varepsilon_i^0$ modeling the assumed changes of the element stiffnesses, while Figure 6.8 (right) plots the computed energy balance and compares it to the reference ANSYS results.

## 6.2.2 Sensitivity Analysis

An important advantage of the VDM-based methods for structural remodeling is the availability of the analytical derivatives of the response with respect to structural parameters, which makes possible precise computations of gradients of response-based objective functions and thus enables effective gradient-based optimization. This subsection considers the derivatives

**Table 6.5** Structural remodeling in elastoplastic structures (numerical example): modifications of material properties

| Element number | $\mu^E$ | $\mu^\rho$ | $\sigma^\star$ (MPa) | $\gamma$ |
|---|---|---|---|---|
| 1 | 1.00 | 1.00 | 294 | 0.01 |
| 2 | 1.00 | 1.00 | 294 | 0.01 |
| 3 | 0.12 | 0.22 | 120 | 0.01 |
| 4 | 0.62 | 1.14 | 210 | 0.01 |
| 5 | 0.62 | 1.14 | 210 | 0.01 |

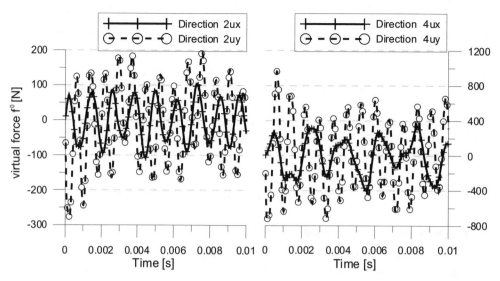

**Figure 6.6**  Structural remodeling in elastoplastic structures, modifications of material properties (numerical example): virtual distortion forces modeling the assumed density modifications

with respect to element modified cross-sections $\hat{A}_i$; derivatives with respect to other types of modifications can be treated in an analogous way.

Direct differentiation of Equation (28) yields the following linear system, which can be used iteratively in successive time steps to compute the derivatives of the virtual distortions:

$$\mathbf{F}^S \frac{\partial \mathbf{x}^0}{\partial \hat{A}_l} = \frac{\partial \mathbf{b}}{\partial \hat{A}_l} - \frac{\partial \mathbf{F}^S}{\partial \hat{A}_l} \mathbf{x}^0 \tag{29}$$

The principal matrix of Equation (29) stays the same as in Equation (28), which simplifies the optimization process. The vector $\partial \mathbf{x}^0 / \partial \hat{A}_l$ collects the derivatives of the virtual

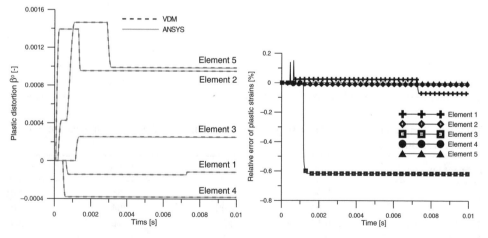

**Figure 6.7**  Structural remodeling in elastoplastic structures, modifications of material properties (numerical example): (left) computed plastic distortions compared to the reference ANSYS results; (right) errors of computed plastic distortions relative to the reference ANSYS results

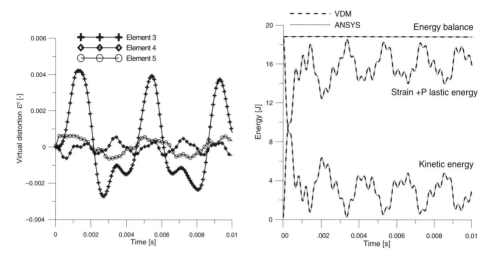

**Figure 6.8** Structural remodeling in elastoplastic structures, modifications of material properties (numerical example): (left) virtual strain distortions modeling assumed modifications of Young's modulus; (right) energy balance of the five-element truss

distortions,

$$\frac{\partial \mathbf{x}^0}{\partial \hat{A}_l} = \left[ \begin{array}{ccc} \dfrac{\partial \varepsilon_j^0(t)}{\partial \hat{A}_l} & \dfrac{\partial f_M^0(t)}{\partial \hat{A}_l} & \dfrac{\partial \Delta \beta_k^0(t)}{\partial \hat{A}_l} \end{array} \right]^{\mathrm{T}}$$

The right-hand side vector is given by

$$\frac{\partial \mathbf{b}}{\partial \hat{A}_l} - \frac{\partial \mathbf{F}^{\mathrm{S}}}{\partial \hat{A}_l} \mathbf{x}^0 = \left[ \begin{array}{c} \left(1 - \mu_{\underline{i}}^{\mathrm{A}}\right) \left[ \dfrac{\partial \varepsilon_{\underline{i}}^{\neq t}(t)}{\partial \hat{A}_l} - \dfrac{\partial \beta_{\underline{i}}^0(t-1)}{\partial \hat{A}_l} \right] - \dfrac{\delta_{il}}{A_{\underline{l}}} [\varepsilon_i(t) - \beta_{\underline{i}}^0(t)] \\[4mm] - \dfrac{\partial \Delta M_{NM}}{\partial \hat{A}_l} \ddot{u}_M(t) - \Delta M_{NM} \dfrac{\partial \ddot{u}_M^{\neq t}(t)}{\partial \hat{A}_l} \\[4mm] E_{\underline{i}} \left[ \dfrac{\partial \varepsilon_{\underline{i}}^{\neq t}(t)}{\partial \hat{A}_l} - \dfrac{\partial \beta_{\underline{i}}^0(t-1)}{\partial \hat{A}_l} \right] - \mathrm{sign}[\sigma_{\underline{i}}^{\mathrm{TR}}(t)] \gamma_{\underline{i}} E_{\underline{i}} \dfrac{\partial \Psi_i(t)}{\partial \hat{A}_l} \end{array} \right]$$

where the derivative of the total plastic strain $\Psi_i(t)$ can be expressed by Equation (37) in Chapter 2 as

$$\frac{\partial \Psi_i(t)}{\partial \hat{A}_l} = \sum_t \mathrm{sign}\left[\Delta \beta_{\underline{i}}^0(t)\right] \frac{\partial \Delta \beta_{\underline{i}}^0(t)}{\partial \hat{A}_l}$$

The derivative of the mass matrix increment $\partial \Delta M_{NM}/\partial \hat{A}_l$ is given by Equation (15) and

$$\frac{\partial \varepsilon_i^{\neq t}(t)}{\partial \hat{A}_l} = \sum_{\tau < t} D_{ij}^\varepsilon(t-\tau) \frac{\partial \varepsilon_j^0(\tau)}{\partial \hat{A}_l} + \sum_{\tau < t} D_{iM}^{\mathrm{f}}(t-\tau) \frac{\partial f_M^0(\tau)}{\partial \hat{A}_l}$$
$$+ \sum_{\tau < t} \sum_{\kappa \leq \tau} D_{ik}^\varepsilon(\tau - \kappa) \frac{\partial \Delta \beta_k^0(\kappa)}{\partial \hat{A}_l} + \sum_{\tau < t} D_{ik}^\varepsilon(t-\tau) \frac{\partial \Delta \beta_k^0(\tau)}{\partial \hat{A}_l}$$

while the derivative of $\ddot{u}_M^{\neq t}(t)$ with respect to $\hat{A}_l$ is analogous.

Given the derivatives of the virtual distortions, the derivatives of the response can be computed either directly by the differentiated versions of Equations (18) and (19) or by

$$\frac{\partial \varepsilon_i(t)}{\partial \hat{A}_l} = \frac{\partial \varepsilon_i^{\neq t}(t)}{\partial \hat{A}_l} + D_{ij}^\varepsilon(0)\frac{\partial \varepsilon_j^0(t)}{\partial \hat{A}_l} + D_{iM}^f(0)\frac{\partial f_M^0(t)}{\partial \hat{A}_l} + D_{ik}^\varepsilon(0)\frac{\partial \Delta\beta_k^0(t)}{\partial \hat{A}_l}$$

or its counterpart for accelerations. The latter should be quicker, since $\partial \varepsilon_i^{\neq t}(t)/\partial \hat{A}_l$ are already computed for use with Equation (29).

### 6.2.2.1 Numerical Testing Example

Consider the following objective function, which describes the total energy dissipated in the process of plastic yielding:

$$\Delta U = \sum_t \Delta U(t) = \sum_t \sigma_i(t)\Delta\beta_i^0(t)l_i A_i \mu_i$$

The derivative of the objective function with respect to the element cross-section can be expressed as follows:

$$\frac{\partial \Delta U}{\partial \hat{A}_l} = \sum_t l_i \left[ \frac{\partial \sigma_i(t)}{\partial \hat{A}_l}\Delta\beta_i^0(t)A_i\mu_i + \sigma_i(t)\frac{\partial \Delta\beta_i^0(t)}{\partial \hat{A}_l}A_i\mu_i + \sigma_i(t)\Delta\beta_i^0(t)\delta_{il} \right]$$

where the component $\partial \sigma_i(t)/\partial \hat{A}_l$ can be determined by differentiating Equation (25a) as

$$\frac{\partial \sigma_i(t)}{\partial \hat{A}_l} = E_i \left( \frac{\partial \varepsilon_i(t)}{\partial \hat{A}_l} - \frac{\partial \beta_i^0(t)}{\partial \hat{A}_l} \right)$$

The sensitivity analysis has been performed for the same testing example as before (see Figure 2.4 and Table 6.4). Table 6.6 lists the numerical values of the computed gradients, reference ANSYS results obtained by the finite differences method (FDM) and the relative discrepancies, which do not exceed $50 \times 10^{-6}$.

## 6.3 Adaptive Structures with Active Elements

The quick reanalysis technique and the possibility of numerically efficient structural modifications presented in the first section, together with accurate sensitivity analysis, provide strong numerical tools to deal with complex optimization problems of elastic structures. However, as demonstrated in the second section, the VDM-based approach to remodeling extends also to the elastoplastic structural behavior. This leads to the main idea of this chapter: the optimum design of *adaptive structures*. If it has been assumed that properly located and actively controlled dissipaters (*structural fuses*) can perform similarly to elastoplastic structural elements, the presented numerical tools can be used for the design of smart structures capable of optimum real-time adaptation to random impact loads. This can be formulated as follows:

1. Assume an initial configuration of a considered structure, which is exposed to a given set of potential impact loads.

**Table 6.6** Sensitivity analysis for material redistribution in elasto-plastic structures (numerical example), validation of gradients of total dissipated energy with respect to element cross-sections: computed (VDM), ANSYS reference results (FDM) and their relative discrepancies

|  | FDM | VDM | Relative discrepancy |
|---|---|---|---|
| $\dfrac{\partial U}{\partial \hat{A}_1}$ | $9.96206 \times 10^3$ | $9.96205 \times 10^3$ | $833.870 \times 10^{-9}$ |
| $\dfrac{\partial U}{\partial \hat{A}_2}$ | $20.4749 \times 10^3$ | $20.4739 \times 10^3$ | $46.7989 \times 10^{-6}$ |
| $\dfrac{\partial U}{\partial \hat{A}_3}$ | $25.9559 \times 10^3$ | $25.9552 \times 10^3$ | $30.0264 \times 10^{-6}$ |
| $\dfrac{\partial U}{\partial \hat{A}_4}$ | $443.956 \times 10^3$ | $443.955 \times 10^3$ | $2.50197 \times 10^{-6}$ |
| $\dfrac{\partial U}{\partial \hat{A}_5}$ | $610.193 \times 10^3$ | $610.192 \times 10^3$ | $1.76812 \times 10^{-6}$ |

2. Redesign the initial structural configuration in such a way that, preserving the same volume of material, a limited number of active elements with controllable plastic-like characteristics can assure the mean-optimum response of the structure to all the considered impact loads. The optimality criterion can be based on nodal displacements, element stresses and overall energy absorption capacities.

A methodology proposed here decomposes the crucial second task into three stages:

1. Redesign the initial structural configuration to obtain *the stiffest elastic structure* by minimizing the displacements of chosen nodes under the considered multiple impact loads. It was observed that the resulting structure has a quasi-isostatic structural topology, with the mass concentrated in a small number of elements; the maximum impact-induced stresses are significantly reduced.
2. Determine the optimum locations for the *structural fuses*, which are elements with controllable elastoplastic properties (e.g. yield stress). In order to minimize the costs, only a limited number of elements should be chosen, which however lead to significant improvement in terms of the structural response and the optimality criterion.
3. Determine the optimum, in terms of energy dissipation, control strategies of the structural fuses, either passive or active. The former assumes no prior information about an imminent impact and is hence a general mean-optimum strategy. The latter allows for optimum control specifically tailored to each of the considered impact cases and requires real-time identification of impact loads [30].

The proposed approach is demonstrated in this section on the example of a planar truss beam structure exposed to impact loads (Figure 6.9). The structure is rigidly supported at both ends and consists of 11 sections 0.1 m × 0.1 m each; the total length is 1.1 m. The elements are made of steel (Young's modulus 210 GPa, density 7800 kg/m$^3$) and initially all 110 mm$^2$ in cross-section. Impact loads are simulated by initial velocities of two masses attached simultaneously to two nodes.

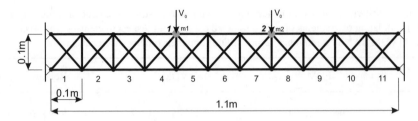

**Figure 6.9**  Two-dimensional truss beam, initial configuration

## 6.3.1 Stiffest Elastic Substructure

### 6.3.1.1 Optimization Problem

The considered structure is to be optimally remodeled to form the stiffest structure with minimum dynamic deflections under the given impact load. The following objective function is thus used:

$$f(\hat{\mathbf{A}}) = \sum_{t,M} u_M^2(t) \tag{30}$$

where $u_M(t)$ denotes the modeled response of the structure with modified element cross-sections $\hat{\mathbf{A}}$, subject to the following constraints:

$$V = A_i l_i \mu_i \leq \tilde{V}$$
$$\hat{A}_i > 0 \tag{31}$$
$$|\sigma_i| < \tilde{\sigma}$$

The index $M$ belongs to the set of the degrees of freedom in which the displacements are being minimized, which in the case considered here are the vertical displacements of the two impacted nodes. The total volume of the material in the initial structural configuration is denoted by $\tilde{V}$ and the maximum admissible stress level by $\tilde{\sigma}$.

In order to simplify the optimization process, the original optimization problem of Equation (30) with the constraints in Equations (31) has been transformed into a corresponding unconstrained problem. Instead of the original constraints, three quadratic penalty functions [31] have been used to form the following modified objective function:

$$f_c(\hat{\mathbf{A}}) = f(\hat{\mathbf{A}}) + c_1 \left(V - \tilde{V}\right)^2 \mathbf{1}_{V > \tilde{V}} + c_2 \sum_i \hat{A}_i^2 \mathbf{1}_{0 > \hat{A}_i}$$
$$+ c_3 \sum_{t,i} \left\{ [\sigma_i(t) + \tilde{\sigma}]^2 \mathbf{1}_{-\tilde{\sigma} > \sigma_i(t)} + [\sigma_i(t) - \tilde{\sigma}]^2 \mathbf{1}_{\tilde{\sigma} < \sigma_i(t)} \right\} \tag{32}$$

where $c_1$, $c_2$ and $c_3$ are suitably chosen constants and

$$\mathbf{1}_{\psi(x)}(x) = \begin{cases} 1 & \text{if } \psi(x) \\ 0 & \text{otherwise} \end{cases}$$

Differentiation of Equation (32) yields the formula for the gradient of the modified objective function:

$$\frac{\partial f_c(\hat{\mathbf{A}})}{\partial \hat{A}_l} = -2 \sum_{t,M} u_M(t) \frac{\partial u_M(t)}{\partial \hat{A}_l} + 2c_1 l_l (V - \tilde{V}) \mathbf{1}_{V > \tilde{V}} + 2c_2 \hat{A}_l \mathbf{1}_{0 > A_l}$$

$$+ 2c_3 \sum_{t,i} \left\{ [\sigma_i(t) + \tilde{\sigma}] \frac{\partial \sigma_i(t)}{\partial \hat{A}_l} \mathbf{1}_{-\tilde{\sigma} > \sigma_i(t)} + [\sigma_i(t) - \tilde{\sigma}] \frac{\partial \sigma_i(t)}{\partial \hat{A}_l} \mathbf{1}_{\tilde{\sigma} < \sigma_i(t)} \right\}$$

where the derivatives of the structural response can be obtained with the VDM-based methods described in this chapter. The Levenberg–Marquardt optimization algorithm used here [31,32] makes use of an easy-to-compute approximation $\tilde{\mathbf{H}}(\hat{\mathbf{A}})$ to the Hessian of the objective function. It is obtained by local linearizations of the displacements $u_M(t)$ and the stresses $\sigma_i(t)$, i.e. by assuming that their second derivatives with respect to element cross-sections vanish:

$$\frac{\partial^2 u_M(t)}{\partial \hat{A}_l \partial \hat{A}_k} \approx 0 \quad \text{and} \quad \frac{\partial^2 \sigma_i(t)}{\partial \hat{A}_l \partial \hat{A}_k} \approx 0$$

which yields the following elements of the approximate Hessian $\tilde{\mathbf{H}}(\hat{\mathbf{A}})$:

$$\frac{\partial^2 f_c(\hat{\mathbf{A}})}{\partial \hat{A}_l \partial \hat{A}_k} \approx 2 \sum_{t,M} \frac{\partial u_M(t)}{\partial \hat{A}_l} \frac{\partial u_M(t)}{\partial \hat{A}_k}$$

$$+ 2c_1 l_l l_k \mathbf{1}_{V > \tilde{V}} + 2c_2 \delta_{lk} \mathbf{1}_{0 > A_l} + 2c_3 \sum_{t,i} \left[ \frac{\partial \sigma_i(t)}{\partial \hat{A}_l} \frac{\partial \sigma_i(t)}{\partial \hat{A}_k} \mathbf{1}_{\tilde{\sigma} < |\sigma_i(t)|} \right]$$

The complete optimization step of the algorithm is listed in Table 6.7. Notice that, in general, the reanalysis algorithm has to take care of skipping the determination of the displacements of a free node, in case all cross-sections $\hat{A}_i$ of the elements related to this node vanish. This can be relatively easily implemented, as the VDM-based methods do not require rebuilding of the global stiffness matrix.

### 6.3.1.2 Asymmetric Single-Impact Load

The impact load was simulated by initial velocities of 5 m/s of two masses $m_1 = 0.1$ kg, $m_2 = 2$ kg simultaneously attached to two nodes of the structure (see Figure 6.9). The subject of minimization is the displacements of the two impacted nodes in the first millisecond after the impact.

The result of the optimum remodeling problem defined by Equations (30) and (31) (or, in fact, its unconstrained counterpart, Equation (32)) is the structure presented in Figure 6.10. The vertical displacements of the two impacted nodes before and after optimization are plotted in Figure 6.11. The objective function was reduced from 172.24 to 80.92. An important side effect of the remodeling process is the significant mass concentration in selected structural elements, which will be used in the next section of this chapter for location of effective shock-absorbers in the adaptive version of the considered structure.

**Table 6.7**    Optimum remodeling example: a single step of the optimization algorithm

1. Derivatives of the response $\partial u_N(t)/\partial \hat{A}_l$ and $\partial \sigma_i(t)/\partial \hat{A}_l$
2. Gradient of the modified objective function $\nabla f_c(\hat{\mathbf{A}})$
3. Approximate Hessian of the modified objective function $\tilde{\mathbf{H}}(\hat{\mathbf{A}})$
4. Adaptive selection of the step length: repeat

    Optimization step $\Delta\hat{\mathbf{A}}$ by solving $[\tilde{\mathbf{H}}(\hat{\mathbf{A}}) + \lambda\mathbf{I}]\Delta\hat{\mathbf{A}} = -\nabla f_c(\hat{\mathbf{A}})$
    Modified objective function at the new point $f_c(\hat{\mathbf{A}} + \Delta\hat{\mathbf{A}})$
    Expected improvement of the modified objective function

$$\frac{1}{2}\Delta\hat{\mathbf{A}}\left[\lambda\mathbf{I}\Delta\hat{\mathbf{A}} - \nabla f_c(\hat{\mathbf{A}})\right]$$

Actual-to-expected improvement ratio

$$\kappa = \frac{f(\hat{A}) - f(\hat{A} + \Delta\hat{A})}{0.5\Delta\hat{\mathbf{A}}\left[\lambda\Delta\hat{\mathbf{A}} - \nabla f_c(\hat{\mathbf{A}})\right]} = \frac{\text{actual improvement}}{\text{expected improvement}}$$

If $\kappa < \kappa_{\min}$, then $\lambda := k\lambda$, else if $\kappa > \kappa_{\max}$, then $\lambda := \lambda/k$
until $\kappa \geq 0$

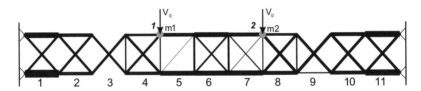

**Figure 6.10**    Optimally remodeled structure (asymmetric impact case)

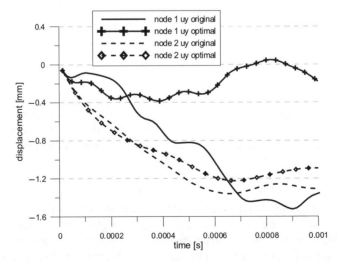

**Figure 6.11**    Original and optimum vertical displacements of impacted nodes (asymmetric impact case)

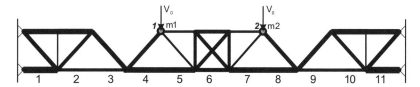

**Figure 6.12** Optimally remodeled 30-element structure (multiple-impact case)

### 6.3.1.3 Multiple-Impact Loads

It is important for further discussion to consider the case of multiple-impact loads. Assume that two impact scenarios are possible: the one just discussed ($m_1 = 0.1$ kg, $m_2 = 2$ kg) and its symmetric counterpart ($m_1 = 2$ kg, $m_2 = 0.1$ kg). The optimization is still based on the algorithm presented in Table 6.7. Since the two considered loads are mutually symmetric, each optimization step should be symmetrized by assuming

$$\Delta \hat{A}_i := \frac{\Delta \hat{A}_i + \Delta \hat{A}_{\bar{\imath}}}{2}$$

where the $\bar{\imath}$th element is located in the structure symmetrically to the $i$th element.

The structure optimally remodeled for the two considered impacts is presented in Figure 6.12, with the original and optimized displacements shown later in Figure 6.14 (left). In terms of the objective function the result is inferior to the previous one, since the optimum configuration has to account for multiple-impact loads. A general optimization algorithm should eliminate free nodes and, in certain cases, replace two collinear elements with one. Since this is not implemented in its present form, these modifications have been done manually. However, an automatic algorithm is also possible. The total number of structural elements is reduced from 54 in Figure 6.9 to 30 in Figure 6.12.

Notice that a similar but significantly simpler solution shown in Figure 6.13 can be obtained by removing two vertical elements between sections 1, 2 and 10, 11, and by replacing each of the two pairs of the remaining upper elements by one with the mean cross-sectional area. The objective function changes only insignificantly as the result (see Figure 6.14, right). This 26-element structure is discussed in the next subsection.

### 6.3.2 Structural Fuses as Active Elements

In the preceding subsection the considered structure has been remodeled to obtain its stiffest elastic substructure (see also Reference [33]). In the next step of the methodology of constructing adaptive structures, outlined at the beginning of this section, controllable elastoplastic

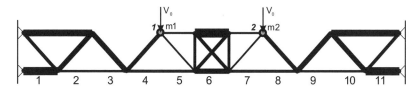

**Figure 6.13** Near-optimum 26-element structure (multiple-impact case)

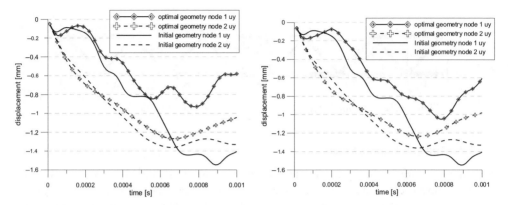

**Figure 6.14**  Original and optimized vertical displacements of impacted nodes (multiple-impact case): (left) 30-element optimum structure; (right) 26-element near-optimum structure

characteristics are attributed to a limited number of chosen elements [34]. These elements are called structural fuses, since they are meant to dissipate impact energy. In the considered optimized truss beam structure (Figure 6.13) it is natural to consider the two elements located in the upper layer of the two section pairs: 1, 2 and 10, 11 (see Figure 6.15). These two structural fuses located in the elements, which accumulate the largest portion of the strain energy in the elastic solution, are good candidates for location of the energy dissipation process. However, in the previously considered case of the single asymmetric impact load, better locations for structural fuses are sections 10, 11 (upper layer) and section 11 (lower layer).

The third step of the methodology considers optimum control strategies for the already allocated structural fuses. Provided a real-time impact identification system is available [30], an active control strategy can be pursued. The optimality is defined in terms of the energy dissipated by the fuses in the process of plastic yielding:

$$\Delta U = \sum_{t \leq T} \sigma_i(t) \Delta \beta_i^0(t) l_i A_i \mu_i \tag{33}$$

which should be maximized for each of the considered impact scenarios. Equation (33) is subject to the following constraints imposed on stresses and plastic distortions (maximum strokes of structural fuses):

$$|\sigma_i(t)| < \tilde{\sigma} \tag{34a}$$

$$|\beta_i^0(t)| < \tilde{\beta} \tag{34b}$$

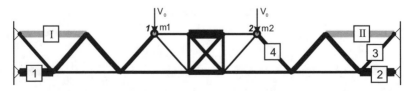

**Figure 6.15**  Location of structural fuses in the 26-element structure

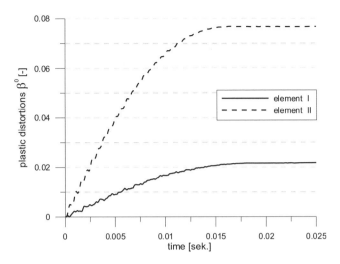

**Figure 6.16**   Development of plastic distortions in adaptive elements I and II (see Figure 6.15)

In Equation (33) the time integration interval $\langle 0, T \rangle$ should satisfy the condition $T_1 < T < T_2$, where $T_1$ and $T_2$ denote respectively the instant of the maximum load peak and the beginning of the elastic springback process. Too small $T$ can cause quick activation of the constraints of Equation (34b) and large residual vibrations after the impact process. On the contrary, too large $T$ can increase the decelerations in the first impact phase.

Assume that the only controllable parameters are the yield stresses $\sigma_i^*$ of the structural fuses. In the active control strategy, their optimum levels should be determined separately for each of the potentially expected impact loads. However, since the two considered impact scenarios are symmetrical, only one of them has been analysed. The optimum yield stress levels were found to be $\sigma_I^* = 2.8$ MPa and $\sigma_{II}^* = 3.6$ MPa. The corresponding developments of the plastic distortions in the structural fuses are plotted in Figure 6.16. The energy balances for the initially optimized elastic and the optimum adaptive structures are shown in Figure 6.17, while Figures 6.18 and 6.19 compare the corresponding stress developments in the four maximally loaded elements, which have been marked 1 to 4 in Figure 6.15. A significant reduction of

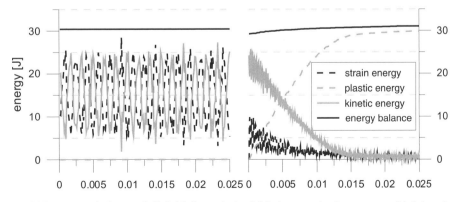

**Figure 6.17**   Energy balance: (left) initially optimized 26-element elastic structure; (right) optimum adaptive structure

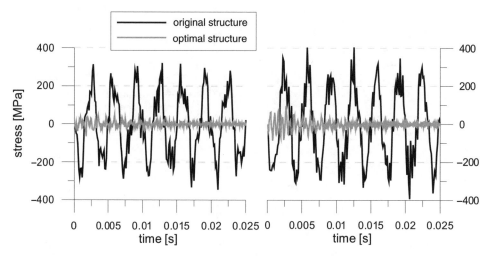

**Figure 6.18**  Development of stresses in maximally loaded elements of the optimum adaptive structure in Figure 6.15: (left) element 1; (right) element 2

stress levels can be observed. The results demonstrate the advantages of the proposed adaptive impact absorption scheme over the purely elastic response of the initially optimized passive structure. Smooth impact energy absorption in the first 15 ms could be achieved. The constraints of Equations (34), defined by $\bar{\beta} = 0.2$ and $\tilde{\sigma} = 200$ MPa, have remained inactive. Only 40 % of the plastic distortion limit $\bar{\beta}$ have been reached, see (Figure 6.16). Therefore, the structure can effectively dissipate impact loads of significantly higher energy.

### 6.3.3  Comments

A VDM-based technique for quick structural dynamic reanalysis has been developed and new algorithms for simulation of coupled structural modifications (mass, stiffness and plastic

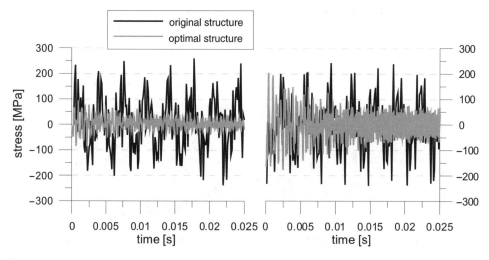

**Figure 6.19**  Development of stresses in maximally loaded elements of the optimum adaptive structure in Figure 6.15: (left) element 3; (right) element 4

zone) have been described, including the corresponding sensitivity analysis. Thereupon, the problem of optimal remodeling for structures exposed to several potential impact loads has been formulated and a methodology to solve it proposed. The assumed procedure selects, via a gradient-based optimization process, the stiffest structure with material concentrated in the overloaded elements and with several elements removed. The resulting structural geometry serves as a basis for the design of the adaptive structure with optimum impact load absorption capacity. The final result of this approach, illustrated in the numerical example of a truss beam structure, demonstrates promising effects of efficient impact energy absorption. The slight residual vibrations observed in Figure 6.17 could be further damped with real-time control of the yield stress levels [35]. The residual post-impact plastic-like distortions of the structural fuses can be released using the pre-stress accumulation release (PAR) vibrorelaxation technique described in Reference [36], which would allow nearly uninterrupted continuous use of the adaptive structure in the case of repetitive impact loads. This technique exploits a special opening and locking strategy applied to the structural fuses in the vibrating structure excited by a shaker. The real-time observability of dynamic loads is a necessary prerequisite for active adaptation strategies, which require the impact (e.g. the impacting mass and its velocity) to be identified in a few milliseconds. Current research [30] proves the feasibility of this challenging problem, at least for some applications. However, large strokes of structural fuses and the resulting geometrical nonlinearity should be avoided.

Another problem (not discussed in this chapter) deals with the distinction between the so-called *fast dynamics* and *slow dynamics*, which are determined by the proportion between the impact velocity and the impacting mass (see Figure 4.1). The same impact energy can be transmitted to the structure by a small mass impacting with a high velocity and by a heavy mass moving with a small velocity. The structural responses are qualitatively different in both cases. Moreover, the discussed optimum remodeling process would lead to different mass distributions. Therefore, when defining the set of potential impact loads, it is important to take into account not only the locations and energies but also the various impact dynamics.

## 6.4 Remodeling of Damped Elastic Structures

All numerical tools considered in the previous sections disregard damping, which is one of the phenomena that play a crucial role in real-world structures. There are many approaches to model damping in mechanical systems: energy methods [37, 38], wave approaches [39, 40], full-scale measurements [41], etc. All of them have advantages and disadvantages, some of them precisely describe the system behavior for particular classes of problems, whereas in other problems other methods excel. This section extends the VDM-based analysis tools to include modeling of both damping itself and of the interactions between the damping and structural parameters, and covers the following stages:

- statement of the damping model;
- statement of the general VDM-based formulation, which includes both the remodeling of damping parameters and the material redistribution in damped structures;
- reformulation of the general statement for three specific problems: damping remodeling, material redistribution and remodeling of material properties, with corresponding sensitivity analyses included to allow gradient-based structural optimization.

Since damping is a strongly frequency-dependent feature, the analysis is pursued in the frequency domain and hence leads to a quasi-static formulation. Only elastic structures are considered.

### 6.4.1 Damping Model

A generalized Rayleigh damping model is used in this section. The standard Rayleigh model decomposes the damping into two components related to environmental and material factors [42]. The damping matrix $\mathbf{C}$ is a linear combination of the mass matrix $\mathbf{M}$ and the stiffness matrix $\mathbf{K}$:

$$C_{NM} = \alpha M_{NM} + \lambda K_{NM}$$

The Rayleigh model is generalized here to allow independent modeling of the environmental damping in chosen degrees of freedom and independent modeling of the material damping in chosen elements. The stiffness matrix is, as in Equation (1), directly related to the diagonal matrix $\mathbf{S}$ of element stiffnesses $E_i A_i$.:

$$K_{NM} = G_{Ni}^{\mathrm{T}} l_i S_{ii} G_{iM} \tag{35}$$

where $\mathbf{G}$ is the geometric matrix. If $\alpha_N$ denotes the coefficient of the environmental damping in the $N$th degree of freedom and $\lambda_i$ denotes the coefficient of the material damping in the $i$th element, then the generalized Rayleigh damping matrix takes the following form:

$$C_{NM} = \alpha_N M_{NM} + G_{Ni}^{\mathrm{T}} \lambda_i l_i S_{ii} G_{iM} \tag{36}$$

### 6.4.2 General VDM Formulation

Assume that the structure is subjected to the harmonic excitation $f_N \mathrm{e}^{i\omega t}$, where $f_N$ is the complex amplitude in the $N$th degree of freedom [43–45]. The standard linear equation of motion of the original unmodified structure is

$$M_{NM} \left( u_M^{\mathrm{L}} \mathrm{e}^{i\omega t} \right)'' + C_{NM} \left( u_M^{\mathrm{L}} \mathrm{e}^{i\omega t} \right)' + K_{NM} u_M^{\mathrm{L}} \mathrm{e}^{i\omega t} = f_N \mathrm{e}^{i\omega t}$$

where $u_M^{\mathrm{L}}$ is the complex amplitude of the original structural response in the $M$th degree of freedom, and yields the following quasi-static formulation:

$$\left( -\omega^2 M_{NM} + i\omega C_{NM} + K_{NM} \right) u_M^{\mathrm{L}} = f_N \tag{37}$$

For the response $u_N$ of the *modified* structure, Equation (37), together with Equations (35) and (36), yields

$$-\omega^2 \hat{M}_{NM} u_M + i\omega \hat{\alpha}_N \hat{M}_{NM} u_M + i\omega G_{Ni}^{\mathrm{T}} \hat{\lambda}_i l_i \hat{S}_{ii} G_{iM} u_M + G_{Ni}^{\mathrm{T}} l_i \hat{S}_{ii} G_{iM} u_M = f_N \tag{38}$$

The virtual distortion method (VDM) models the modifications of mass, environmental damping, material damping and stiffness by four respective virtual distortions, $f_N^0$, $d_N^0$, $\phi_i^0$ and $\varepsilon_i^0$, which are the complex amplitudes of the corresponding oscillating force distortions related to nodes (for modifications of mass and environmental damping) and strain distortions related to elements (for modifications of material damping and stiffness). Therefore, the quasi-static formulation of the equation of motion of the *modeled* structure takes the form

$$\begin{aligned}
\left(-\omega^2 M_{NM} u_M - f_N^0\right) &+ \left(i\omega\alpha_N M_{NM} u_M - d_N^0\right) + \left[G_{Ni}^T l_i S_{ii} \left(i\omega\lambda_i G_{iM} u_M - \phi_i^0\right)\right] \\
&+ \left[G_{Ni}^T l_i S_{ii} \left(G_{iM} u_M - \varepsilon_i^0\right)\right] = f_N
\end{aligned} \tag{39}$$

Each distortion models exactly one type of modification and the responses and the member forces in the modified and the modeled structure have to be equal. Hence Equations (38) and (39) together yield the following system of four equations similar to Equation (45) in Chapter 2 and Equation (6):

$$\begin{aligned}
-\omega^2 \Delta M_{NM} u_M &= -f_N^0 \\
i\omega \left(\Delta\alpha M\right)_{NM} u_M &= -d_N^0 \\
i\omega \left(\Delta\lambda S\right)_{ii} \varepsilon_{\underline{i}} &= -S_{\underline{ii}} \phi_{\underline{i}}^0 \\
\Delta S_{\underline{ii}} \varepsilon_{\underline{i}} &= -S_{\underline{ii}} \varepsilon_{\underline{i}}^0
\end{aligned} \tag{40}$$

where, for brevity and similar to Equation (46) in Chapter 2

$$\Delta M_{NM} := \hat{M}_{NM} - M_{NM} = \left(\mu_i^A - 1\right) M_{NM}^i \tag{41a}$$

$$\left(\Delta\alpha M\right)_{NM} := \hat{\alpha}_{\underline{N}} \hat{M}_{NM} - \alpha_{\underline{N}} M_{NM} = \left(\hat{\alpha}_{\underline{N}} \mu_i^A - \alpha_{\underline{N}}\right) M_{NM}^i \tag{41b}$$

$$\left(\Delta\lambda S\right)_{\underline{ii}} := \hat{\lambda}_{\underline{i}} \hat{S}_{\underline{ii}} - \lambda_{\underline{i}} S_{\underline{ii}} = \left(\hat{\lambda}_{\underline{i}} \mu_{\underline{i}}^A - \lambda_{\underline{i}}\right) S_{\underline{ii}} \tag{41c}$$

$$\Delta S_{\underline{ii}} := \hat{S}_{\underline{ii}} - S_{\underline{ii}} = \left(\mu_{\underline{i}}^A - 1\right) S_{\underline{ii}} \tag{41d}$$

In Equations (41) the modification parameter $\mu_i^A$ of element cross-section have been used, which is the formulation for the problem of material redistribution. If the element's material properties (the densities $\rho_i$ and the stiffnesses $E_i$) are directly and independently modified, then instead of $\mu_i^A$ the two corresponding modification parameters have to be used: $\mu_i^\rho$ in Equations (41a) and (41b) and $\mu_i^E$ in Equations (41c) and (41d). The general system of Equation (43) given later should also be modified accordingly.

Equation (39), rewritten in the form

$$\begin{aligned}
\left[-\omega^2 M_{NM} + i\omega \left(\alpha_N M_{NM} + G_{Ni}^T \lambda_i l_i S_{ii} G_{iM}\right) + G_{Ni}^T l_i S_{ii} G_{iM}\right] u_M \\
= f_N + f_N^0 + d_N^0 + G_{Ni}^T l_i S_{ii} \phi_i^0 + G_{Ni}^T l_i S_{ii} \varepsilon_i^0
\end{aligned}$$

proves that the response $u_N$ and $\varepsilon_i$ of the *modeled* structure depends linearly on the virtual distortions and hence

$$\begin{aligned}
u_N &= u_N^L + B_{NM}^f f_M^0 + B_{NM}^f d_M^0 + B_{Ni}^\varepsilon \phi_i^0 + B_{Ni}^\varepsilon \varepsilon_i^0 \\
\varepsilon_i &= \varepsilon_i^L + D_{iM}^f f_M^0 + D_{iM}^f d_M^0 + D_{ij}^\varepsilon \phi_j^0 + D_{ij}^\varepsilon \varepsilon_j^0
\end{aligned} \tag{42}$$

where $B_{NM}^f$, $B_{Nj}^\varepsilon$, $D_{iM}^f$ and $D_{ij}^\varepsilon$ are quasi-static influence matrices, which describe in the frequency domain the response of the structure to the unit harmonic force in the $M$th degree of freedom ($B_{NM}^f$ and $D_{iM}^f$) and to the unit harmonic distortion of the $j$th element ($B_{Nj}^\varepsilon$ and $D_{ij}^\varepsilon$).

Equations (42) can be substituted in to Equations (40) to yield the following linear system:

$$\mathbf{F}^{\omega}\mathbf{x}^0 = \mathbf{b}^{\omega} \tag{43}$$

where the principal matrix $\mathbf{F}^{\omega}$ is given by

$$
\mathbf{F}^{\omega} =
\begin{bmatrix}
-\omega^2 \Delta M_{NM} B_{MK}^{\mathrm{f}} + \delta_{KN} & -\omega^2 \Delta M_{NM} B_{MK}^{\mathrm{f}} \\
i\omega(\Delta\alpha M)_{NM} B_{MK}^{\mathrm{f}} & i\omega(\Delta\alpha M)_{NM} B_{MK}^{\mathrm{f}} + \delta_{KN} \\
i\omega(\Delta\lambda S)_{\underline{i}\underline{i}} D_{\underline{i}K}^{\mathrm{f}} & i\omega(\Delta\lambda S)_{\underline{i}\underline{i}} D_{\underline{i}K}^{\mathrm{f}} \\
\Delta S_{\underline{i}\underline{i}} D_{\underline{i}K}^{\mathrm{f}} & \Delta S_{\underline{i}\underline{i}} D_{\underline{i}K}^{\mathrm{f}}
\end{bmatrix}
$$

$$
\begin{matrix}
-\omega^2 \Delta M_{NM} B_{Mk}^{\varepsilon} & -\omega^2 \Delta M_{NM} B_{Mk}^{\varepsilon} \\
i\omega(\Delta\alpha M)_{NM} B_{Mk}^{\varepsilon} & i\omega(\Delta\alpha M)_{NM} B_{Mk}^{\varepsilon} \\
i\omega(\Delta\lambda S)_{\underline{i}\underline{i}} D_{\underline{i}k}^{\varepsilon} + \delta_{ik} S_{\underline{i}\underline{i}} & i\omega(\Delta\lambda S)_{\underline{i}\underline{i}} D_{\underline{i}k}^{\varepsilon} \\
\Delta S_{\underline{i}\underline{i}} D_{\underline{i}k}^{\varepsilon} & \Delta S_{\underline{i}\underline{i}} D_{\underline{i}k}^{\varepsilon} + \delta_{ik} S_{\underline{i}\underline{i}}
\end{matrix}
$$

the vector $\mathbf{x}^0$ collects the virtual distortions

$$\mathbf{x}^0 = \begin{bmatrix} f_K^0 & d_K^0 & \phi_k^0 & \varepsilon_k^0 \end{bmatrix}^T$$

and the right-hand side vector

$$
\mathbf{b}^{\omega} =
\begin{bmatrix}
\omega^2 \Delta M_{NM} u_M^{\mathrm{L}} \\
-i\omega (\Delta\alpha M)_{NM} u_M^{\mathrm{L}} \\
-i\omega (\Delta\lambda S)_{\underline{i}\underline{i}} \varepsilon_{\underline{i}}^{\mathrm{L}} \\
-\Delta S_{\underline{i}\underline{i}} \varepsilon_{\underline{i}}^{\mathrm{L}}
\end{bmatrix}
$$

In practice, the dimensions of the linear system of Equation (43) (or rather its problem-specific versions, see the next subsection) are to be compared with the dimensions of Equation (38): if the structure is remodeled in a limited number of locations, the presented VDM-based approach is numerically less costly in comparison to the direct solution of the quasi-static problem by Equation (38), since then there are fewer unknown virtual distortions being sought than all the degrees of freedom. The same remark applies to the sensitivity analysis.

### 6.4.3 Specific Formulations and Sensitivity Analysis

Structural optimization is usually based on an objective function, which expresses the desired structural characteristics. Although other formulations are possible, the objective function $f$ considered here measures the discrepancy between the structural response $u_N$ and a desired response $u_N^{\mathrm{D}}$:

$$f := \left(u_N - u_N^{\mathrm{D}}\right)\left(\bar{u}_N - \bar{u}_N^{\mathrm{D}}\right) = \sum_N \left| u_N - u_N^{\mathrm{D}} \right|^2$$

and depends on the structural variables being optimized via the virtual distortions, which model the corresponding structural modifications. Notice that, although the virtual distortions are complex quantities, both the objective function and the structural variables (coefficients of damping, element cross-sections, material densities and Young's moduli) are real. Therefore, standard gradient-based and steepest-descent optimization techniques are directly applicable. The derivative of the objective function with respect to any structural variable $\hat{x}$ is expressed as

$$\frac{\partial f}{\partial \hat{x}} := 2\Re \left[ \frac{\partial u_N}{\partial \hat{x}} \left( \bar{u}_N - \bar{u}_N^{\mathrm{D}} \right) \right]$$

where

$$\frac{\partial u_N}{\partial \hat{x}} = B_{NM}^{\mathrm{f}} \frac{\partial f_M^0}{\partial \hat{x}} + B_{NM}^{\mathrm{f}} \frac{\partial d_M^0}{\partial \hat{x}} + B_{Ni}^{\varepsilon} \frac{\partial \phi_i^0}{\partial \hat{x}} + B_{Ni}^{\varepsilon} \frac{\partial \varepsilon_i^0}{\partial \hat{x}}$$

$$\frac{\partial \varepsilon_i}{\partial \hat{x}} = D_{NM}^{\mathrm{f}} \frac{\partial f_M^0}{\partial \hat{x}} + D_{NM}^{\mathrm{f}} \frac{\partial d_M^0}{\partial \hat{x}} + D_{Ni}^{\varepsilon} \frac{\partial \phi_i^0}{\partial \hat{x}} + D_{Ni}^{\varepsilon} \frac{\partial \varepsilon_i^0}{\partial \hat{x}}$$

(44)

where the second formula is necessary if the objective function takes into account the strains.

Therefore, the structural response and the derivatives of the response are necessary for optimization purposes, in order to compute the objective function and its gradient. In the following, the general approach of the previous subsection is reformulated for three specific practical remodeling problems; the sensitivity analyses of the corresponding responses are included.

### 6.4.3.1 Remodeling of Damping

It is assumed that the original structural configuration is preserved, cross-sections, masses and stiffnesses of elements are not subject to modifications and the only variables are the environmental and material damping coefficients of the modified structure, $\hat{\alpha}_L$ and $\hat{\lambda}_l$, respectively. Consequently, $\mu_i^{\mathrm{A}} = 1$, $\Delta M_{NM} = 0$, $\Delta S_{ii} = 0$ and thus $f_N^0 = 0$ and $\varepsilon_i^0 = 0$. The principal matrix $\mathbf{F}^{\omega}$ therefore takes the following simplified form:

$$\mathbf{F}_1^{\omega} = \begin{bmatrix} (\Delta\alpha M)_{NM} B_{MK}^{\mathrm{f}} + \delta_{KN} & (\Delta\alpha M)_{NM} B_{Mk}^{\varepsilon} \\ (\Delta\lambda S)_{\underline{ii}} D_{\underline{i}K}^{\mathrm{f}} & (\Delta\lambda S)_{\underline{ii}} D_{\underline{i}k}^{\varepsilon} + \delta_{ik} S_{\underline{ii}} \end{bmatrix}$$

(45)

and the general system of Equation (43) becomes

$$\mathbf{F}_1^{\omega} \begin{bmatrix} d_K^0 \\ \phi_k^0 \end{bmatrix} = \begin{bmatrix} -\Delta(\alpha M)_{NM} u_M^{\mathrm{L}} \\ -\Delta(\lambda S)_{\underline{ii}} \varepsilon_{\underline{i}}^{\mathrm{L}} \end{bmatrix}$$

This simplified system can be used to compute the nonvanishing virtual distortions and, by Equations (42), the response of the modified structure.

The respective sensitivity analysis is simplified by the fact that the differentiation of Equations (41) with respect to the damping variables yields only two nonvanishing results:

$$\frac{\partial \Delta M_{NM}}{\partial \hat{\alpha}_L} = \frac{\partial (\Delta\lambda S)_{ii}}{\partial \hat{\alpha}_L} = \frac{\partial \Delta S_{ii}}{\partial \hat{\alpha}_L} = 0$$

$$\frac{\partial \Delta M_{NM}}{\partial \hat{\lambda}_l} = \frac{\partial (\Delta\alpha M)_{NM}}{\partial \hat{\lambda}_l} = \frac{\partial \Delta S_{ii}}{\partial \hat{\lambda}_l} = 0$$

$$\frac{\partial (\Delta\alpha M)_{NM}}{\partial \hat{\alpha}_L} = \delta_{L\underline{N}} M_{\underline{N}M}$$

$$\frac{\partial (\Delta\lambda S)_{ii}}{\partial \hat{\lambda}_l} = \delta_{l\underline{i}} S_{\underline{ii}}$$

which, substituted into the simplified system of Equation (45) differentiated with respect to both variables, yield the two following linear systems:

$$\mathbf{F}_1^\omega \begin{bmatrix} \dfrac{\partial d_K^0}{\partial \hat{\alpha}_L} \\[2mm] \dfrac{\partial \phi_k^0}{\partial \hat{\alpha}_L} \end{bmatrix} = \begin{bmatrix} -\delta_{LN} M_{NM} u_M \\ 0 \end{bmatrix}, \qquad \mathbf{F}_1^\omega \begin{bmatrix} \dfrac{\partial d_K^0}{\partial \hat{\lambda}_l} \\[2mm] \dfrac{\partial \phi_k^0}{\partial \hat{\lambda}_l} \end{bmatrix} = \begin{bmatrix} 0 \\ -\delta_{li} S_{ii} \varepsilon_i \end{bmatrix}$$

These systems can be used to obtain the corresponding derivatives of the nonvanishing virtual distortions and, by Equations (44), of the response.

### 6.4.3.2 Redistribution of Material

In this case, the problem is to redistribute the material between the elements while preserving at the same time the damping characteristics of the structure. The damping coefficients are hence assumed to be constant, $\hat{\alpha}_N = \alpha_N$ and $\hat{\lambda}_i = \lambda_i$, and the only variables are the cross-sectional areas $\hat{A}_l$ of the elements of the modified structure. Since the element masses and stiffnesses are modified, the damping-related virtual distortions also change, even though the damping coefficients remain constant. Consequently, all four virtual distortions are nonvanishing and the general form of the system of Equation (43) is retained, besides the minor computational simplifications related to the fact that $(\Delta\alpha M)_{NM} = \alpha_N \Delta M_{NM}$ and $(\Delta\lambda S)_{ii} = \lambda_i \Delta S_{ii}$. The response of the modified structure can be computed by solving the general system and by substituting the resulting virtual distortions into Equations (42).

As the first step of the corresponding sensitivity analysis, Equations (41) are differentiated with respect to the modified cross-sectional areas $\hat{A}_l$ of elements,

$$\begin{aligned} \frac{\partial \Delta M_{NM}}{\partial \hat{A}_l} &= \frac{M_{NM}^l}{A_l}, & \frac{\partial (\Delta\lambda S)_{ii}}{\partial \hat{A}_l} &= \delta_{il} \lambda_i E_i \\[3mm] \frac{\partial (\Delta\alpha M)_{NM}}{\partial \hat{A}_l} &= \alpha_N \frac{M_{NM}^l}{A_l}, & \frac{\partial \Delta S_{ii}}{\partial \hat{A}_l} &= \delta_{il} E_i \end{aligned} \tag{46}$$

Differentiation of the original system of Equation (43) yields the following linear system:

$$\mathbf{F}^\omega \begin{bmatrix} \dfrac{\partial f_K^0}{\partial \hat{A}_l} \\[2mm] \dfrac{\partial d_K^0}{\partial \hat{A}_l} \\[2mm] \dfrac{\partial \phi_k^0}{\partial \hat{A}_l} \\[2mm] \dfrac{\partial \varepsilon_k^0}{\partial \hat{A}_l} \end{bmatrix} = \begin{bmatrix} \omega^2 \dfrac{\partial \Delta M_{NM}}{\partial \hat{A}_l} u_M \\[2mm] -i\omega \dfrac{\partial (\Delta\alpha M)_{NM}}{\partial \hat{A}_l} u_M \\[2mm] -i\omega \dfrac{\partial (\Delta\lambda S)_{ii}}{\partial \hat{A}_l} \varepsilon_i \\[2mm] -\dfrac{\partial \Delta S_{ii}}{\partial \hat{A}_l} \varepsilon_i \end{bmatrix}$$

After substitution of Equations (46), this system can be used to obtain the derivatives of all virtual distortions and, by Equations (44), the derivatives of the response.

### 6.4.3.3 Remodeling of Material Properties

Modification of the element cross-section influences both its stiffness and mass at the same time. However, they are also related to the properties of the material, which is often modified directly by exchanging structural elements while retaining the same element cross-sections. In this case, the density and Young's modulus modification parameters, $\mu_i^\rho$ and $\mu_i^E$, have to be used in Equations (41) instead of $\mu_i^A$. As before, the damping coefficients are assumed to remain constant, $\hat{\alpha}_N = \alpha_N$ and $\hat{\lambda}_i = \lambda_i$; the variables are the densities $\hat{\rho}_l$ and Young's moduli $\hat{E}_i$ of the elements of the modified structure. As in the previously considered case of material redistribution, all four virtual distortions are nonvanishing and the general form of the system of Equation (43) is retained, besides the same minor computational simplifications, which are related to the fact that $(\Delta \alpha M)_{NM} = \alpha_N \Delta M_{NM}$ and $(\Delta \lambda S)_{ii} = \lambda_i \Delta S_{ii}$. The response of the modified structure can be computed by solving the general system and by substituting the resulting virtual distortions into Equations (42).

Differentiations of Equations (41) with respect to the densities and Young's moduli of the elements of the modified structure yield

$$
\frac{\partial \Delta M_{NM}}{\partial \hat{\rho}_l} = \frac{M_{NM}^l}{\hat{\rho}_l}, \qquad \frac{\partial \Delta M_{NM}}{\partial \hat{E}_l} = \frac{\partial (\Delta \alpha M)_{NM}}{\partial \hat{E}_l} = 0
$$

$$
\frac{\partial (\Delta \alpha M)_{NM}}{\partial \hat{\rho}_l} = \alpha_N \frac{M_{NM}^l}{\hat{\rho}_l} \qquad \frac{\partial (\Delta \lambda S)_{ii}}{\partial \hat{E}_l} = \delta_{il} \lambda_i A_i \tag{47}
$$

$$
\frac{\partial (\Delta \lambda S)_{ii}}{\partial \hat{\rho}_l} = \frac{\partial \Delta S_{ii}}{\partial \hat{\rho}_l} = 0, \qquad \frac{\partial \Delta S_{ii}}{\partial \hat{E}_l} = \delta_{il} A_i
$$

Differentiations of the original system of Equation (43) yield the two following linear systems:

$$
\mathbf{F}^\omega \begin{bmatrix} \dfrac{\partial f_K^0}{\partial \hat{\rho}_l} \\[2mm] \dfrac{\partial d_K^0}{\partial \hat{\rho}_l} \\[2mm] \dfrac{\partial \phi_k^0}{\partial \hat{\rho}_l} \\[2mm] \dfrac{\partial \varepsilon_k^0}{\partial \hat{\rho}_l} \end{bmatrix} = \begin{bmatrix} \omega^2 \dfrac{\partial \Delta M_{NM}}{\partial \hat{\rho}_l} u_M \\[2mm] -i\omega \dfrac{\partial (\Delta \alpha M)_{NM}}{\partial \hat{\rho}_l} u_M \\[2mm] 0 \\[2mm] 0 \end{bmatrix}, \qquad \mathbf{F}^\omega \begin{bmatrix} \dfrac{\partial f_K^0}{\partial \hat{E}_l} \\[2mm] \dfrac{\partial d_K^0}{\partial \hat{E}_l} \\[2mm] \dfrac{\partial \phi_k^0}{\partial \hat{E}_l} \\[2mm] \dfrac{\partial \varepsilon_k^0}{\partial \hat{E}_l} \end{bmatrix} = \begin{bmatrix} 0 \\[2mm] 0 \\[2mm] -i\omega \dfrac{\partial (\Delta \lambda S)_{ii}}{\partial \hat{E}_l} \varepsilon_i \\[2mm] -\dfrac{\partial \Delta S_{ii}}{\partial \hat{E}_l} \varepsilon_i \end{bmatrix}
$$

which, after substitution of Equations (47), can be used to obtain the derivatives of the virtual distortions and, by Equations (44), yield the derivatives of the response.

## References

1. N. Kikuchi and M. P. Bendsøe, Generating optimal topologies in structural design using homogenization method, *Computer Methods in Applied Mechanics and Engineering*, **71**(2), November 1988, 197–224.
2. A. R. Diaz and M. P. Bendsøe, Shape optimization of multipurpose structures by homogenization method, *Structural Optimization*, 4, 1992, 17–22.

3. M. P. Bendsøe and J. E. Taylor, An interpretation for min–max structural design problems including a method for relaxing constraints, *International Journal of Solids and Structures*, **20**(4), 1984, 301–314.

4. V. Tvergaard, On the optimum shape of a fillet in a flat bar with restrictions, in *procedings of IUTAM Symposium on Optimization in Structural Design (Warsaw 1973)*, Springer-Verlag, Berlin, 1975, pp. 181–195.

5. R. Steinhauser and G. Kreisselmeier, Systematic control design by optimizing a vector performance index, in *IFAC Symposium on Computer Aided Design of Control Systems* (ed. M. A. Cuenod), Zurich, Switzerland, 1979, Pergamon Press, Oxford, 1979, pp. 113–117.

6. R. A. Scott, R. R. Mayer and N. Kikuchi, Application of topological optimization techniques to structural crashworthiness, *International Journal for Numerical Methods in Engineering*, **39**(8), 1996, 1383–1403.

7. Z. Marzec and J. Holnicki-Szulc, Adaptive barriers with maximal impact energy absorption, in *Proceedings of the 3rd World Congress of Structural and Multidisciplinary Optimization*, Buffalo, May 1999.

8. J. Holnicki-Szulc and L. Knap, Optimal design of adaptive structures for the best crashworthiness, in *Proceedings of the 3rd World Congress of Structural and Multidisciplinary Optimization*, Buffalo, May 1999.

9. A. R. Diaz and C. A. Soto, Lattice models for crash resistant design and optimization, in *Proceedings of the 3rd World Congress of Structural and Multidisciplinary Optimization*, Buffalo, May 1999.

10. N. Kikuchi, K. Yuge and N. Iwai, Topology optimization algorithm for plates and shells subjected to plastic deformations, in *Proceedings of ASME Design Engineering Technical Conference*, Atlanta, September 1998.

11. K. Maute, S. Schwartz and E. Ramm, Adaptive topology optimization for elastoplastic structures, *Structural Optimization*, **15**(2), April 1998, 81–91.

12. C. A. Soto, Application of structural topology optimization in the automotive industry: past, present and future, in *Proceedings of the 5th World Congress of Computational Mechanics*, Vienna, July 2002.

13. M. M. Kamal, Analysis and simulation of vehicle-to-barrier impact, Technical Report 700414, SAE, 1970.

14. M. M. Kamal and J. A. Wolf Jr, *Modern Automotive Structural Analysis*, Van Nostrand Reinhold Company, New York, February 1982.

15. J. T. Wang, J. A. Bennet and R. V. Lust, Optimal design strategies in crashworthiness and occupant protection, in *Crashworthiness and Occupant Protection in Transport Systems*, AMD Vol. 126, ASME, New York, 1991.

16. J. O. Song and C. M. Ni, Computer-aided design analysis methods for vehicle structural crashworthiness, in *Proceedings of Symposium on Vehicle Crashworthiness Including Impact Biomechanics*, AMD Vol. 79, ASME, New York, 1986, pp. 125–139.

17. M. A. El-Bkaily, J. Y. Bakkar, H. F. Mahmood and D. G. Wheatley, On the front end design of automotive vehicle crashworthiness, in *Crashworthiness and Occupant Protection in Transport Systems*, AMD Vol. 169, ASME, New York, 1993.

18. A. R. Mijar, J. S. Arora and C. H. Kim, Simplified models for automotive crash simulation and design optimization, in *Proceedings of the 3rd World Congress of Structural and Multidisciplinary Optimization*, Buffalo, May 1999.

19. C. H. Kim, A. R. Mijar and J. S. Arora, Development of simplified models for design and optimization of automotive structures for crashworthiness, *Structural and Multidisciplinary Optimization*, **22**(4), November 2001, 307–321.

20. C. B. W. Pedersen, Topology optimization of 2D-frame structures with path-dependent response, *International Journal for Numerical Methods in Engineering*, **57**(10), 2003, 1471–1501.

21. C. B. W. Pedersen, Topology optimization for crashworthiness of frame structures, *International Journal of Crashworthiness*, **8**(1), January 2003, 29–39.

22. C. B. W. Pedersen, *On Topology Design of Frame Structures for Crashworthiness*, PhD thesis, Technical University of Denmark, July 2002.
23. C. L. Ignatovich and A. R. Diaz, Lattices, tuning, and space mapping in approximations of structures under impact loads, in *3rd ISSMO/AIAA Internet Conference on Approximations in Optimization*, (eds R. Haftka and F. von Keulen), October 2002.
24. C. L. Ignatovich, A. R. Diaz and C. A. Soto, On improving the accuracy of lattice models in crashworthiness analysis, in *Proceedings of ASME DETC'2002 Design Engineering Technical Conference*, Baltimore, September 2002.
25. J. Holnicki-Szulc and J.T. Gierliński, *Structural Analysis, Design and Control by the Virtual Distortion Method*, John Wiley & Sons, Ltd, Chichester, 1995.
26. P. Kolakowski, T. G. Zieliński and J. Holnicki-Szulc, Damage identification by the dynamic virtual distortion method, *Journal of Intelligent Material Systems and Structures*, 15(6), 2004, 479–493.
27. P. Kolakowski, L. E. Mujica and J. Vehi, Two approaches to structural damage identification: model updating vs. soft computing, *Journal of Intelligent Material Systems and Structures*, 17(1), 2006, 63–79.
28. A. Swiercz, P. Kolakowski and J. Holnicki-Szulc, Structural damage identification using low frequency non-resonance harmonic excitation, *Key Engineering Materials*, 347, 2007, 427–432.
29. Ł. Jankowski, M. Wiklo and J. Holnicki-Szulc, Robust post-accident reconstruction of loading forces, *Key Engineering Materials*, 347, 2007, 659–664.
30. K. Sekula and J. Holnicki-Szulc, On-line impact load identification, *International Journal of Impact Engineering*, January 2008 (submitted).
31. J. Nocedal and S. J. Wright, *Numerical Optimization*, Springer Series in Operations Research, Springer, New York, 1999.
32. Ł. Jankowski, Programming, Numerics and Optimisation, Smart-Tech Expert Course, Autumn 2006, http://smart.ippt.gov.pl.
33. M. Wiklo and J. Holnicki-Szulc, Optimal design of adaptive structures, Part I. Remodeling for impact reception, *Structural and Multidisciplinary Optimization*, 2008 (in press).
34. M. Wiklo and J. Holnicki-Szulc, Optimal design of adaptive structures, Part II. Adaptation to impact loads, *Structural and Multidisciplinary Optimization*, 2008 (in press).
35. G. Mikulowski and J. Holnicki-Szulc, Adaptive landing gear concept – feedback control validation, *Smart Materials and Structures*, 16(6), 2007, 2146–2158.
36. J. Holnicki-Szulc and P. Pawlowski, Adaptive structures under extreme loads – impact detection, self-adaptation, self-repairing, in *Proceedings of the 3rd European Conference on Structural Control, 3ECSC*, Vienna, July 2004.
37. R. H. Lyon and R. G. DeJong, *Theory and Application of Statistical Energy Analysis*, Butterworth-Heinemann, Oxford, 1995.
38. A. J. Keane and W. G. Price, *Statistical Energy Analysis – An Overview, with Applications in Structural Dynamics*, Cambridge University Press, Cambridge, 1994.
39. J. G. McDaniel, P. Dupont and L. Salvino, A wave approach to estimating frequency-dependent damping under transient loading, *Journal of Sound and Vibration*, 231(2), 2000, 433–449.
40. J. Esteban and C. A. Rogers, Wave localization due to material damping, *Computer Methods in Applied Mechanics and Engineering*, 177, 1999, 93–107.
41. M. J. Glanville, K. C. S. Kwok and R. O. Denoon, Full-scale damping measurements of structures in Australia, *Journal of Wind Engineering and Industrial Aerodynamics*, 59, 1996, 349–364.
42. J. L. Humar, *Dynamics of Structures*, Prentice-Hall, Englewood Cliffs, New Jersey, 1990.
43. A. D. Nashif, D. I. G. Jones and J. P. Henderson, *Vibration Damping*, John Wiley & Sons, Ltd, Chichester, 1985.
44. R. E. D. Bishop and D. C. Jonson, *The Mechanics of Vibration*, Cambridge University Press, Cambridge, 1979.
45. S. S. Rao, *Mechanical Vibrations*, Addison-Wesley Publishing Company, Reading, Massachusetts, 1986.

# 7

# Adaptive Damping of Vibration by the Prestress Accumulation/Release Strategy

Arkadiusz Mróz, Anita Orłowska and Jan Holnicki-Szulc

## 7.1 Introduction

The majority of the recently discussed active damping of vibration techniques in engineering structures are based on magnetorheological fluids (semi-active and applicable to real structures; e.g. see Reference [1]) and piezoelectric devices (fully active, but applicable mostly to lab-scale demonstrators, e.g. see Reference [2]).

The prestress accumulation/release (PAR) is a method to convert the strain energy of a vibrating system into kinetic energy, which is then released from the system by means of a dissipative device. The process is fully semi-active; it does not require adding any substantial amount of energy into the system. As a consequence, it creates the possibility of applications to real structures. The semi-active methods are popular because of their high efficiency and relatively low cost, compared to passive and active damping of vibrations [3,4]. Some interesting investigations of semi-active techniques have been published in References [5] and [6].

The first formulation of the PAR concept for a truss–beam system and double-layered cantilever beam was presented in References [7] and [8]. Based on the use of an on–off joint connections control concept for energy dissipation in a flexible truss-beam structure, a similar formulation was presented in Reference [9], whereas Reference [10] investigates the vibration suppression in a mass–two-spring system due to controlled detaching and reattaching of a spring.

The dissipating process discussed in the present chapter consists of two phases. In the first phase, some kinematic constraints imposed on the system are released at the instant when the maximum strain energy can be converted to kinetic energy. It is usually manifested by local, high-frequency vibrations. In the second phase, kinematic constraints are reimposed, which leads to conversion of a part of the kinetic energy into another, nonmechanical form, e.g. heating-up of the actuator device.

*Smart Technologies for Safety Engineering*   Edited by J. Holnicki-Szulc
© 2008 John Wiley & Sons, Ltd

First, the proposed approach is described theoretically on a simple spring–mass system in order to demonstrate the idea of response mitigation and to show the energy balance of the system. Second, the numerical studies are presented for a layered beam simulating a pedestrian bridge, where the control is based on disconnecting (for a very short instant of time) and then sticking back two layers (delamination effect).

Finally, the experimental results are presented. A laboratory-scale set-up was built to verify the effectiveness of the PAR strategy on a cantilever beam demonstrator. A controllable de-lamination effect was obtained by means of piezoelectric actuators. The control was carried out as a closed-loop feedback system.

## 7.2 Mass–Spring System

### 7.2.1 The Concept

A simple 'mass–two-spring' system is considered, as shown in Figure 7.1(a). One of the springs is active in the sense that it can be detached and then reattached to the mass anywhere along the spring length. During the free vibration of the system, the active spring can be detached, in particular at the point of maximum displacement of the mass, and reattached as it comes to its free end (cf. Figure 7.1(b)). In the following phase of vibration, a force that opposes further movement is introduced, proportional to the displacement of the active spring from its new equilibrium position. Thus, a new equilibrium of the whole system is established (dotted line in Figure 7.1(c)). Then the active spring can be detached and reattached again, resulting in returning to the initial configuration. During the following vibration, the system behaves the same as before introducing the control, but a considerable part of the total vibration amplitude vanishes.

### 7.2.2 Analytical Solution

#### 7.2.2.1 Equation of Motion

In this section the system shown in Figure 7.1 is analysed in detail. If natural damping is not considered and no force excitation is used, then the motion in the first phase of vibration is governed by the equation:

$$m\,\ddot{x}(t) + (k_1 + k_2)\,x(t) = 0 \tag{1}$$

where $m$ is the moving mass and $k_i = E_i A_i / L_i$ is the stiffness of a spring with cross-section $A_i$, Young's modulus $E_i$ and length $L_i$. The solution, under the given initial conditions

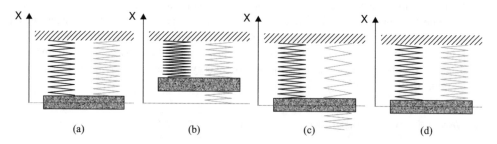

(a)                    (b)                    (c)                    (d)

**Figure 7.1**   Mass–two-spring system

$x(t = 0) = -\epsilon$ and $\dot{x}(t = 0) = 0$, takes the form:

$$x(t) = -\epsilon \cos\left(\sqrt{\frac{k_1 + k_2}{m}}\, t\right) \tag{2}$$

At the time instant of maximum displacement, $t = t^1$, the active spring is detached and reattached as it comes to its equilibrium position (cf. Figure 7.1(b)). At this point it is assumed that the inertia of springs is not taken into account. Now the equation of motion is given by

$$m\,\ddot{x}(t) + k_1\, x(t) + k_2'\, x_2(t) = 0 \tag{3}$$

where $k_2' = E_2 A_2/L_2'$ with new active spring length $L_2' = L_2 - \epsilon$ and $x_2$ is the active spring displacement in the second phase: $x_2(t) = x(t) - \epsilon$. The system has still a single degree of freedom, namely the displacement of the mass $m$. Equation (3) can be rewritten as

$$m\,\ddot{x}(t) + (k_1 + k_2')\, x(t) = k_2'\, \epsilon \tag{4}$$

It can be seen that in the second phase of the process, the governing equation is nonhomogeneous with a term $k_2'\,\epsilon$, which can be viewed as an additional constant force applied to the system. Now the solution takes the form:

$$x(t) = C_1 \cos\left(\sqrt{\frac{k_1 + k_2'}{m}}\, t\right) + C_2 \sin\left(\sqrt{\frac{k_1 + k_2'}{m}}\, t\right) + \frac{k_2'\,\epsilon}{k_1 + k_2'} \tag{5}$$

with constants $C_1$ and $C_2$ calculated from the initial conditions: $x(t = t^1) = \epsilon$ and $\dot{x}(t = t^1) = 0$. This operation can be repeated several times in order to enhance the amplitude mitigation effect. If the desired effectiveness is reached, then the operation of 'detaching/reattaching' can be repeated again near the initial equilibrium position, in order for the active spring to return to its initial length.

### 7.2.2.2 Energy Balance

The potential energy of a spring is equal to the work done by the elastic force along the displacement direction:

$$E_{\text{pot}} = \int_0^{x_k} k_i\, x\, \mathrm{d}x = \tfrac{1}{2}\, k_i\, x_k^2 \tag{6}$$

The potential energy of the system at the instant before activation of the control $t = t^1 - \mathrm{d}t$ is

$$E_{\text{pot}} = \tfrac{1}{2} k_1\, \epsilon^2 + \tfrac{1}{2} k_2\, \epsilon^2 \tag{7}$$

The potential energy of the system at the instant after activation of control $t = t^1 + \mathrm{d}t$ is

$$E_{\text{pot}} = \tfrac{1}{2} k_1\, \epsilon^2 + \tfrac{1}{2} k_2'\, (\epsilon - \epsilon)^2 = \tfrac{1}{2} k_1\, \epsilon^2 \tag{8}$$

The control is activated at the point of maximum displacement, where $\dot{x} = 0$, thus at this point the kinetic energy vanishes. If the inertia of springs was not neglected then some energy would

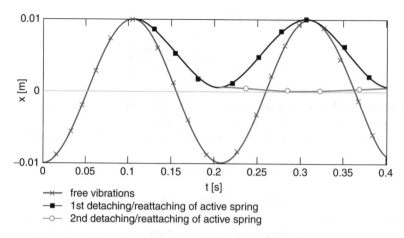

— free vibrations
— 1st detaching/reattaching of active spring
— 2nd detaching/reattaching of active spring

**Figure 7.2**  Resulting mass displacement; the control is triggered twice

be transferred to this part of the active spring, which can vibrate freely after reattachment. This situation will be discussed in further sections.

### 7.2.2.3 Numerical Example

Figure 7.2 depicts the resulting displacement of the mass if the following data were used for calculations (both springs equal):

$$m = 20\,\text{kg}, \quad L = 0.1\,\text{m}, \quad E = 6 \times 10^{10}\,\text{Pa}, \quad A = 1.54 \times 10^{-8}\,\text{m}^2, \quad \epsilon = 0.01\,\text{m}$$

After performing detaching/reattaching of the active spring twice, vibrations of the mass have ceased almost completely. A slight change in frequency of the controlled response, as compared with the reference case, is due to a small change in the spring stiffness, which affects the frequency and is caused by a change in the spring length. It can also be observed that after activating the control, the system oscillates about the new equilibrium position (cf. Figure 7.2).

Activating control takes place at the maximum value of the accumulated strain energy and results in an instant decrease in the strain energy (cf. Figure 7.3 and Equations (7) and (8)). At

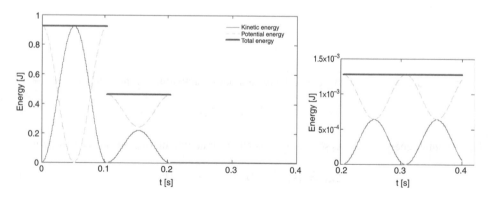

**Figure 7.3**  Energy balance

the end of the process, almost all the energy is released from the system. In a more practical approach, part of the energy would be transferred into higher frequency vibrations of the detached part of the active spring and another part would be dissipated in the process, which is here idealized as imposing some initial values. The last phase of the process is shown in Figure 7.3(b).

### 7.2.3 Case with Inertia of the Active Spring Considered

#### 7.2.3.1 Introduction

As stated before, detaching the active spring results in converting the accumulated strain energy into kinetic energy, which can be dissipated from the system during reattachment of the spring. This whole process was idealized in the previous section by imposing proper initial conditions, which resulted in an instant decrease in energy of the system. In practice, part of the released strain energy is dissipated by a device that reattaches the spring and the remaining part introduces higher frequency vibrations, which can be easily suppressed by natural damping of the system.

In the present analysis, the control device is idealized by imposing/releasing local constraints between the geometrical point of mass $m$ and any point along the active spring. The mass of active spring is concentrated at its full length and in the middle (cf. Figure 7.4). Slight natural damping is also introduced into the equations. The Rayleigh damping coefficients introduce little damping (1 or 0.2 %) around the first natural frequency and relatively much higher damping of higher frequencies. All remaining parameters do not change. (Simulations were performed using the Abaqus/Standard code.)

#### 7.2.3.2 Results

Displacements of mass $m$ and the tip of the spring are depicted in Figure 7.5. Detaching an active spring and reattaching it at some point along its length introduces higher frequency vibrations of the spring loose end. Releasing/reimposing the constraint again results in returning the system to the initial configuration, but with the amplitude of vibrations decreased.

First, the spring is detached at the point of maximum displacement, i.e. where there is the maximum stress accumulated in the spring. Then the spring is reattached when its end passes the equilibrium position, which means that the length of the spring is decreased by about 10 mm. This corresponds to Figure 7.4(b). If the control is stopped there, the system would oscillate about a new equilibrium position. The time instant for the next spring detachment has to be chosen properly, so that it can be reattached as the mass $m$ is as close to the initial equilibrium position as possible. The whole procedure can be repeated several times, as needed. If, for instance, the active spring stiffness is considerably smaller compared to the passive one, then the desired mitigation effect would have to be obtained in more steps. In the analysed

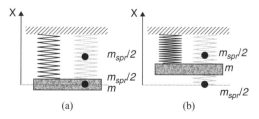

(a)                                      (b)

**Figure 7.4**  Dynamic model with inertia of the active spring included

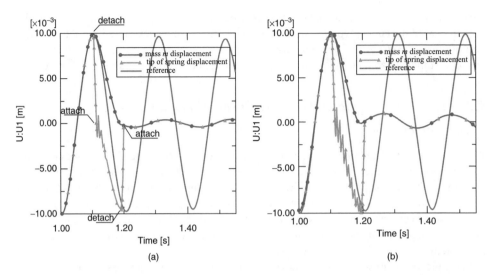

**Figure 7.5**    Response of the system: (a) 1 % of critical damping, (b) 0.2 % of critical damping

example, after the described sequence of activating the control, the amplitude of the mass *m* displacement was decreased by 96 % and 92 % for 1 % and 0.2 % of critical damping, respectively. The procedure is very sensitive to the time instant of detaching the spring.

Reimposing the constraints causes higher frequency vibrations of the mass located in the middle of the spring. These vibrations, however, are effectively damped out by the natural damping of the system. The typical behavior of the middle mass is shown in Figure 7.6.

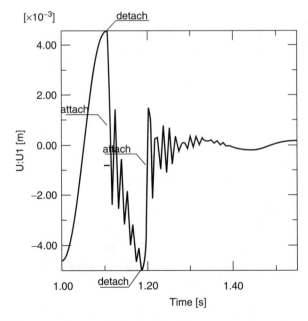

**Figure 7.6**    Middle mass response with indicated points corresponding to the activating control

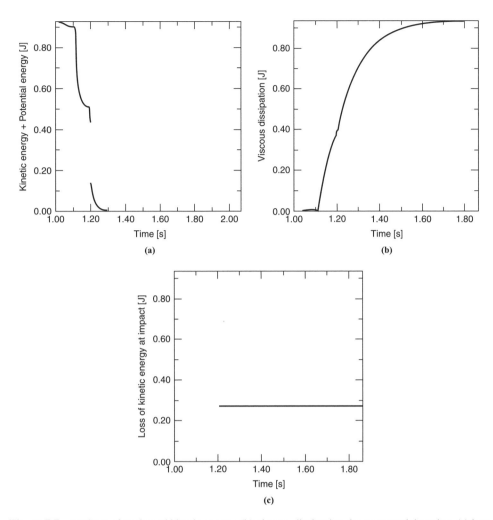

**Figure 7.7**   (a) Sum of strain and kinetic energy, (b) viscous dissipation due to natural damping, (c) loss of kinetic energy while imposing constraints

Steep exponential decline in the sum of a potential and kinetic energy graph is due to viscous dissipation, which increases with an increase in the vibration velocity. Vibration velocity, in turn, is at its highest when the local, higher frequency vibrations are introduced due to imposing/releasing the constraints. The viscous dissipation is due to natural damping of the system.

Discontinuity of the graph is caused by the loss of kinetic energy at the instant when constraints are reimposed. The size of this gap indicates the maximum amount of energy that can be dissipated by the active device. All nonzero forms of energy can be seen in Figure 7.7.

## 7.3  Delamination of a Layered Beam

### 7.3.1  PAR Strategy for Layered Beams

The strategy of releasing the accumulated strain energy in order to dissipate it can in theory be effectively used for various types of structures. If a layered beam is considered as shown

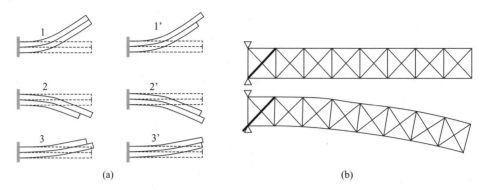

(a)                                                              (b)

**Figure 7.8**   PAR strategy for a cantilever beam

in Figure 7.8, the idea of adaptation would be as follows. First, at the point of maximum deflection two layers are disconnected, resulting in almost instant dislocation of layers (1' in Figure 7.8(a)). This dislocation can then be frozen if the layers are reconnected again. This yields an introduction of the elastic force that opposes further vibration of the cantilever (2 in Figure 7.8(a)). Then, near the equilibrium position the layers are disconnected/reconnected again in order to return to the initial configuration. The whole sequence can be repeated until the desired effect is obtained.

A similar effect of response mitigation can be obtained if a truss structure is considered with a detachable element (cf. Figure 7.8(b)). Applying the same methodology for control, the axial strain accumulated in the active element can be released as the element is disconnected at one of its ends. It is worth mentioning that in both cases only one active member is required.

## 7.3.2 Numerical Example of a Simply Supported Beam

### 7.3.2.1 Numerical Model

A simply supported, two-layered beam with the span of 15.6 m is considered. The bending stiffness of each layer is $EI = 2.218 \times 10^6 \, \text{N m}^2$. Material damping of 1% of critical damping is assumed around the first natural frequency. Layers are permanently connected together at the left support. It is assumed that there is a device at the right support, capable of instantly disconnecting or sticking the layers. Along the beam length the distance between the layers remains the same, whereas the frictionless, relative movement of layers is possible in the direction parallel to the beam axis. The considered beam model is depicted in Figure 7.9.

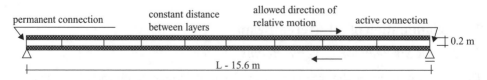

**Figure 7.9**   Assumed model of the layered beam

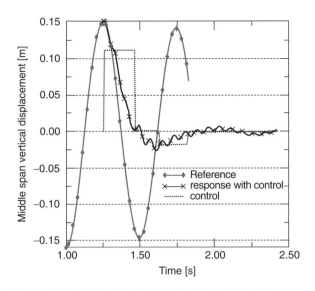

**Figure 7.10**  Vertical displacement of the middle of the beam

### 7.3.2.2 Results

In the first step, an initial displacement of 16 cm was applied to the model. In the following steps, the dynamic analysis procedure was used to calculate the free vibrations of the system as in the reference case. Then the calculations were repeated with control procedures added. The vertical displacement of the middle of the beam is shown in Figure 7.10.

It can be seen that ca. 95 % of the vibration amplitude is damped out after two cycles of vibration. The control signal, i.e. the relative displacement between layer ends shown in the picture, is magnified 10 times. After only two cycles of disconnection/reconnection of layers, the vibration of the first mode is considerably mitigated, while higher modes of vibrations are introduced.The second sequence of control activation is triggered close to the maximum accumulated strain, at the point where the deformation shape is of the first mode (cf. Figure 7.11(a)). It is worth mentioning that if the control is triggered at the local peak, shortly after the first sequence, it would not give the desired effect, since the deformation shape at that point is of the third mode of vibration (cf. Figure 7.11(b)). The performed control does not affect higher modes, which can, however, be mitigated by the natural damping. Appropriate modification of the control strategy could also make PAR a useful tool in higher mode mitigation.

The control in the numerical simulations was applied by instant releasing of the available component of relative motion between the layers or freezing it at some value. This resulted in a very fast dislocation of layers in the direction parallel to the beam axis. Of course, a drawback to such an approach is that between the subsequent disconnecting and reconnecting of layers, the beam stiffness is decreased. However, the time intervals when this is the case are very short,

**Figure 7.11**  Desired (a) and undesired (b) deformation shapes for activating the control

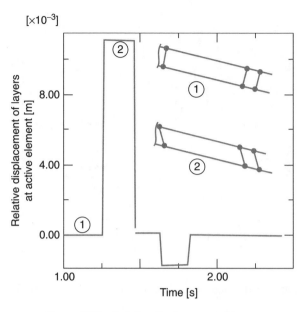

**Figure 7.12**   Relative displacement of layers

as can be seen in Figure 7.12. It can also be observed that at the end of the process, relative displacement between the layers is zero, which means that the structure has returned to the initial configuration.

## 7.3.3  PAR – the VDM Formulation

The VDM-based model of a composite beam with the contact layer presented in Chapter 3 (Section 3.3.2) can be effectively used to describe the PAR process. Assuming feasibility of the full delamination/adhesion actions (performed by smart actuators), the following formula can be used to determine strains in the diagonal truss elements A and B (cf. Figure 3.16) of the contact layer:

$$\varepsilon_i\,(t > t_1) = \varepsilon_i^{\mathrm{L}}\,(t) + \sum_{\tau=0}^{t} D_{ij}\,(t - t_1 - \tau)\,\varepsilon_j^0\,(\tau) \tag{9}$$

where $\varepsilon_i^{\mathrm{L}}$ denotes the response of the fully bonded two-layer beam, $t_1$ denotes the time instant of introducing delamination (readhesion is taking place immediately afterwards at $t_1 + \delta t$) and $\varepsilon_j^0$ is constant and equal to the strain in the diagonal element A (or B) at the time instant $t_1$. The virtual distortions introduced at the time instant $t_1$ have to be removed at the time instant $t_2$, when the maximum rebound of the two bonded (by actuators) layers is taking place. This means that an extra component has to be added to the strain formula:

$$\varepsilon_i(t) = \varepsilon_i^{\mathrm{L}}\,(t) + \sum_{\tau=0}^{t} D_{ij}\,(t - t_1 - \tau)\,\varepsilon_j^0\,(\tau) - \sum_{\tau=0}^{t} D_{ij}\,(t - t_2 - \tau)\,\varepsilon_j^0\,(\tau) \tag{10}$$

where $\varepsilon_j^0$ is constant over the time period $\langle t_1, t_2 \rangle$. The formula (10) specifies the strains related to the accumulated and released parts of the energy.

If the damping process is less effective, the delamination/adhesion actions have to be repeated several times and the corresponding virtual distortions, modeling the prestress accumulation and release at the contact interface, can be written as follows:

$$\varepsilon_i(t) = \varepsilon_i^L(t) + \sum_{\tau=0}^{t} \sum_{\alpha} D_{ij}\left(t - t_1^\alpha - \tau\right) \varepsilon_j^0(\tau, \alpha)$$

$$-\sum_{\tau=0}^{t} \sum_{\alpha} D_{ij}\left(t - t_2^\alpha - \tau\right) \varepsilon_j^0(\tau, \alpha) \tag{11}$$

where $\alpha$ denotes the number of successive cycles of accumulating and releasing the energy in the time intervals $\langle t_1^\alpha, t_2^\alpha \rangle$. Each cycle (characterized by constant distortion $\varepsilon_j^0$ starts with activation of the actuator (adhesion) and finishes with its deactivation (delamination), see Figure 7.13). Note that the cycles occur continuously one after another, i.e. $t_2^\alpha = t_1^{\alpha+1}$. Thus, the algorithm of the PAR strategy simulation consists of the following stages:

1. Initialize – influence the matrix, linear response of the structure.
2. For every time step:
   - track the instant in which the potential energy of the structure is maximal;
   - update the $\alpha$ value and take a new $\varepsilon_j^0(\tau, \alpha)$;
   - solve the set (11) for the structural response.

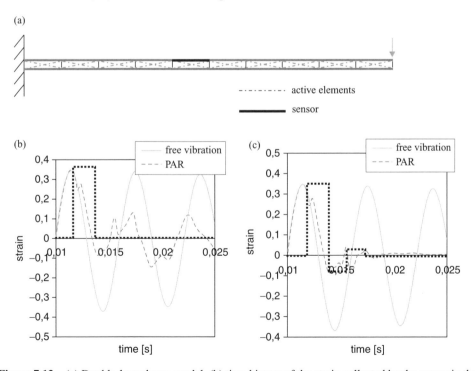

**Figure 7.13** (a) Double-layer beam model, (b) time history of the strain collected by the sensor in the case of a single delamination/adhesion action, (c) time history of the strain collected by the sensor in the case of a few delamination/adhesion actions

The presented algorithm was applied to a double-layer cantilever beam (the material properties of the structural elements were $E = 80 \times 10^9\,\mathrm{Pa}$, $A = 1 \times 10^{-5}\,\mathrm{m}^2$). The structure was vibrating at the fourth eigenfrequency. The virtual distortions were applied in these diagonal elements, belonging to the contact layer, which met the following condition:

$$\varepsilon_{\underline{i}}^0 \left( \varepsilon_{\underline{i}} - \varepsilon_{\underline{i}}^0 \right) \geq 0 \qquad\qquad (12)$$

meaning that the elastic energy was accumulated in an adaptive way (without supplying the energy to the system – the distortion must not exceed the strain). Figure 7.13 shows the history of the strain for an element belonging to the upper beam (sensor). The vibration was successfully damped after applying the control action four times ($\alpha = 4$).

## 7.4  Experimental Verification

### 7.4.1  Experimental Set-up

In order to validate the efficiency of the PAR strategy in vibration mitigation, a cantilever two-layered beam has been chosen as a demonstration. The experimental beam was submitted to free vibrations. The idea of the experiment was to control the delamination between layers. As described in the numerical model section (Section 7.3.2.1), the time during which the layers are disconnected should be short as compared to the full oscillation period. Results were compared with a reference case in which the layers were pressed together with a maximum force.

The experimental set-up consists of two adjacent aluminum bars clamped at the root. Bars are not connected with each other along the length, except at the tip, where they can be pressed together with the actuator device. The actuator used for the experiment consists of piezo-stacks and a mechanical displacement amplifier [11]. The resulting stroke can reach 0.23 mm and the maximum force is about 1300 N. Using this device, the layers can be pressed together with enough force to hold them together during the deformation process.

The actuator state is updated, based on a piezoelectric sensor reading placed close to the cantilever root. A general view of the set-up is shown in Figure 7.14.

### 7.4.2  Control Procedure

The process of adaptation of the system during free vibrations is based on a closed loop between the strain sensor reading and the voltage applied to the piezo-stacks of the actuator device. The process starts with the maximum voltage applied to piezo-stacks, i.e. with maximum force holding the layers together. The control algorithm recognizes a displacement peak at which the layers are disconnected for a time interval of 100 ms. The time interval at which the force is completely removed is significantly smaller in practice due to the internal constraints of the amplifier. In fact, the 100 ms interval was possibly the smallest at which the actual force could decrease to zero. The choice of the amplifier was a compromise, where on the one hand it was possibly the fastest reaction time and on the other hand it was the safety of piezo-stacks, which could be destroyed if too high current peaks were applied. After this, the layers were connected again until the next displacement peak is recognized; cf. the numerical simulations section (Section 7.3.2) for the expected behavior of the layers during the activation control. The control loop was updated with the sampling frequency of 100 Hz.

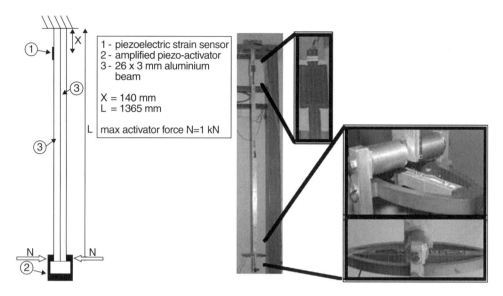

**Figure 7.14**   Experimental set-up: detailed view of the sensor and actuator

## 7.4.3 Results

### 7.4.3.1 Initial Tests

In the initial stage the system was submitted to free vibrations without applying any control procedure. At first no voltage was applied to the piezo-stacks, and then the maximum voltage was applied. The fundamental natural frequency of both systems was identified. The resulting Fast Fourier Transform (FFT) spectra are shown in Figure 7.15. A 22 % change in the fundamental frequency of the structure can be observed, which means that the actuator state can significantly influence the response of the system.

### 7.4.3.2 Semi-active Control

Free oscillations were introduced, with the initial displacement of 10 cm applied at the tip of the cantilever. After a few oscillations the control procedure was enabled. Then, after the delamination had been allowed in a few displacement extremes (in this particular case, in three consecutive peaks), the control procedure was disabled again. For the reference case the control procedure was disabled during the entire process. Disabling of the control procedure means that a constant maximum voltage is applied to piezo-stacks regardless of the sensor reading. Results are shown in Figure 7.16. Triggering the delamination in three consecutive peaks causes the vibration amplitude to be mitigated by over 80 %. The remaining control triggering occurs when the vibration is stopped almost completely and is of no importance. It can be seen that only a threefold delamination has a substantial effect on the structural vibrations. The remaining amplitude could be further mitigated if the control had been activated for longer periods of time.

It has to be emphasized that the actual delamination time is much shorter than the assumed 100 ms due the amplifier constraints. This means that the delamination, or 'weakening' of the structure, is allowed during only a few percent of the total oscillation period.

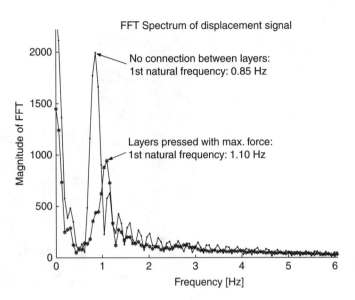

**Figure 7.15**   First natural frequency shift between the disconnected and connected layers

The system under investigation has quite high structural damping, which can be seen in Figure 7.16. Figure 7.17, in turn, illustrates the relative difference in amplitudes of the first 10 vibration peaks, as related to the reference case.

In order to release possibly most of the strain energy, the delamination has to coincide with the deformation peaks. Therefore the PAR efficiency is very sensitive to the time instant of the delamination. If it occurs too long before or after the maximum deflection, then the

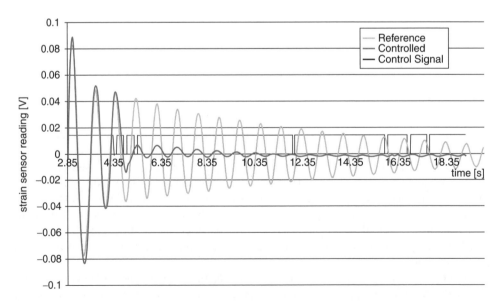

**Figure 7.16**   Time history response of the systems

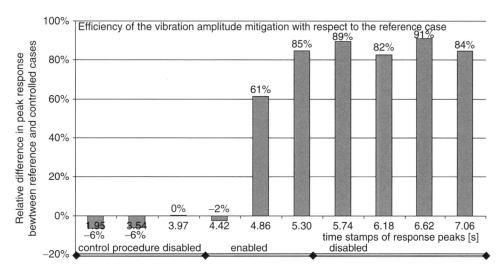

**Figure 7.17**   Relative difference in amplitudes as related to the reference case

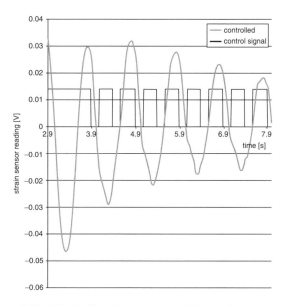

**Figure 7.18**   Weak effect due to a wrong delamination trigger time

obtained response mitigation is less significant. If, in the worst case, the delamination occurs around the equilibrium position of the cantilever, then there is no effect at all since there are no deformation incompatibilities between the layers that could be 'frozen'. Figure 7.18 shows no damping effect due to a wrong delamination trigger time.

## 7.5 Possible Applications

Among possible applications are:

- pedestrian bridges,
- pipelines,
- truss structures,
- small-scale mechanical systems.

The two-layered beam discussed in Section 7.2 has the span and stiffness of an experimental, lightweight pedestrian bridge located in EMPA Laboratory, Switzerland. Cables sustaining the span, which completely change the dynamic behavior of the structure, were not modeled in the numerical example. In order to apply the PAR strategy effectively to this type of structure, mitigation of higher modes has to be accounted for. For this purpose, only one active member at the support is not sufficient. The detachable spring which was discussed in Section 7.3 could be used at the supports of pipeline systems in order to accumulate the deformation energy and convert it to kinetic energy, which can then be dissipated.

## References

1. A. Ruangrassamee and K. Kawashima, Semi-active control of bridges with use of magnetorheological damper, in *Proceedings of the 12th European Conference on Earthquake Engineering*, London, September 2002.
2. F. dell'Isola and S. Vidoli, Damping of bending waves in truss beams by electrical transmission lines with pzt actuators, *Archive of Applied Mechanics*, **96**(2), 1998, 619–626.
3. G.C. Lee, Z. Liang and M. Tong, Development of a semi-active structural control system. in research progress and accomplishments 1997–1999, *Multidisciplinary Center for Earthquake Engineering Research MCEER*, July 1999.
4. M.D. Symans and M.C. Constantinou, Semi-active control systems for seismic protection of structures: a state-of-the-art review, *Engineering Structures*, **21**(6), 1999, 469–487.
5. S.J. Dyke, B.F. Spencer Jr, M.K. Sain and J.D. Carlson, Experimental verification of semi-active structural control strategies using acceleration feedback, in *Proceedings of the 3rd International Conference on Motion and Vibration Control*, Chiba, Japan, 1–6 September 1996.
6. D.C. Kamopp, M.J. Crosby and R.A. Harwood. Vibration control using semi active force generation, *Journal of Engineering for Industry*, **96**(2), 1974, 619–626.
7. J. Holnicki-Szulc and Z. Marzec, Adaptive structures with semi-active interfaces, in *Proceedings of the EUROMECH 373 Colloquium on Modelling and Control of Adaptive Mechanica Structures*, Magdeburg, Germany, 1998.
8. Z. Marzec, J. Holnicki-Szulc and F. López-Almansa, Strategy of impulse release of strain energy for damping of vibration, in *Proceedings NATO ARW Smart Structures'98*, Pultusk, Poland, 1998.
9. L. Gaul, R. Nitsche and D. Sachau, Semi-active vibration control of flexible structures. in *Proceedings of the EUROMECH 373 Colloquium on Modelling and Control of Adaptive Mechanical Structure*, 1998.

10. D.F. Ledezma-Ramirez, N.S. Ferguson and M.J. Brennan, Vibration decay using on–off stiffness control, in *Proceedings of the ISMA International Conference on Noise and Vibration Engineering*, Leuven, Belgium, 2006.
11. H. Bruneau, R. Le Letty, F. Barillet and N. Lhermet, Application of a new amplified piezoelectric actuator to semi-active control of vibration, in *Proceedings of the 2nd International Conference on Active Control in Mechanical Engineering*, 1997.

# 8

# Modeling and Analysis of Smart Technologies in Vibroacoustics

Tomasz G. Zieliński

## 8.1 Introduction

### 8.1.1 Smart Hybrid Approach in Vibroacoustics

Effective noise reduction is nowadays a very important topic, relevant for many applications. Certainly, noise has become a crucial factor in the design of automotive vehicles and aircraft. Noise reduction techniques can be traditionally divided into two groups, namely:

(1) *passive techniques*: porous liners, screens with air-gaps, multilayered panels, etc.;
(2) *active techniques*: the active noise control (ANC) and active structural acoustic control (ASAC).

The first approach is suitable for attenuation of the high-frequency contributions of noise and vibrations, while the active control technologies appear to be the only way to reduce the low-frequency components. As a matter of fact, the active noise control and active structural acoustic control have become classic approaches to cope with low-frequency noise. In the ANC, the noise is cancelled in some areas by the addition of a secondary noise field, which is an inverse replica of the first noise field [1] (and, in practice, causes a noise augmentation in some other areas), while in the ASAC the vibrations of noise radiating surfaces (plates, beams, shells) are actively controlled to reduce the generation of low-frequency noise [2]. These classic solutions have drawbacks and practical limitations and a commonly drawn conclusion is that combined solutions should be the most suited to cover the entire frequency range. Thus, a *smart hybrid approach* was proposed [3–8] that is relevant especially for barriers limiting the transmission of acoustic waves and, in general, for attenuators and dissipative materials for noise insulation and absorption. In such applications porous liners and multilayered panels (usually with a core of porous material and thin elastic faceplates) are widely used, but since they are passive, their efficiency is limited only to high and medium frequencies. The hybrid approach is also termed

*Smart Technologies for Safety Engineering*   Edited by J. Holnicki-Szulc
© 2008 John Wiley & Sons, Ltd

the *hybrid active–passive approach* [6, 8, 9] since it proposes an active control as a remedy for the lack of performance at low frequency while in the high- and medium-frequency ranges an excellent passive acoustic absorption should be guaranteed due to the inherited absorbing properties of well-chosen porous components.

Serious investigations concerning such smart hybrid approach in vibroacoustics started a decade ago in the USA [3, 4] and in France, where a few designs of hybrid noise absorbers were developed and tested for complex situations such as coupled vibroacoustic cavities [5], flow ducts [6, 10] or panels limiting the transmission of acoustic waves [7, 8, 11]. For example, a hybrid active-passive liner prototype was proposed in the form of a square cell (55 mm wide and less than 30 mm thick) composed of a porous layer with an air-gap backed by an elastic plate with a piezoceramic actuator, suggested to perform a pressure reduction at the rear face of a porous layer in order to enhance absorption at low frequencies [6]. The first series of tests confirmed the ability of such active absorbers to achieve noise reduction in flow ducts. The proposed prototype is small and larger active surfaces are obtained by juxtaposing the active cells.

The complete design and optimization of such active absorbers is complicated and first mathematical models were extremely simplified. The most substantial simplification consisted in describing the behavior of a porous layer using the fluid-equivalent model by Johnson and Allard [12]. Moreover, analytical solutions were used, which are valid for plane acoustic waves, i.e. for a one-dimensional analysis of multilayered acoustic media. This very approximative approach allowed, however, some very important behavior to be predicted, while experimental investigations confirmed the efficiency of the hybrid approach [6, 10].

## 8.1.2 A Concept of an Active Composite Noise Absorber

A novel approach in modeling hybrid noise attenuators (absorbers and insulators) has recently been developed. It applies Biot's theory of poroelasticity for porous media and involves an accurate three-dimensional, finite element modeling of combined poroelastic, acoustic, elastic and piezoelectric components. A complete theoretical background as well as the corresponding Galerkin finite element model are presented in Reference [13]. Accurate predictive tools developed on the basis of this model can be used for exact modelling of smart acoustic panels and should allow even the most innovative ideas to be verified, such as the concept of an active *composite* noise attenuator made of porous layers (foams, etc.) with some solid implants: passive inclusions (e.g. small distributed masses [14]) and active elements (e.g. patches of piezoceramic PZT, pieces of PVDF foil, piezo-fibres, etc.). Figure 8.1 shows a model configuration of such a hybrid active–passive noise attenuator. The widespread design of such smart mufflers (composites, liners, panels) is still an open topic and should be addressed with accurate predictive tools. Moreover, very often the interaction of mufflers with air-gaps or a fragment of the surrounding acoustical medium (the air) should also be taken into account in the modeling. The first sections of this chapter (to Section 8.7) intend to provide a complete theoretical basis necessary for the development of such tools using the finite element method, which allows complex geometries to be modeled. Modeling and analysis of smart multilayered panels as well as of porous layers with mass inclusions improving the acoustic absorption are presented in the second part of the chapter (Sections 8.8 to 8.10).

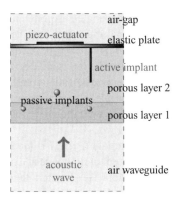

**Figure 8.1** A model configuration of a hybrid active–passive noise attenuator (a smart single-plate composite panel).

### 8.1.3 Physical Problems Involved and Relevant Theories

Accurate modeling of active elastoporoelastic noise attenuators (liners, panels or composites) means a multiphysics approach involving the finite element method to cope with a complex geometry. To this end the following theories will be used (relevant to the physical problems involved):

- Biot's theory of poroelasticity, to model the vibroacoustic transmission and passive dissipation of acoustic waves in porous layers;
- linear acoustics, to model the propagation of acoustic waves in the surrounding air and in air-gaps;
- linear elasticity, to model the vibrations of elastic faceplates (and implants);
- the theory of piezoelectricity, to model the piezo-actuators and active control of low-frequency vibrations.

Moreover, a mutual interaction of all these various physical problems is of the utmost importance, so the relevant couplings must be thoroughly investigated and taken into account in modeling.

### 8.1.4 General Assumptions and Some Remarks on Notation

The considered vibroacoustic application allows perfectly linear theories to be used, so the superposition principle holds and may be effectively used. Consequently, the frequency analysis may be used as an efficient and sufficient tool for design and testing of the active liners, panels or composites. Therefore, apart from Equations (1) and (2) below, all other expressions (for all the problems involved) will be formulated for the case of harmonic oscillations with angular frequency $\omega = 2\pi f$, where $f$ is the frequency of oscillations. Then, all the time-dependent quantities $q(x, t)$ have the following (complex) form: $q(x, t) = q(x)\exp(j\,\omega\,t)$ (here $x$ and $t$ are the independent variables denoting the position and time, and j is the imaginary unit, $j = \sqrt{-1}$), where *physically meaningful* is only the real part of the complex value. After performing the time differentiation for occurring velocities, $\dot{q}(x, t) = j\,\omega\,q(x)\exp(j\,\omega\,t)$, and accelerations,

$\ddot{q}(\mathbf{x}, t) = -\omega^2 q(\mathbf{x}) \exp(j \omega t)$, the time-dependent exponential term $\exp(j \omega t)$ can be easily disposed from the equations, so all the quantities are in fact their (complex) spatial amplitudes. Notice that these amplitudes are moreover frequency-dependent, i.e. $q(\mathbf{x}) = q(\mathbf{x}, \omega)$.

The equations of poroelasticity presented below assume no body forces acting on the poroelastic material. Consequently, the problems of elasticity and piezoelectricity are considered with zero body forces. Moreover, in the piezoelectricity problem there is no body electric charge applied. These assumptions comply with the modeling requirements of hybrid piezo–elasto–poroelastic noise attenuators.

Two sets of subscripts will be used, namely $i, j, k, l \in \{1, 2, 3\}$ to denote vector and tensor components in the three-dimensional system of reference and $m, n \in \{1, \ldots, N_{\mathrm{DOF}}\}$ to number the degrees of freedom of a discrete model ($N_{\mathrm{DOF}}$ is the total number of degrees of freedom). The summation convention is in use for both types of subscripts. The (invariant) differentiation symbol is used which, in the Cartesian coordinate system, simply reads $(.)_{|i} = \partial(.)/\partial x_i$.

For brevity, the symbols $d\Omega$ and $d\Gamma$ are skipped in all the integrals presented below since it is obvious that integration takes place on the specified domain or boundary. Furthermore, the following notation rule for the symbol of variation (or test function) is used: $\delta(v\,w) = v\,\delta w + w\,\delta v$, where $u$ and $v$ are two dependent variables (fields) and $\delta v$ and $\delta w$ are their admissible variations.

## 8.2 Biot's Theory of Poroelasticity

### 8.2.1 Isotropic Poroelasticity and the Two Formulations

Certainly, a vital component of an acoustic panel, liner or composite is a layer (or layers) of porous material. Porous materials are quite often modeled in acoustics by using the so-called fluid-equivalent approach (see, for example, Reference [12]). This is acceptable for most of the porous media in some applications, especially in the higher frequency range, where the vibrations of the skeleton can be completely neglected; then the so-called models of the porous materials with a rigid frame are valid. There are, however, many applications where the contribution of elastic frame vibrations is very significant, particularly in lower frequencies. This is relevant to sandwich panels with a poroelastic core (layer) and certainly it is the very case for porous composite noise absorbers. Here, the simple fluid-equivalent modeling is no longer valid and, instead, a more complicated theory should be used since it is necessary to take into consideration the vibrations of the elastic skeleton and their coupling to the wave propagating in the fluid in pores. There are two main theories that permit an adequate and thorough description of such problems: Biot's theory of poroelasticity [12, 15, 16] and the so-called theory of porous media [17]; the latter one has been essentially established quite recently and is more general. An excellent work by de Boer [17] provides a current state of the theory of porous media, also offering highlights in the historical development and a comparison to Biot's poroelasticity. Biot's theory [15, 16] allows materials to be modeled that are made up of solid elastic skeleton (matrix, frame) with the pores filled up with a compressible fluid. Without doubt, within the framework of geometrically and physically linear theory, it gives good results for a wide range of practical problems – in particular for dynamic ones. A large number of applications have been worked out using this theory: starting from acoustics [12, 18] (and vibroacoustics) up to bio- and geomechanics [19]. In this theory a *biphasic* approach is applied where the so-called *solid phase* is used to describe the behavior of the elastic skeleton, while the so-called *fluid phase* pertains to the fluid in the pores. Both phases are, in fact, coupled homogenized continua of the 'smeared' skeleton and pore fluid. This homogenization (or, rather, averaging) uses the concept of the average volume element (AVE) and works very well if the shortest length of waves propagating in a porous medium is significantly greater than the characteristic dimension of pores.

In practical applications the most frequently used is Biot's *isotropic* theory of poroelasticity [12, 15, 16, 18]. In this approach, both phases are isotropic. Moreover, the fluid is modeled as perfect (i.e. inviscid), though viscous forces are taken into account, but only when modeling the interaction between the fluid and the frame.

Two formulations of Biot's isotropic poroelasticity may be distinguished:

- the classical *displacement formulation* proposed by Biot, where the unknowns are the solid and fluid phase displacements, which yields 6 DOF in every node of a three-dimensional numerical model;
- the *mixed displacement–pressure formulation*, where the dependent variables are the solid phase displacements and the pressure of the fluid in pores. Therefore, there are only 4 DOF in a three-dimensional model.

The second formulation is valid only for harmonic motion and was presented by Atalla *et al.* [20]. Debergue *et al.* [21] discussed a very important subject of the boundary and interface–coupling conditions for this formulation.

### 8.2.2 The Classical Displacement Formulation

As mentioned above, in the classical formulation [12, 15, 16] a state of poroelastic medium is unequivocally described by the displacements of the solid phase, $u = \{u_i\}$, and the fluid phase, $U = \{U_i\}$. Therefore, this is often referred to as the displacement–displacement, or $\{u, U\}$ formulation. Biot's equations for a local dynamic equilibrium state of poroelastic medium link partial stress tensors associated with the skeleton particle ($\sigma_{ij}^s$) and the macroscopic fluid particle ($\sigma_{ij}^f$) with the solid and fluid macroscopic displacements:

$$\sigma_{ij|j}^s = \varrho_{ss} \ddot{u}_i + \varrho_{sf} \ddot{U}_i + \tilde{b}(\dot{U}_i - \dot{u}_i) \tag{1}$$

$$\sigma_{ij|j}^f = \varrho_{ff} \ddot{U}_i + \varrho_{sf} \ddot{u}_i + \tilde{b}(\dot{u}_i - \dot{U}_i) \tag{2}$$

where $\tilde{b}$ is the *viscous drag coefficient* and $\varrho_{ss}$, $\varrho_{ff}$, $\varrho_{sf}$ are the so-called *effective densities*. The first of these *equilibrium equations* refers to the solid phase and the second one to the fluid phase. Nevertheless, it is easy to notice that both equations are strongly coupled by the inertial and viscous coupling terms: the viscous drag coefficient pertains to the traction between the interstitial fluid and the solid skeleton (the fluid by itself is inviscid, i.e. in the sense that there are no viscous forces between the fluid particles), whereas the last of the effective densities, $\varrho_{sf}$, is the so-called *mass coupling coefficient*, responsible for consideration of the inertial interaction forces that occur between the solid skeleton and the fluid. The effective densities are expressed as follows:

$$\varrho_{ss} = (1 - \phi)\varrho_s - \varrho_{sf}, \qquad \varrho_{ff} = \phi \varrho_f - \varrho_{sf}, \qquad \varrho_{sf} = -(\alpha_\infty - 1)\phi \varrho_f \tag{3}$$

They depend on the porosity, $\phi$, the tortuosity of pores, $\alpha_\infty$, the density of the material of the skeleton, $\varrho_s$, and the density of the saturating fluid, $\varrho_f$.

Consider now the case of harmonic motion (with the angular frequency $\omega$). Then, as explained above, the time-dependent exponential term $\exp(\mathrm{j}\,\omega\,t)$ is reduced from the equations and all the relevant quantities (like excitations or state variables, i.e. $\sigma_{ij}^{s}$, $\sigma_{ij}^{f}$, $u_i$ and $U_i$) become frequency-dependent complex amplitudes. Remembering this, the equilibrium equations (1) and (2) read as follows:

$$\sigma_{ij|j}^{s} + \omega^2 \tilde{\varrho}_{ss}\, u_i + \omega^2 \tilde{\varrho}_{sf}\, U_i = 0 \tag{4}$$

$$\sigma_{ij|j}^{f} + \omega^2 \tilde{\varrho}_{ff}\, U_i + \omega^2 \tilde{\varrho}_{sf}\, u_i = 0 \tag{5}$$

where the so-called *frequency-dependent effective densities* are introduced:

$$\tilde{\varrho}_{ss} = \varrho_{ss} + \frac{\tilde{b}}{\mathrm{j}\,\omega}, \qquad \tilde{\varrho}_{ff} = \varrho_{ff} + \frac{\tilde{b}}{\mathrm{j}\,\omega}, \qquad \tilde{\varrho}_{sf} = \varrho_{sf} - \frac{\tilde{b}}{\mathrm{j}\,\omega} \tag{6}$$

As a matter of fact, these densities are responsible not only for the inertia of solid or fluid phase particles but also for the combined inertial and viscous coupling (interaction) of both phases.

The partial solid and fluid stress tensors are linearly related to the partial strain tensors prevailing in the skeleton and the interstitial fluid. This is given by the following linear and isotropic *constitutive equations* of Biot's theory of poroelasticity (where linear kinematic relations have already been used to replace the strain tensors with the gradients of displacements):

$$\sigma_{ij}^{s} = \mu_s\, (u_{i|j} + u_{j|i}) + \left(\tilde{\lambda}_s\, u_{k|k} + \tilde{\lambda}_{sf}\, U_{k|k}\right)\delta_{ij} \tag{7}$$

$$\sigma_{ij}^{f} = \left(\tilde{\lambda}_f\, U_{k|k} + \tilde{\lambda}_{sf}\, u_{k|k}\right)\delta_{ij} \tag{8}$$

(In these equations and further below, $\delta_{ij}$ is the Kronecker's delta symbol). It can clearly be seen that in this modeling both phases are isotropic. Four material constants are involved here, namely $\mu_s$, $\tilde{\lambda}_s$, $\tilde{\lambda}_f$ and $\tilde{\lambda}_{sf}$. The first two are similar to the two Lamé coefficients of isotropic elasticity. Moreover, $\mu_s$ is the shear modulus of the poroelastic material and, consequently, the shear modulus of the frame, since the fluid does not contribute to the shear restoring force. The three dilatational constants, $\tilde{\lambda}_s$, $\tilde{\lambda}_f$ and $\tilde{\lambda}_{sf}$, are frequency-dependent and are functions of $K_b$, $K_s$ and $\tilde{K}_a$ ($\tilde{\lambda}_s$ depends also on $\mu_s$), where $K_b$ is the bulk modulus of the frame at constant pressure in the fluid, $K_s$ is the bulk modulus of the elastic solid from which the frame is made and $\tilde{K}_a$ is the effective bulk modulus of fluid in the porous material. Adequate exact formulas to compute the poroelastic material constants can be found in Reference [12]. Section 8.3 will discuss the material parameters for rigid-frame and poroelastic models of porous media, and the necessary formulas for poroelastic material constants will be given for the case of poroelastic materials with an incompressible skeleton (i.e. the elastic skeleton can vibrate and deform, but its volume does not change, even locally). This simplification is valid if the elastic material of the skeleton is much stiffer than the poroelastic medium itself, so it is very important in practice and it provides much simpler expressions for $\tilde{\lambda}_s$, $\tilde{\lambda}_f$ and $\tilde{\lambda}_{sf}$ by taking the infinite limit for the bulk modulus of the elastic solid from which the frame is made, i.e. $K_s \to \infty$.

Notice that only one material constant, namely the *constitutive coupling coefficient*, $\tilde{\lambda}_{sf}$, is responsible for a multiphysical coupling occurring between the constitutive equations of both phases. However, the reader should be reminded of the visco–inertial coupling present in the equations of equilibrium; thus, the interaction of the solid skeleton with the fluid in the pores is very well represented in this biphasic approach.

The equations of equilibrium (1) and (2), or (4) and (5) in the case of harmonic motion, form with the constitutive relations (7) and (8) the *displacement formulation* of linear isotropic poroelasticity. Finally, total quantities are defined for this biphasic model, namely the *total stress tensor* as a simple sum of the partial, i.e. phasic, stress tensors:

$$\sigma_{ij}^t = \sigma_{ij}^s + \sigma_{ij}^f = \mu_s \left(u_{i|j} + u_{j|i}\right) + \left[\left(\tilde{\lambda}_s + \tilde{\lambda}_{sf}\right)u_{k|k} + \left(\tilde{\lambda}_f + \tilde{\lambda}_{sf}\right)U_{k|k}\right]\delta_{ij} \tag{9}$$

and the *total displacement vector* of the poroelastic medium, which reads

$$u_i^t = (1 - \phi)u_i + \phi U_i \tag{10}$$

where the porosity-dependent contributions of the displacements of both phases are involved.

### 8.2.3 The Mixed Displacement–Pressure Formulation

Finite element models based on the displacement formulation of Biot's poroelasticity have been used to predict the acoustical and structural behavior of porous multilayer structures [22–25]. Since these models, while accurate, lead to large frequency-dependent matrices for three-dimensional problems, Atalla *et al.* proposed in Reference [20] a novel exact mixed displacement–pressure formulation derived directly from Biot's poroelasticity equations. The boundary conditions for this formulation were extensively discussed in Reference [21].

Equations for the mixed formulation will be derived below to show that it has the form of a classical coupled fluid–structure problem, involving the dynamic equations of the skeleton *in vacuo* and the equivalent fluid in the rigid skeleton limit. The most important boundary conditions will be discussed later on in Section 8.5, which is devoted to the boundary integrals of the weak form of poroelasticity.

The fluid phase stress tensor can be expressed as follows:

$$\sigma_{ij}^f = -\phi \, p \, \delta_{ij} \tag{11}$$

where $p$ is the pressure of fluid in the pores (the so-called *porefluid pressure*; it should not be mistaken for the partial pressure of the fluid phase which equals $\phi p$). Using this relation for the constitutive equation of fluid phase (8) leads to the following expressions:

$$p = -\frac{\tilde{\lambda}_f}{\phi} U_{k|k} - \frac{\tilde{\lambda}_{sf}}{\phi} u_{k|k} \qquad \text{or} \qquad U_{k|k} = -\frac{\phi}{\tilde{\lambda}_f} p - \frac{\tilde{\lambda}_{sf}}{\tilde{\lambda}_f} u_{k|k} \tag{12}$$

They are valid for the general case since only a constitutive equation has been used. Now, however, interest is restricted to the harmonic oscillations (with the angular frequency $\omega$). In this case, by using Equation (11) in the harmonic equilibrium equation of fluid phase (5), the fluid phase displacements can be expressed as a function of the pressure in the pores and the solid phase displacements:

$$U_i = \frac{\phi}{\omega^2 \tilde{\varrho}_{\text{ff}}} p_{|i} - \frac{\tilde{\varrho}_{\text{sf}}}{\tilde{\varrho}_{\text{ff}}} u_i \tag{13}$$

This in turn can be used for Equation (4); so now the harmonic equilibrium for the solid phase can be expressed as follows:

$$\sigma_{ij|j}^{\text{ss}} + \omega^2 \tilde{\varrho} \, u_i + \phi \left( \frac{\tilde{\varrho}_{\text{sf}}}{\tilde{\varrho}_{\text{ff}}} - \frac{\tilde{\lambda}_{\text{sf}}}{\tilde{\lambda}_{\text{f}}} \right) p_{|i} = 0 \quad \text{where } \tilde{\varrho} = \tilde{\varrho}_{\text{ss}} - \frac{\tilde{\varrho}_{\text{sf}}^2}{\tilde{\varrho}_{\text{ff}}} \tag{14}$$

Here, a new stress tensor is introduced

$$\sigma_{ij}^{\text{ss}} = \mu_{\text{s}} (u_{i|j} + u_{j|i}) + \tilde{\lambda}_{\text{ss}} u_{k|k} \delta_{ij} \quad \text{where } \tilde{\lambda}_{\text{ss}} = \tilde{\lambda}_{\text{s}} - \frac{\tilde{\lambda}_{\text{sf}}^2}{\tilde{\lambda}_{\text{f}}} \tag{15}$$

This tensor depends only on the solid phase displacements and has an interesting physical interpretation: it is called the stress tensor of the skeleton *in vacuo* because it describes the stresses in the skeleton when there is no fluid in the pores or when at least the pressure of fluid is constant in the pores. This can be easily noticed when putting $p(x) = $ constant in Equation (14); then, the last term (which couples this equation with its fluid phase counterpart) vanishes and so the remaining terms clearly describe the behavior of the skeleton of poroelastic medium filled with a fluid under the same pressure everywhere. The new stress tensor is related to the solid phase stress tensor in the following way:

$$\sigma_{ij}^{\text{s}} = \sigma_{ij}^{\text{ss}} - \phi \frac{\tilde{\lambda}_{\text{sf}}}{\tilde{\lambda}_{\text{f}}} p \, \delta_{ij} \tag{16}$$

Now, the fluid phase displacements are to be eliminated from the fluid phase harmonic equilibrium equations (5). To this end, Equations (11) and the second equation in (12) are used for Equation (5) to obtain (after multiplication by $-\phi/(\omega^2 \tilde{\varrho}_{\text{ff}})$)

$$\frac{\phi^2}{\omega^2 \tilde{\varrho}_{\text{ff}}} p_{|ii} + \frac{\phi^2}{\tilde{\lambda}_{\text{f}}} p - \phi \left( \frac{\tilde{\varrho}_{\text{sf}}}{\tilde{\varrho}_{\text{ff}}} - \frac{\tilde{\lambda}_{\text{sf}}}{\tilde{\lambda}_{\text{f}}} \right) u_{i|i} = 0 \tag{17}$$

This equation pertains to the fluid phase but the last term couples it with the solid phase equation (14). This term vanishes for the rigid body motion of the skeleton (i.e. when $u_i = $ constant). That means that the main terms describe the behavior of the fluid when the skeleton is motionless or rigid. Notice also that the expression that stands by $u_{i|i}$ in the coupling term is similar to the one standing by $p_{|i}$ in the coupling term of the solid phase equation (14). This feature is quite important when constructing the weak variational formulation (and it justifies presenting Equation (17) in such a form), since it permits the handling of some coupling conditions at the interface between two different poroelastic media to be simplified.

Equations (14) and (17) together with the constitutive relation (15) constitute the *mixed displacement–pressure formulation* of harmonic isotropic poroelasticity. For completeness, the total stresses and total displacements in terms of the fluid pressure and solid phase displacements are given here:

$$\sigma_{ij}^{t} = \sigma_{ij}^{ss} - \phi\left(1 + \frac{\tilde{\lambda}_{sf}}{\tilde{\lambda}_{f}}\right) p\,\delta_{ij} = \mu_{s}\left(u_{i|j} + u_{j|i}\right) + \left[\tilde{\lambda}_{ss}\,u_{k|k} - \phi\left(1 + \frac{\tilde{\lambda}_{sf}}{\tilde{\lambda}_{f}}\right) p\right]\delta_{ij} \qquad (18)$$

$$u_{i}^{t} = \left[1 - \phi\left(1 + \frac{\tilde{\varrho}_{sf}}{\tilde{\varrho}_{ff}}\right)\right]u_{i} + \frac{\phi^{2}}{\omega^{2}\tilde{\varrho}_{ff}}\,p_{|i} \qquad (19)$$

## 8.3 Porous and Poroelastic Material Data and Coefficients

### 8.3.1 Porous Materials with a Rigid Frame

In many applications of porous media and for a wide frequency range the solid skeleton vibrations can often be completely neglected. In such cases the frame of porous medium is considered to be rigid and the so-called *fluid-equivalent model* by Johnson and Allard is applied which uses the classical Helmholtz equation of linear acoustics. The parameters characterizing an acoustic medium (i.e. inviscid perfectly elastic fluid) are the density and bulk modulus (or, equivalently, the speed of sound). In the case of the fluid-equivalent model of porous material, the only difference is that these quantities are frequency-dependent. Therefore, they are termed the *effective* density and bulk modulus. All (quite cumbersome) formulas to calculate these quantities will be given in this section. Notice that the same effective bulk modulus of the fluid in a porous material is also used by the theory of poroelasticity (see Section 8.3.2).

#### 8.3.1.1 Effective Density

The frequency-dependent effective density of a porous material with a rigid frame is computed using the following formula:

$$\tilde{\varrho}_{a} = \alpha_{\infty}\varrho_{f} + \frac{\tilde{b}}{\phi\,j\,\omega} \qquad (20)$$

where the frequency-dependent viscous drag coefficient reads

$$\tilde{b} = \sigma\,\phi^{2}\tilde{G}(\omega) \quad \text{where } \tilde{G}(\omega) = \sqrt{1 + \frac{4\eta\,\alpha_{\infty}^{2}\varrho_{f}}{\sigma^{2}\,\Lambda^{2}\,\phi^{2}}\,j\,\omega} \qquad (21)$$

This coefficient will also be used in the poroelastic model. In the above expressions the following six material parameters appear:

- $\varrho_{f}$ (kg/m$^{3}$) is the density of the saturating fluid;
- $\eta$ (Pa s) is the fluid viscosity;

these two parameters pertain to the fluid that saturates the pores (which is usually the air), whereas the remaining parameters pertain to the porous material itself:

- $\sigma$ (Pa s/m$^{2}$) is the flow resistivity;
- $\phi$ is the porosity;

- $\alpha_\infty$ is the tortuosity of pores ($\alpha_\infty = \phi\, r_s/r_f$, where $r_s$ and $r_f$ are the resistivity of the material saturated with a conducting fluid and the resistivity of the saturating fluid, respectively);
- $\Lambda$ (m) is the characteristic dimension of pores for viscous forces.

### 8.3.1.2 Bulk Modulus of the Fluid in the Porous Material

The effective bulk modulus of porous material with a rigid skeleton is in fact the bulk modulus of the saturating fluid in a porous material (but not the specific bulk modulus of this fluid). This is a frequency-dependent quantity computed as follows:

$$\tilde{K}_a = \frac{\gamma\, P_a}{\gamma - (\gamma - 1)/[1 + \tilde{H}(\omega)]} \quad \text{where } \tilde{H}(\omega) = \frac{\sqrt{64\eta^2 + 4\eta\, \Lambda'^2 B^2 \varrho_f j \omega}}{\Lambda'^2 B^2 \varrho_f j \omega} \tag{22}$$

Here, apart from the saturating fluid density, $\varrho_f$, and viscosity, $\eta$, another three saturating fluid parameters appear, namely:

- $\gamma$ is the ratio of specific heats ($\gamma = c_v/c_p$, where $c_v$ and $c_p$ are the specific heats per unit mass at constant volume and pressure, respectively);
- $P_a$ (Pa) is the ambient mean pressure;
- $B^2$ is the Prandtl number ($B^2 = \kappa/(\varrho_f c_v)$, where $\kappa$ is the thermal conductivity);

yet another porous material property is also involved, namely:

- $\Lambda'$ (m) is the characteristic dimension of pores for the bulk modulus of saturating fluid.

It is as well to mention here that the bulk modulus of fluid in the porous material is also used in the calculation of the 'dilatational' coefficients of the poroelastic medium (see Section 8.3.2), i.e. when the assumption of rigidity of the skeleton does not hold anylonger.

## 8.3.2 Poroelastic Materials

There are only three *additional* parameters to describe a porous material with an elastic frame, namely:

- $\varrho_{sph}$ (kg/m$^3$) is the density of the solid phase (notice that for the material with open pores $\varrho_{sph} = (1 - \phi)\varrho_s$, where $\varrho_s$ is the density of the material of the skeleton);
- $\mu_s$ (Pa) is the shear modulus of the material of skeleton;
- $\nu_b$ is the bulk Poisson's ratio of the poroelastic material.

The other necessary parameters are identical to the ones used for a porous media with a rigid skeleton. As a matter of fact, the effective bulk modulus of the fluid in a porous material is computed in exactly the same way for the porous media with a rigid or elastic frame.

### 8.3.2.1 Effective Densities

Equations (3) define two effective densities of the poroelastic medium, $\varrho_{ss}$ and $\varrho_{ff}$, and the (negative) inertial coupling coefficient $\varrho_{sf}$. Their frequency-dependent counterparts are defined by formulas (6) involving the viscous drag coefficient $\tilde{b}$ calculated using Equation (21).

### 8.3.2.2 Material Constants

There are four Lamé-like coefficients in the displacement equations of poroelastic media, namely:

- the shear modulus of the porous material, which consequently is the shear modulus of the skeleton, $\mu_s$, and
- the three 'dilatational' coefficients,[1]

$$\tilde{\lambda}_s = -\frac{2}{3}\mu_s + K_b + \frac{(1-\phi)^2}{\phi}\tilde{K}_a, \qquad \tilde{\lambda}_{sf} = (1-\phi)\tilde{K}_a, \qquad \tilde{\lambda}_f = \phi\,\tilde{K}_a, \qquad (23)$$

which depend on the porosity, $\phi$, and the bulk modulus of the fluid in the porous material, $\tilde{K}_a$ (see Equation (22)). The first of the 'dilatational' coefficients depends also on the shear modulus, $\mu_s$, and the bulk modulus of the skeleton at constant pressure, $K_b$. In vacuum (i.e. no pressure) this bulk modulus is expressed as

$$K_b = \frac{2(1+\nu_b)}{3(1-2\nu_b)}\mu_s \qquad (24)$$

## 8.4 Weak Forms of Poroelasticity, Elasticity, Piezoelectricity and Acoustics

### 8.4.1 Weak Form of the Mixed Formulation of Poroelasticity

The weak integral form of the mixed formulation of Biot's poroelasticity was presented by Atalla *et al.* [20]. An enhanced version of this weak formulation was proposed in Reference [26]. Here, the enhanced version will be used since the enhancement allows some boundary and interface–coupling conditions to be handled easily. This matter was extensively discussed in References [26] and [27]. Finite element models based on the enhanced weak form of the mixed poroelasticity problem involving coupling to elastic and acoustic media are presented in References [26] to [28]. In Reference [28] the convergence of the model using hierarchical elements was investigated.

Let $\Omega_p$ be a domain of poroelastic material and $\Gamma_p$ its boundary, with $n_i$ being the components of the vector normal to the boundary and pointing outside the domain. The harmonic poroelasticity problem can be described in this domain by the mixed formulation equations (14) and (17). Both sides of these equations are multiplied by the so-called test (or weighting) functions, $\delta u_i$ and $\delta p$, for the solid phase equations and the fluid phase equation, respectively, and then integrated in the whole domain $\Omega_p$ and summed up to one integral equation. Integrating by parts some of the terms and using the divergence theorem yields the weak form for the harmonic poroelasticity problem, valid for any arbitrary yet admissible virtual displacements,

---

[1] Here, simplified expressions for the 'dilatational' coefficients are presented. They are valid for every poroelastic medium, which is much less stiff than the elastic material of its skeleton, so that the volume of the skeleton can be assumed constant (while its shape changes). In practice, it means that these expressions can be used for most of the sound-absorbing porous materials. They are obtained directly from the general ones (given, for example, in Reference [12]) by taking an infinite limit for the bulk modulus of the elastic solid from which the frame is made, i.e. $K_s \to \infty$. Another practical remark is that most of the authors tend to use (after Biot [15, 16] or Allard [12]) the following denotation for the poroelastic material constants: $P = \tilde{\lambda}_s + 2\mu_s$, $Q = \tilde{\lambda}_{sf}$, $R = \tilde{\lambda}_f$ and $N = \mu_s$.

$\delta u_i$, and pressure, $\delta p$. This form can be expressed as follows:

$$
\begin{aligned}
\mathcal{WF}_{\mathrm{p}} = &-\int_{\Omega_{\mathrm{p}}} \sigma_{ij}^{\mathrm{ss}}\, \delta u_{i|j} + \int_{\Omega_{\mathrm{p}}} \omega^2 \tilde{\varrho}\, u_i\, \delta u_i - \int_{\Omega_{\mathrm{p}}} \frac{\phi^2}{\omega^2 \tilde{\varrho}_{\mathrm{ff}}}\, p_{|i}\, \delta p_{|i} + \int_{\Omega_{\mathrm{p}}} \frac{\phi^2}{\tilde{\lambda}_{\mathrm{f}}}\, p\, \delta p \\
&+ \int_{\Omega_{\mathrm{p}}} \phi\left(1 + \frac{\tilde{\varrho}_{\mathrm{sf}}}{\tilde{\varrho}_{\mathrm{ff}}}\right) \delta(p_{|i}\, u_i) + \int_{\Omega_{\mathrm{p}}} \phi\left(1 + \frac{\tilde{\lambda}_{\mathrm{sf}}}{\tilde{\lambda}_{\mathrm{f}}}\right) \delta(p\, u_{i|i}) \\
&+ \int_{\Gamma_{\mathrm{p}}} \sigma_{ij}^{\mathrm{t}}\, n_j\, \delta u_i + \int_{\Gamma_{\mathrm{p}}} \phi\, (U_i - u_i)\, n_i\, \delta p = 0
\end{aligned}
\tag{25}
$$

Here, $\sigma_{ij}^{\mathrm{ss}} = \sigma_{ij}^{\mathrm{ss}}(\boldsymbol{u})$ is a function of solid phase displacements according to Equation (15). The total stresses, $\sigma_{ij}^{\mathrm{t}} = \sigma_{ij}^{\mathrm{t}}(\boldsymbol{u}, p)$, and the fluid phase displacements, $U_i = U_i(\boldsymbol{u}, p)$, may also be formally considered as functions of solid phase displacements and fluid phase pressure (see Equations (18) and (13)), but they appear only in the boundary integrals and will be reduced or replaced by specific prescribed values when considering the boundary or interface coupling conditions. These integrals of (Neumann) boundary conditions are in the last line of Equation (25) while the second line contains the coupling terms. Owing to the proposed weak formulation, the boundary and interface conditions are naturally handled for rigid piston displacements and when coupling to an elastic medium or to a layer of another poroelastic material. They are also adequately simple when imposing a pressure field and in the case of coupling to an acoustic medium. Since the issue of boundary and coupling interface conditions is not a simple one in the case of a double-phase modeling, this matter will be extensively discussed later on in Sections 8.5 and 8.6.

### 8.4.2 Weak Form for an Elastic Solid

The theory of (linear) elasticity and the derivation of the weak form used by the finite element method (FEM) and other variational methods can be found in many textbooks (e.g. Reference [29]). Below, the weak integral of the principle of virtual work for a harmonic elastic-body system is given (without derivation), and the natural and essential boundary conditions are briefly discussed.

Let $\Omega_{\mathrm{e}}$ be an elastic solid domain with mass density $\varrho_{\mathrm{e}}$ and boundary $\Gamma_{\mathrm{e}}$, and $n_i^{\mathrm{e}}$ the components of the vector normal to the boundary and pointing outside the domain. Assuming zero body forces and the case of harmonic oscillations, the weak variational form of the problem of elasticity expressing the principle of virtual work reads

$$
\mathcal{WF}_{\mathrm{e}} = -\int_{\Omega_{\mathrm{e}}} \sigma_{ij}^{\mathrm{e}}\, \delta u_{i|j}^{\mathrm{e}} + \int_{\Omega_{\mathrm{e}}} \omega^2 \varrho_{\mathrm{e}}\, u_i^{\mathrm{e}}\, \delta u_i^{\mathrm{e}} + \int_{\Gamma_{\mathrm{e}}} \sigma_{ij}^{\mathrm{e}}\, n_j^{\mathrm{e}}\, \delta u_i^{\mathrm{e}} = 0
\tag{26}
$$

where $u_i^{\mathrm{e}}$ are the elastic solid displacements and $\delta u_i^{\mathrm{e}}$ are their arbitral yet admissible variations; the elastic stress tensor $\sigma_{ij}^{\mathrm{e}} = \sigma_{ij}^{\mathrm{e}}(\boldsymbol{u}^{\mathrm{e}})$ is a substitute here for a linear function of elastic displacements. Generally, in the anisotropic case this equals

$$
\sigma_{ij}^{\mathrm{e}} = C_{ijkl}^{\mathrm{e}}\, \frac{u_{k|l}^{\mathrm{e}} + u_{l|k}^{\mathrm{e}}}{2}
\tag{27}
$$

where $C^e_{ijkl}$ is the fourth-order tensor of linear elasticity. It can be seen that the linear kinematic relations between the elastic strain tensor and the elastic displacements, $\varepsilon^e_{ij} = \frac{1}{2}(u^e_{i|j} + u^e_{j|i})$, have already been used in Equation (27). Let it also be remembered that thanks to the symmetrical properties, only 21 of the total number of 81 elastic tensor components are independent (since $C^e_{ijkl} = C^e_{ijlk}$, $C^e_{ijkl} = C^e_{jikl}$ and $C^e_{ijkl} = C^e_{klij}$). Moreover, in the case of an orthotropic material there are nine nonzero independent material constants, while in the case of transversal isotropy there one only five; finally, every isotropic elastic material is described completely by only two material constants. In this latter case the constitutive equation (of linear isotropic elasticity) can be expressed as follows:

$$\sigma^e_{ij} = \mu_e \left( u^e_{i|j} + u^e_{j|i} \right) + \lambda_e \, u^e_{k|k} \, \delta_{ij} \tag{28}$$

where the well-known Lamé coefficients: the shear modulus, $\mu_e$, and the dilatational constant, $\lambda_e$, appear. They are related to the material's Young modulus, $E_e$, and Poisson's coefficient, $\nu_e$, in the following way:

$$\mu_e = \frac{E_e}{2(1 + \nu_e)}, \qquad \lambda_e = \frac{\nu_e \, E_e}{(1 + \nu_e)(1 - 2\nu_e)} \tag{29}$$

### 8.4.2.1 Boundary Conditions

Two kinds of boundary conditions will be discussed here, namely Neumann's and Dirichlet's, although they may be combined into the third specific type, the so-called Robin (or generalized Dirichlet) boundary condition. For the sake of brevity, the latter type will not be considered; remember only that, in practice, the well-known technique of Lagrange multipliers is usually involved when applying it. The Neumann (or natural) boundary conditions describe the case when forces $\hat{t}^e_i$ are applied on a boundary, i.e.

$$\sigma^e_{ij} \, n^e_j = \hat{t}^e_i \text{ on } \Gamma^t_e \tag{30}$$

whereas the displacements $\hat{u}^e_i$ are prescribed by the Dirichlet (or essential) boundary conditions

$$u^e_i = \hat{u}^e_i \text{ on } \Gamma^u_e \tag{31}$$

According to these conditions the boundary is divided into two (directionally disjoint) parts, i.e. $\Gamma_e = \Gamma^t_e \cup \Gamma^u_e$. There is an essential difference between the two kinds of conditions. The displacement constraints form the kinematic requirements for the trial functions while the imposed forces appear in the weak form; thus, the boundary integral, i.e. the last left-hand-side term of Equation (26), equals

$$\mathcal{BI}_e = \int_{\Gamma_e} \sigma^e_{ij} \, n^e_j \, \delta u^e_i = \int_{\Gamma^t_e} \hat{t}^e_i \, \delta u^e_i \tag{32}$$

Here, the property $\delta u^e_i = 0$ on $\Gamma^u_e$ has been used.

### 8.4.3 Weak Form of Piezoelectricity

The theory of piezoelectricity is extensively discussed, for example, in References [30] and [31]. More or less brief recapitulations of the linear theory of piezoelectricity may also be found in many papers and books on active vibration control and piezoelectric actuators and sensors (e.g. References [2], [32] and [33]). A very good survey of the advances and trends in finite element modeling of piezoelectricity was presented by Benjeddou [34]. In this paper the basic theoretical considerations and equations of linear piezoelectricity as well as the variational piezoelectric equations are also given.

Piezoelectric elements of the proposed active composites, liners and panels are to be modeled using the linear theory. It is adequate enough and, moreover, it is a very accurate model when compared it to some frequently used approximations like the so-called thermal analogy approach. Here, a variational form of linear piezoelectricity will be presented as being the most used one for piezoelectric finite element formulations. This form should be regarded as the sum of the conventional principle of virtual mechanical displacements and the principle of virtual electric potential.

Let $\Omega_{\mathrm{pz}}$ be a domain of piezoelectric material, $\varrho_{\mathrm{pz}}$ its mass density and $\Gamma_{\mathrm{pz}}$ its boundary. The unit boundary-normal vector, $n_i^{\mathrm{pz}}$, points outside the domain. The dependent variables of piezoelectric medium are the mechanical displacements, $u_i^{\mathrm{pz}}$, and electric potential, $V^{\mathrm{pz}}$. The case of harmonic oscillations (with the angular frequency $\omega$) with no mechanical body forces and electric body charge is considered. Then, for arbitrary yet admissible virtual displacements, $\delta u_i^{\mathrm{pz}}$, and virtual electric potential, $\delta V^{\mathrm{pz}}$, the variational formulation of the piezoelectricity problem can be given as

$$\mathcal{WF}_{\mathrm{pz}} = -\int_{\Omega_{\mathrm{pz}}} \sigma_{ij}^{\mathrm{pz}} \, \delta u_{i|j}^{\mathrm{pz}} + \int_{\Omega_{\mathrm{pz}}} \omega^2 \varrho_{\mathrm{pz}} u_i^{\mathrm{pz}} \, \delta u_i^{\mathrm{pz}} + \int_{\Gamma_{\mathrm{pz}}} \sigma_{ij}^{\mathrm{pz}} \, n_j^{\mathrm{pz}} \, \delta u_i^{\mathrm{pz}}$$
$$- \int_{\Omega_{\mathrm{pz}}} D_i^{\mathrm{pz}} \, \delta V_{|i}^{\mathrm{pz}} + \int_{\Gamma_{\mathrm{pz}}} D_i^{\mathrm{pz}} \, n_i^{\mathrm{pz}} \, \delta V^{\mathrm{pz}} = 0 \tag{33}$$

where $\sigma_{ij}^{\mathrm{pz}} = \sigma_{ij}^{\mathrm{pz}}(\boldsymbol{u}^{\mathrm{pz}}, V^{\mathrm{pz}})$ and $D_i^{\mathrm{pz}} = D_i^{\mathrm{pz}}(\boldsymbol{u}^{\mathrm{pz}}, V^{\mathrm{pz}})$ are expressions of mechanical displacements and electric potential. Obviously, from the physical point of view they represent the mechanical stress tensor and the electric displacement vector, respectively. As a matter of fact, these expressions are the so-called *stress-charge form* of the constitutive relations of piezoelectricity. They are given below for the case of linear anisotropic piezoelectricity:

$$\sigma_{ij}^{\mathrm{pz}} = C_{ijkl}^{\mathrm{pz}} \frac{u_{k|l}^{\mathrm{pz}} + u_{l|k}^{\mathrm{pz}}}{2} - e_{kij}^{\mathrm{pz}} V_{|k}^{\mathrm{pz}}, \qquad D_i^{\mathrm{pz}} = e_{ikl}^{\mathrm{pz}} \frac{u_{k|l}^{\mathrm{pz}} + u_{l|k}^{\mathrm{pz}}}{2} + \epsilon_{ik}^{\mathrm{pz}} V_{|k}^{\mathrm{pz}} \tag{34}$$

Here, $C_{ijkl}^{\mathrm{pz}}$, $e_{ikl}^{\mathrm{pz}}$ and $\epsilon_{ik}^{\mathrm{pz}}$ denote (the components of) the fourth-order tensor of elastic material constants, the third-order tensor of piezoelectric material constants and the second-order tensor of dielectric material constants, respectively. These three tensors of material constants characterize completely any piezoelectric material, i.e. its elastic, piezoelectric and dielectric properties. Only one of these tensors is responsible for the piezoelectric effects. Therefore, piezoelectricity can be viewed as a multiphysics problem where in one domain of a piezo-electric medium the problems of elasticity and electricity are coupled by the piezoelectric material constants present in (additional) coupling terms in the constitutive relations. Note that the (linear) kinematic relations, $\varepsilon_{ij}^{\mathrm{pz}} = (u_{k|l}^{\mathrm{pz}} + u_{k|l}^{\mathrm{pz}})/2$, linking mechanical strain ($\varepsilon_{ij}^{\mathrm{pz}}$) and displacements ($u_i^{\mathrm{pz}}$), and the Maxwell's law for electrostatics, $E_i^{\mathrm{pz}} = -V_{|i}^{\mathrm{pz}}$, relating the electric field ($E_i^{\mathrm{pz}}$) with its potential ($V^{\mathrm{pz}}$), have been explicitly used in Equations (34).

The constitutive equations (34) are given for a general case of anisotropic piezoelectricity. However, piezoelectric materials are usually treated as orthotropic or even transversally isotropic. This is certainly valid for thin piezoelectric patches with through-thickness polarization, which are often used in smart structures as fixed-on-surface piezoelectric actuators and sensors. In this case, assuming that the thickness (and therefore, the polarization) of the piezo-patch is directed along the $x_3$ axis, the orthotropic constitutive equations of the piezoelectric material can be expressed as follows, where the formulas for stresses are

$$
\begin{aligned}
\sigma_{11}^{\mathrm{pz}} &= \hat{C}_{11}^{\mathrm{pz}}\, u_{1|1}^{\mathrm{pz}} + \hat{C}_{12}^{\mathrm{pz}}\, u_{2|2}^{\mathrm{pz}} + \hat{C}_{13}^{\mathrm{pz}}\, u_{3|3}^{\mathrm{pz}} - \hat{e}_{31}^{\mathrm{pz}}\, V_{|3}^{\mathrm{pz}} \\
\sigma_{22}^{\mathrm{pz}} &= \hat{C}_{12}^{\mathrm{pz}}\, u_{1|1}^{\mathrm{pz}} + \hat{C}_{22}^{\mathrm{pz}}\, u_{2|2}^{\mathrm{pz}} + \hat{C}_{23}^{\mathrm{pz}}\, u_{3|3}^{\mathrm{pz}} - \hat{e}_{32}^{\mathrm{pz}}\, V_{|3}^{\mathrm{pz}} \\
\sigma_{33}^{\mathrm{pz}} &= \hat{C}_{13}^{\mathrm{pz}}\, u_{1|1}^{\mathrm{pz}} + \hat{C}_{23}^{\mathrm{pz}}\, u_{2|2}^{\mathrm{pz}} + \hat{C}_{33}^{\mathrm{pz}}\, u_{3|3}^{\mathrm{pz}} - \hat{e}_{33}^{\mathrm{pz}}\, V_{|3}^{\mathrm{pz}} \\
\sigma_{23}^{\mathrm{pz}} &= \hat{C}_{44}^{\mathrm{pz}}\, (u_{2|3}^{\mathrm{pz}} + u_{3|2}^{\mathrm{pz}}) - \hat{e}_{24}^{\mathrm{pz}}\, V_{|2}^{\mathrm{pz}} \\
\sigma_{13}^{\mathrm{pz}} &= \hat{C}_{55}^{\mathrm{pz}}\, (u_{1|3}^{\mathrm{pz}} + u_{3|1}^{\mathrm{pz}}) - \hat{e}_{15}^{\mathrm{pz}}\, V_{|1}^{\mathrm{pz}} \\
\sigma_{12}^{\mathrm{pz}} &= \hat{C}_{66}^{\mathrm{pz}}\, (u_{1|2}^{\mathrm{pz}} + u_{2|1}^{\mathrm{pz}})
\end{aligned}
\tag{35}
$$

and for the electric displacements are

$$
\begin{aligned}
D_1^{\mathrm{pz}} &= \hat{e}_{15}^{\mathrm{pz}}\, (u_{1|3}^{\mathrm{pz}} + u_{3|1}^{\mathrm{pz}}) + \epsilon_{11}^{\mathrm{pz}}\, V_{|1}^{\mathrm{pz}} \\
D_2^{\mathrm{pz}} &= \hat{e}_{24}^{\mathrm{pz}}\, (u_{2|3}^{\mathrm{pz}} + u_{3|2}^{\mathrm{pz}}) + \epsilon_{22}^{\mathrm{pz}}\, V_{|2}^{\mathrm{pz}} \\
D_3^{\mathrm{pz}} &= \hat{e}_{31}^{\mathrm{pz}}\, u_{1|1}^{\mathrm{pz}} + \hat{e}_{32}^{\mathrm{pz}}\, u_{2|2}^{\mathrm{pz}} + \hat{e}_{33}^{\mathrm{pz}}\, u_{3|3}^{\mathrm{pz}} + \epsilon_{33}^{\mathrm{pz}}\, V_{|3}^{\mathrm{pz}}
\end{aligned}
\tag{36}
$$

Moreover, for transversal isotropy the following relations are satsfied by the elastic material constants:

$$
\hat{C}_{22}^{\mathrm{pz}} = \hat{C}_{11}^{\mathrm{pz}}, \qquad \hat{C}_{23}^{\mathrm{pz}} = \hat{C}_{13}^{\mathrm{pz}}, \qquad \hat{C}_{55}^{\mathrm{pz}} = \hat{C}_{44}^{\mathrm{pz}}, \qquad \hat{C}_{66}^{\mathrm{pz}} = \frac{\hat{C}_{11}^{\mathrm{pz}} - \hat{C}_{12}^{\mathrm{pz}}}{2}
\tag{37}
$$

by the piezoelectric constants:

$$
\hat{e}_{24}^{\mathrm{pz}} = \hat{e}_{15}^{\mathrm{pz}}, \qquad \hat{e}_{32}^{\mathrm{pz}} = \hat{e}_{33}^{\mathrm{pz}}
\tag{38}
$$

and by the dielectric constants:

$$
\epsilon_{22}^{\mathrm{pz}} = \epsilon_{11}^{\mathrm{pz}}
\tag{39}
$$

In the above relations, some pairs of indices have been replaced by new subscripts as given by the following rule of change (known from the so-called Kelvin–Voigt notation): $11 \mapsto 1$, $22 \mapsto 2$, $33 \mapsto 3$, $23 \mapsto 4$, $13 \mapsto 5$ and $12 \mapsto 6$. Thus, $C_{ijkl}^{\mathrm{pz}}$ is represented as $\hat{C}_{IJ}^{\mathrm{pz}}$ and $e_{kij}^{\mathrm{pz}}$ as $\hat{e}_{kI}^{\mathrm{pz}}$, where $i, j, k, l = 1, 2, 3$ and $I, J = 1, \ldots, 6$. This contracted notation allows 21 independent anisotropic elastic material constants to be grouped in a $6 \times 6$ symmetrical matrix and the third-order tensor of piezoelectric effects to be represented as a $3 \times 6$ matrix. Notice again that in the case of orthotropic elasticity there are only nine independent nonzero elastic matrix components, namely $\hat{C}_{11}^{pz}$, $\hat{C}_{12}^{pz}$, $\hat{C}_{13}^{pz}$, $\hat{C}_{22}^{pz}$, $\hat{C}_{23}^{pz}$, $\hat{C}_{33}^{pz}$, $\hat{C}_{44}^{pz}$, $\hat{C}_{55}^{pz}$ and $\hat{C}_{66}^{pz}$. Moreover, in the case of transversal isotropy only five of them are independent in view of the relations (37). The number of nonzero components of the tensor of piezoelectric constants is five, which are $\hat{e}_{15}^{pz}$, $\hat{e}_{24}^{pz}$, $\hat{e}_{31}^{pz}$, $\hat{e}_{32}^{pz}$ and $\hat{e}_{33}^{pz}$, but only three of them are independent, as stated by Equations (38), in the case of transversal isotropy. The tensor of dielectric constants has a diagonal matrix representation, but only two independent components, $\epsilon_{11}^{pz}$ and $\epsilon_{33}^{pz}$, in the transversal isotropic case where $\epsilon_{22}^{pz} = \epsilon_{11}^{pz}$. All the mentioned features have to be used in the constitutive equations (35) and (36). Finally, instead of the dielectric constants, $\epsilon_{ik}^{pz}$, there is often a preference to use nondimensional constants, the so-called relative permittivities, $\hat{e}_{ik}^{pz}$, which are defined relative to the electric permittivity of a vacuum. Then, the following relation is used:

$$\epsilon_{ik}^{pz} = \epsilon_0 \, \hat{e}_{ik}^{pz} \tag{40}$$

where $\epsilon_0 = 8.854 \times 10^{-12}$ F/m is the permittivity of the vacuum.

### 8.4.3.1 Boundary Conditions

In piezoelectricity the boundary conditions are divided into two groups: mechanical conditions (referring to the elasticity problem) and electrical conditions (referring to the electricity). Consequently, the boundary of the piezoelectric domain can be subdivided as follows:

$$\Gamma_{pz} = \Gamma_{pz}^t \cup \Gamma_{pz}^u \quad \text{and} \quad \Gamma_{pz} = \Gamma_{pz}^Q \cup \Gamma_{pz}^V \tag{41}$$

The parts belonging to the same group of subdivision are disjoint and both subdivisions are completely independent. Here, $\Gamma_{pz}^t$ and $\Gamma_{pz}^Q$ pertain to the Neumann conditions for surface-applied mechanical forces and electric charge, respectively, while $\Gamma_{pz}^u$ and $\Gamma_{pz}^V$ refer to the Dirichlet conditions on imposed mechanical displacements and electric potential, respectively. The third possibility of the Robin boundary condition is skipped; however, it would involve another parts, one in the mechanical and one in the electric subdivision of the boundary.

First, consider the mechanical boundary conditions. The forces, $\hat{t}_i^{pz}$, applied to a boundary are expressed by the Neumann (or natural) condition

$$\sigma_{ij}^{pz} \, n_j^{pz} = \hat{t}_i^{pz} \text{ on } \Gamma_{pz}^t \tag{42}$$

whereas the imposed displacements, $\hat{u}_i^{pz}$, will appear in the Dirichlet (i.e. essential) boundary condition

$$u_i^{pz} = \hat{u}_i^{pz} \text{ on } \Gamma_{pz}^u \tag{43}$$

The Dirichlet condition must be *a priori* explicitly met by the trial functions while the Neumann condition (42) is used for the mechanical boundary integral, i.e. the third term in Equation (33), which equals

$$\mathcal{BI}_{pz}^{mech} = \int_{\Gamma_{pz}} \sigma_{ij}^{pz} \, n_j^{pz} \, \delta u_i^{pz} = \int_{\Gamma_{pz}^t} \hat{t}_i^{pz} \, \delta u_i^{pz} \tag{44}$$

since $\delta u_i^{pz} = 0$ on $\Gamma_{pz}^u$.

The electric boundary condition of the Neumann kind serves for a surface electric charge $\hat{Q}^{\mathrm{pz}}$ applied on a boundary:

$$-D_i^{\mathrm{pz}} \, n_i^{\mathrm{pz}} = \hat{Q}^{\mathrm{pz}} \text{ on } \Gamma_{\mathrm{pz}}^Q \tag{45}$$

whereas the Dirichlet condition allows the electric potential $\hat{V}^{\mathrm{pz}}$ to be prescribed on a boundary:

$$V^{\mathrm{pz}} = \hat{V}^{\mathrm{pz}} \text{ on } \Gamma_{\mathrm{pz}}^V \tag{46}$$

The electric boundary integral, i.e. the last term in Equation (33), equals

$$\mathcal{BI}_{\mathrm{pz}}^{\mathrm{elec}} = -\int_{\Gamma_{\mathrm{pz}}} D_i^{\mathrm{pz}} \, n_i^{\mathrm{pz}} \, \delta V^{\mathrm{pz}} = \int_{\Gamma_{\mathrm{pz}}^Q} \hat{Q}^{\mathrm{pz}} \, \delta V^{\mathrm{pz}} \tag{47}$$

Here, the Neumann condition for electric charge (45) has been used together with the condition for voltage variation, $\delta V^{\mathrm{pz}} = 0$ on $\Gamma_{\mathrm{pz}}^V$.

By summing up the mechanical and electrical boundary integrals (44) and (47), the following total mechanical–electric boundary integral results:

$$\mathcal{BI}_{\mathrm{pz}} = \mathcal{BI}_{\mathrm{pz}}^{\mathrm{mech}} + \mathcal{BI}_{\mathrm{pz}}^{\mathrm{elec}} = \int_{\Gamma_{\mathrm{pz}}^t} \hat{t}_i^{\mathrm{pz}} \, \delta u_i^{\mathrm{pz}} + \int_{\Gamma_{\mathrm{pz}}^Q} \hat{Q}^{\mathrm{pz}} \, \delta V^{\mathrm{pz}} \tag{48}$$

### 8.4.4 Weak Form for an Acoustic Medium

Classical acoustic media are homogeneous inviscid perfectly elastic fluids where compressional acoustic waves propagate, with velocity being the material property of the medium termed as the speed of sound. The classical linear time-harmonic acoustics is governed by the Helmholtz equation. The derivation of this equation may be found in many textbooks, e.g. in Reference [35]. Finite (and infinite) element methods for time-harmonic acoustics are reviewed in References [36] and [37]. Below, the weak integral form (used by the FEM) of harmonic acoustics is given and the relevant natural and essential boundary conditions are briefly discussed.

Let $\Omega_{\mathrm{a}}$ be an acoustic medium domain and $\Gamma_{\mathrm{a}}$ its boundary with $n_i^{\mathrm{a}}$ being the components of unit normal vector pointing outside the domain. The dependent variable of the acoustical medium is the acoustic pressure, $p^{\mathrm{a}}$. For harmonic motion with the angular frequency $\omega$, the following weak form should be used:

$$\mathcal{WF}_{\mathrm{a}} = -\int_{\Omega_{\mathrm{a}}} \frac{1}{\omega^2 \varrho_{\mathrm{a}}} \, p_{|i}^{\mathrm{a}} \, \delta p_{|i}^{\mathrm{a}} + \int_{\Omega_{\mathrm{a}}} \frac{1}{K_{\mathrm{a}}} \, p^{\mathrm{a}} \delta p^{\mathrm{a}} + \int_{\Gamma_{\mathrm{a}}} \frac{1}{\omega^2 \varrho_{\mathrm{a}}} \, p_{|i}^{\mathrm{a}} \, n_i^{\mathrm{a}} \, \delta p^{\mathrm{a}} = 0 \tag{49}$$

where $\varrho_{\mathrm{a}}$ and $K_{\mathrm{a}}$ are the acoustic medium mass density and bulk modulus, respectively. In the case of fluids (especially gases), usually the given data describe how fast a sound wave propagates in the medium. Therefore, the bulk modulus can always be replaced by $K_{\mathrm{a}} = \varrho_{\mathrm{a}} c_{\mathrm{a}}^2$, where $c_{\mathrm{a}}$ is the speed of sound. However, in the case of fluid-equivalent models of porous materials (with rigid frame) there is often a preference to use the bulk modulus which, together with the (now effective) density, is a frequency-dependent quantity, i.e. $K_{\mathrm{a}} = \tilde{K}_{\mathrm{a}}(\omega)$ and $\varrho_{\mathrm{a}} = \tilde{\varrho}_{\mathrm{a}}(\omega)$. Knowing the acoustic pressure, the (complex amplitudes of) displacements, velocities and accelerations of a fluid particle can always be determined using the following formulas:

$$u_i^a = \frac{1}{\omega^2 \varrho_a} p_{|i}^a, \qquad v_i^a = j \omega u_i^a = -\frac{1}{j \omega \varrho_a} p_{|i}^a, \qquad a_i^a = -\omega^2 u_i^a = -\frac{1}{\varrho_a} p_{|i}^a \qquad (50)$$

### 8.4.4.1 Boundary Conditions

Two kinds of boundary conditions will be considered: the Neumann condition when a rigid piston of known accaleration, $\hat{a}_i^a$, is imposed on a boundary and the Dirichlet condition when a value of acoustic pressure, $\hat{p}^a$, is prescribed. In the harmonic case, $\hat{a}_i^a = -\omega^2 \hat{u}_i^a$ with $\hat{u}_i^a$ being the (complex) amplitude of displacements, and the Neumann condition reads

$$\frac{1}{\omega^2 \varrho_a} p_{|i}^a = \hat{u}_i^a \text{ on } \Gamma_a^u \qquad (51)$$

The Dirichlet boundary condition simply states that

$$p^a = \hat{p}^a \text{ on } \Gamma_a^p \qquad (52)$$

As in the case of poroelastic, elastic and piezoelectric media, the third (i.e. Robin's) kind of boundary conditions is skipped in the present discussion.

Now, using the Neumann condition (51) and the condition for pressure variation, $\delta p^a = 0$ on $\Gamma_a^p$, the boundary integral, i.e. the last term in Equation (49), can be written as follows:

$$\mathcal{BI}_a = \int_{\Gamma_a} \frac{1}{\omega^2 \varrho_a} p_{|i}^a n_i^a \delta p^a = \int_{\Gamma_a^u} u_i^a n_i^a \delta p^a \qquad (53)$$

## 8.5 Boundary Conditions for Poroelastic Medium

### 8.5.1 The Boundary Integral

The boundary integral in the weak variational form of the mixed formulation of poroelasticity (25) has the following form:

$$\mathcal{BI}_p = \int_{\Gamma_p} \sigma_{ij}^t n_j \delta u_i + \int_{\Gamma_p} \phi (U_i - u_i) n_i \delta p \qquad (54)$$

Here, two types of boundary conditions that may occur at the boundary of the poroelastic medium will be discussed. Although some other conditions might be formally applied, these two are the most representative and important in practice. In other words, skipped will be, for example, the mixed conditions that independently prescribe in the same point of the boundary different fields to both phases.

### 8.5.2 Imposed Displacement Field

A displacement field, $\hat{u}_i$, applied on a boundary of the poroelastic medium describes, for example, the case of a piston in motion acting on the surface of the medium. Here, it is assumed that the solid skeleton is fixed to the surface of the piston while the fluid obviously cannot penetrate into the piston. Therefore,

$$u_i = \hat{u}_i, \qquad (U_i - u_i) n_i = 0 \tag{55}$$

The first condition expresses the continuity between the imposed displacement vector and the solid phase displacement vector. The second equation expresses the continuity of normal displacements between the solid phase and the fluid phase. Using these conditions and the fact that the variations of the known solid displacements are zero ($\delta u_i = 0$), the boundary integral reduces to zero [26]:

$$\mathcal{BI}_\mathrm{p} = 0 \tag{56}$$

Notice that this result also holds when the poroelastic medium is not glued but only adherent to the rigid piston, providing that there is no friction or any imposed tangential forces at the interface between the piston and the poroelastic medium. In that case, the second one of the boundary conditions (55) holds (and so the second term of the boundary integral of Equation (54) disappears) whereas, instead of the three constraints for the solid displacements given by the first one of the conditions (55), there is one requirement for the *normal* solid displacement $u_i n_i = \hat{u}$ (where $\hat{u}$ is the prescribed normal displacement of piston), and two additional requirements about the total stress vector ($\sigma_{ij}^\mathrm{t} n_j$), which state that the components tangential to the surface of the piston are zero. This assumption together with the fact that the variation of the prescribed normal component of the solid displacement must be zero ($\delta u_i n_i = 0$) make the first term of the boundary integral (54) vanish, and so the result (56) is valid. Remember, however, that this result cannot be used if the friction occurs between the piston and the poroelastic medium, or if any tangential forces are imposed. In this latter (rather academic) case the prescribed tangential forces would appear in the boundary integral. The case of friction can be important in practice and will yield a nonlinear boundary condition.

### 8.5.3 Imposed Pressure Field

A harmonic pressure field of amplitude $\hat{p}$ is imposed on the boundary of the poroelastic domain, which means that it affects at the same time the fluid in the pores and the solid skeleton. Therefore, the following boundary conditions must be met:

$$p = \hat{p}, \qquad \sigma_{ij}^\mathrm{t} n_j = -\hat{p} n_i \tag{57}$$

The first condition is of the Dirichlet type and must be applied explicitly. It describes the continuity of pressure in the fluid. It also means that the pressure variation is zero ($\delta p = 0$) at the boundary. The second condition expresses the continuity of the total normal stress. All this, when used for Equation (54), leads to the following boundary integral [26, 27]:

$$\mathcal{BI}_\mathrm{p} = -\int_{\Gamma_\mathrm{p}} \hat{p} \, n_i \, \delta u_i \tag{58}$$

Now, consider an important case when there is no pressure (nor any displacement field) applied on the boundary of a poroelastic medium. In spite of appearances, this is not identical with, but can only be approximated by, the case when the pressure at the boundary is kept at zero

($\hat{p} = 0$). Then, the boundary integral vanishes, $\mathit{BI}_p = 0$, and only the Dirichlet boundary condition, $p = 0$, must be applied.

## 8.6 Interface Coupling Conditions for Poroelastic and Other Media

### 8.6.1 Poroelastic–Poroelastic Coupling

To begin with, consider the coupling conditions between two different poroelastic media (domains) fixed to one another. The superscripts 1 and 2 (put in parentheses) denote which domain the superscripted quantity belongs to. Let $\Gamma_{(1)\text{-}(2)}$ be an interface between the two media and let $n_i^{(1)}$ be the components of the unit vector normal to the interface and pointing outside medium 1 (and into medium 2), while $n_i^{(2)}$ are the components of the unit normal vector pointing outside medium 2 (into medium 1), which means that at every point of the interface $n_i^{(2)} = -n_i^{(1)}$. The coupling integral terms (given at the interface) are a combination of the boundary integrals resulting from the weak variational forms (25) obtained for both poroelastic domains, i.e.

$$\mathit{CI}_{(1)\text{-}(2)} = \int_{\Gamma_{(1)\text{-}(2)}} \sigma_{ij}^{t(1)} n_j^{(1)} \delta u_i^{(1)} + \int_{\Gamma_{(1)\text{-}(2)}} \phi_{(1)}(U_i^{(1)} - u_i^{(1)}) n_i^{(1)} \delta p^{(1)}$$

$$+ \int_{\Gamma_{(1)\text{-}(2)}} \sigma_{ij}^{t(2)} n_j^{(2)} \delta u_i^{(2)} + \int_{\Gamma_{(1)\text{-}(2)}} \phi_{(2)}(U_i^{(2)} - u_i^{(2)}) n_i^{(2)} \delta p^{(2)} \tag{59}$$

It will be demonstrated that this coupling integral (resulting from the weak form (25) of the mixed formulation of harmonic poroelasticity) equals zero, which means that the coupling conditions are *naturally handled* [26,27] at the interface between two domains made of poroelastic materials.

At the interface between two poroelastic media, the following coupling conditions must be met:

$$\sigma_{ij}^{t(1)} n_j^{(1)} = \sigma^{t(2)} n_j^{(1)}, \qquad \phi_{(1)}(U_i^{(1)} - u_i^{(1)}) n_i^{(1)} = \phi_{(2)}(U_i^{(2)} - u_i^{(2)}) n_i^{(1)}$$

$$u_i^{(1)} = u_i^{(2)}, \qquad p^{(1)} = p^{(2)} \tag{60}$$

The first condition ensures the continuity of total stresses while the second one ensures the continuity of the relative mass flux across the interface. The two last conditions express the continuity of the solid phase displacements and of the pressure of porefluids, respectively. This also indicates that the appropriate variations are the same (i.e. $\delta u_i^{(1)} = \delta u_i^{(2)}$ and $\delta p^{(1)} = \delta p^{(2)}$). Now, applying the coupling conditions for Equation (59) and taking into account that $n_i^{(2)} = -n_i^{(1)}$, it is easy to obtain the following result:

$$\mathit{CI}_{(1)\text{-}(2)} = 0 \tag{61}$$

which means that the coupling conditions between two poroelastic media are indeed naturally handled [26, 27].

### 8.6.2 Poroelastic–Elastic Coupling

Let $\Gamma_{p\text{-}e}$ be an interface between poroelastic and elastic media. Let $n_i$ be the components of the unit vector normal to the interface and pointing outside the poroelastic domain into the elastic one. The coupling integral combines the boundary integral terms resulting from both poroelastic and elastic weak forms (Equations (25) and (26), respectively):

$$CI_{\text{p-e}} = \int_{\Gamma_{\text{p-e}}} \sigma_{ij}^{\text{t}} \, n_j \, \delta u_i + \int_{\Gamma_{\text{p-e}}} \phi \, (U_i - u_i) \, n_i \, \delta p + \int_{\Gamma_{\text{p-e}}} \sigma_{ij}^{\text{e}} \, n_j^{\text{e}} \, \delta u_i^{\text{e}} \qquad (62)$$

where $n_i^{\text{e}} = -n_i$ are the components of the unit normal vector pointing outside the elastic domain (and into the poroelastic medium). Now, the following coupling conditions must be met at the interface:

$$\sigma_{ij}^{\text{t}} \, n_j = \sigma_{ij}^{\text{e}} \, n_j \,, \qquad (U_i - u_i) \, n_i = 0 \,, \qquad u_i = u_i^{\text{e}} \qquad (63)$$

The first condition states the continuity of the total stress tensor, the second one expresses that there is no mass flux across the interface and the last one assumes the continuity of the solid displacements. The last condition also involves the equality of the variations of displacements, $\delta u_i = \delta u_i^{\text{e}}$. Now, applying the coupling conditions for the coupling integral (62) results in

$$CI_{\text{p-e}} = 0 \qquad (64)$$

This is similar to the result obtained for coupling between two poroelastic domains: the coupling between poroelastic and elastic media is also *naturally handled* [26, 27].

### 8.6.3 Poroelastic–Acoustic Coupling

Now, the coupling between poroelastic and acoustic media will be discussed. Let $\Gamma_{\text{p-e}}$ be an interface between a poroelastic material and an acoustic medium, with $n_i$ being the components of the unit vector normal to the interface and pointing outside the poroelastic domain (and into the acoustic medium), whereas $n_i^{\text{a}}$ are the components of the similar unit normal vector pointing in the opposite direction; therefore, in every point of the interface $n_i^{\text{a}} = -n_i$. The coupling integral is a combination of the boundary integral terms from the poroelastic weak form (25) and the acoustic weak form (49):

$$CI_{\text{p-a}} = \int_{\Gamma_{\text{p-a}}} \sigma_{ij}^{\text{t}} \, n_j \, \delta u_i + \int_{\Gamma_{\text{p-a}}} \phi \, (U_i - u_i) \, n_i \, \delta p + \int_{\Gamma_{\text{p-a}}} \frac{1}{\omega^2 \varrho_{\text{a}}} \, p_{|i}^{\text{a}} \, n_i^{\text{a}} \, \delta p^{\text{a}} \qquad (65)$$

The coupling conditions between the two media express the continuity of (total) stresses, (total) normal displacements and pressure, respectively:

$$\sigma_{ij}^{\text{t}} \, n_j = -p \, n_i \,, \qquad \frac{1}{\omega^2 \varrho_{\text{a}}} \, p_{|i}^{\text{a}} \, n_i^{\text{a}} = u_i^{\text{t}} \, n_i^{\text{a}} \,, \qquad p = p^{\text{a}} \qquad (66)$$

Now, using these conditions and the expression for the total displacements of the poroelastic medium, $u_i^t = (1 - \phi) u_i + \phi U_i$, the coupling integral (65) [26, 27] simplifies to

$$CI_{\text{p-a}} = -\int_{\Gamma_{\text{p-a}}} p\, n_i\, \delta u_i + \int_{\Gamma_{\text{p-a}}} \phi\, (U_i - u_i)\, n_i\, \delta p - \int_{\Gamma_{\text{p-a}}} \left[(1 - \phi)\, u_i + \phi\, U_i\right] n_i\, \delta p$$

$$= -\int_{\Gamma_{\text{p-a}}} \left(p\, n_i\, \delta u_i + u_i\, n_i\, \delta p\right) = -\int_{\Gamma_{\text{p-a}}} \delta(p\, u_i\, n_i) \tag{67}$$

### 8.6.4 Acoustic–Elastic Coupling

The coupling integral on an interface $\Gamma_{\text{a-e}}$ between elastic and acoustic subdomains reads as follows:

$$CI_{\text{a-e}} = \int_{\Gamma_{\text{a-e}}} \frac{1}{\omega^2 \varrho_a}\, p_{|i}^a\, n_i^a\, \delta p^a + \int_{\Gamma_{\text{a-e}}} \sigma_{ij}^e\, n_j^e\, \delta u_i^e \tag{68}$$

On the interface, the conditions of continuity of the displacements normal to the interface and normal stresses must be satisfied, i.e.

$$\frac{1}{\omega^2 \varrho_a}\, p_{|i}^a\, n_i^a = u_i^e\, n_i^a, \qquad \sigma_{ij}^e\, n_j^e = -p^a\, n_i^e \tag{69}$$

These conditions are used for the integral and since on the interface the two unit normal vectors are in the opposite direction to one another, i.e. $n_i^a = -n_i^e$, the interface coupling integral (68) simplifies to

$$CI_{\text{a-e}} = \int_{\Gamma_{\text{a-e}}} \left(p^a\, n_i^a\, \delta u_i^e + u_i^e\, n_i^a\, \delta p^a\right) = \int_{\Gamma_{\text{a-e}}} \delta(p^a\, u_i^e\, n_i^a) \tag{70}$$

Obviously, this result is also valid and, moreover, complete in the case of a piezoelectric medium in contact with an acoustic one, since the interface coupling occurs explicitly only between the acoustic problem and its mechanical (i.e. elastic) counterpart in the piezoelectric subdomain. To be formal, $\Gamma_{\text{a-e}}$ should only be changed to $\Gamma_{\text{a-pz}}$ and $u_i^e$ to $u_i^{\text{pz}}$ in the formulas given above.

## 8.7 Galerkin Finite Element Model of a Coupled System of Piezoelectric, Elastic, Poroelastic and Acoustic Media

### 8.7.1 A Coupled Multiphysics System

Consider a coupled multiphysics system made up of $N_{\text{pz}}$ piezoelectric, $N_e$ elastic, $N_p$ poroelastic and $N_a$ acoustic subdomains; the subdomains are denoted by, respectively, $\Omega_{\text{pz}(N)}$ where $N \in \{1, \ldots, N_{\text{pz}}\}$, $\Omega_{e(N)}$ where $N \in \{1, \ldots, N_e\}$, $\Omega_{p(N)}$ where $N \in \{1, \ldots, N_p\}$ and $\Omega_{a(N)}$ where $N \in \{1, \ldots, N_a\}$. In practice, the system can describe an active composite noise absorber or insulator, in particular, a panel composed of a couple of poroelastic layers fixed to an elastic faceplate (or faceplates) to which some patches of piezoelectric transducers (actuators or sensors) are glued, the whole assembly being immersed in an acoustic medium. The Galerkin method will be used to approach the problem by means of finite elements. For simplicity, the analysis will be reduced to the case of consistent finite element meshes where the nodes of two contacting domains are coincident on the interface. In the general case of nonconsistent meshes, a standard method of Lagrange multipliers must be used to deal with noncontiguous nodes.

The assumption of consistent meshes means that the essential interface continuity conditions can be easily met at the discretized interfaces since the corresponding degrees of freedom are at contiguous nodes. This concerns the field continuity conditions at the interfaces between subdomains of the same kind (for example, $u_i^{(1)} = u_i^{(2)}$ on $\Gamma_{p(1)\text{-}p(2)}$, $u_i^{e(1)} = u_i^{e(2)}$ on $\Gamma_{e(1)\text{-}e(2)}$ or $p^{a(1)} = p^{a(2)}$ on $\Gamma_{a(1)\text{-}a(2)}$, etc., assuming that the exemplary interfaces exist) as well as the displacement continuity conditions on the interfaces between poroelastic, elastic and piezoelectric subdomains (for example, $u_i^{(1)} = u_i^{e(1)}$ on $\Gamma_{p(1)\text{-}e(1)}$ or $u_i^{e(1)} = u_i^{pz(1)}$ on $\Gamma_{e(1)\text{-}pz(1)}$, etc.) and the pressure continuity conditions on the interfaces between poroelastic and acoustic subdomains (for example, $p^{(1)} = p^{a(1)}$ on $\Gamma_{p(1)\text{-}a(1)}$, etc.). As a matter of fact, in the assumed notation these interface continuity conditions will be implicitly imposed by choosing the same variables referring to the similar fields in different subdomains. Therefore, the same field of displacements, $u_i$, will be used defined continuously in poroelastic, $\Omega_p$, elastic, $\Omega_e$, and piezoelectric, $\Omega_{pz}$, domains, as well as the field of pressure, $p$, continuous through poroelastic, $\Omega_p$, and acoustic, $\Omega_a$, domains, and the field of electric potential continuous in a whole piezoelectric domain, $\Omega_{pz}$. The mentioned poroelastic, acoustic, elastic and piezoelectric domains are unions of all subdomains of the same kind; i.e. they are formally defined as follows (see also the diagram in Figure 8.2, page 299):

$$\Omega_p = \bigcup_{N=1}^{N_p} \Omega_{p(N)}\,, \qquad \Omega_a = \bigcup_{N=1}^{N_a} \Omega_{a(N)}\,, \qquad \Omega_e = \bigcup_{N=1}^{N_e} \Omega_{e(N)}\,, \qquad \Omega_{pz} = \bigcup_{N=1}^{N_{pz}} \Omega_{pz(N)} \qquad (71)$$

Obviously, the fields $u_i$ and $p$, although being continuous functions of position $x$ in $\Omega_p \cup \Omega_e \cup \Omega_{pz}$ and in $\Omega_p \cup \Omega_a$, respectively, they have, however, a slightly different meaning in different domains; namely $u_i$ are the solid phase displacements in $\Omega_p$ and the elastic displacements in $\Omega_e$ or in $\Omega_{pz}$, whereas $p$ is the pore pressure in $\Omega_p$ and the acoustic pressure in $\Omega_a$. Nevertheless, when knowing $u_i$ and $p$, any of the fields defined in a particular subdomain can be specified as follows:

$$u_i^{(N)} = u_i \text{ in } \Omega_{p(N)}, \qquad p^{(N)} = p \text{ in } \Omega_{p(N)}\,, \qquad N \in \{1, \dots, N_p\} \qquad (72)$$

$$p^{a(N)} = p \text{ in } \Omega_{a(N)}\,, \qquad N \in \{1, \dots, N_a\} \qquad (73)$$

$$u_i^{e(N)} = u_i \text{ in } \Omega_{e(N)}\,, \qquad N \in \{1, \dots, N_e\} \qquad (74)$$

$$u_i^{pz(N)} = u_i \text{ in } \Omega_{pz(N)}\,, \qquad V^{pz(N)} = V \text{ in } \Omega_{pz(N)}\,, \qquad N \in \{1, \dots, N_{pz}\} \qquad (75)$$

and then other relevant quantities (like strains or stresses, or electric displacements) can be computed in the particular subdomain.

Consequently, boundaries are defined as adequate unions of boundaries of the subdomains of the same kind. Let it point out here that the term *boundary* refers only to these bounds of any subdomain where boundary conditions (constraints or excitations) are applied, and not to the interfaces with other subdomains. Therefore, for the poroelastic domain the total boundary,

$\Gamma_p$, is an assembly of the boundaries of all poroelastic subdamains, $\Gamma_{p(N)}$ ($N \in \{1, \dots, N_p\}$), i.e.

$$\Gamma_p = \bigcup_{N=1}^{N_p} \Gamma_{p(N)} = \bigcup_{N=1}^{N_p} \left( \Gamma_{p(N)}^u \cup \Gamma_{p(N)}^p \right) = \bigcup_{N=1}^{N_p} \Gamma_{p(N)}^u \cup \bigcup_{N=1}^{N_p} \Gamma_{p(N)}^p = \Gamma_p^u \cup \Gamma_p^p \qquad (76)$$

and, as shown here, this can be divided into two kinds, namely where the value of displacement or the value of pressure is prescribed, $\Gamma_p^u$ or $\Gamma_p^p$, respectively. In fact, two other types could have been distinguished but, like previously, they are left out because of their low practical importance. In other words, only piston displacement or acoustic pressure can be imposed on to the boundaries of poroelastic media.

Similarly, the boundary of the acoustic domain, $\Gamma_a$, incorporates all boundaries of acoustic subdomains, $\Gamma_{a(N)}$ ($N \in \{1, \dots, N_a\}$), i.e.

$$\Gamma_a = \bigcup_{N=1}^{N_a} \Gamma_{a(N)} = \bigcup_{N=1}^{N_a} \left( \Gamma_{a(N)}^p \cup \Gamma_{a(N)}^u \right) = \bigcup_{N=1}^{N_a} \Gamma_{a(N)}^p \cup \bigcup_{N=1}^{N_a} \Gamma_{a(N)}^u = \Gamma_a^p \cup \Gamma_a^u \qquad (77)$$

Here, as well, two parts are distinguished where pressure or displacement boundary conditions are applied, $\Gamma_a^p$ or $\Gamma_a^u$, respectively. This time they are clearly the Dirichlet (essential) and Neumann (natural) boundary conditions.

The boundaries of elastic subdomains, $\Gamma_{a(N)}$ ($N \in \{1, \dots, N_a\}$), make up the total boundary of elastic domain, i.e.

$$\Gamma_e = \bigcup_{N=1}^{N_e} \Gamma_{e(N)} = \bigcup_{N=1}^{N_e} \left( \Gamma_{e(N)}^u \cup \Gamma_{e(N)}^t \right) = \bigcup_{N=1}^{N_e} \Gamma_{e(N)}^u \cup \bigcup_{N=1}^{N_e} \Gamma_{e(N)}^t = \Gamma_e^u \cup \Gamma_e^t \qquad (78)$$

where, again, a part referring to the prescribed displacement (the Dirichlet condition), $\Gamma_e^u$, and a part referring to the prescribed load (the Neumann condition), $\Gamma_e^t$, are discriminated.

The piezoelectric domain represents a slightly more complicated case. The total boundary of the piezoelectric domain, $\Gamma_{pz}$, consists of the boundaries of all piezoelectric subdomains, $\Gamma_{pz(N)}$ ($N \in \{1, \dots, N_{pz}\}$); this time, however, there is a need to distinguish the two independent subdivisions relevant for the mechanical and electrical conditions:

$$\Gamma_{pz} = \bigcup_{N=1}^{N_{pz}} \Gamma_{pz(N)} = \begin{cases} \bigcup_{N=1}^{N_{pz}} \left( \Gamma_{pz(N)}^u \cup \Gamma_{pz(N)}^t \right) = \bigcup_{N=1}^{N_{pz}} \Gamma_{pz(N)}^u \cup \bigcup_{N=1}^{N_{pz}} \Gamma_{pz(N)}^t = \Gamma_{pz}^u \cup \Gamma_{pz}^t \\ \bigcup_{N=1}^{N_{pz}} \left( \Gamma_{pz(N)}^V \cup \Gamma_{pz(N)}^Q \right) = \bigcup_{N=1}^{N_{pz}} \Gamma_{pz(N)}^V \cup \bigcup_{N=1}^{N_{pz}} \Gamma_{pz(N)}^Q = \Gamma_{pz}^V \cup \Gamma_{pz}^Q \end{cases} \qquad (79)$$

Eventually, the essential boundary conditions can formally be written for the coupled system as follows:

$$u_i = \begin{cases} \hat{u}_i \text{ on } \Gamma_p^u, \\ \hat{u}_i^e \text{ on } \Gamma_e^u, \\ \hat{u}_i^{pz} \text{ on } \Gamma_{pz}^u, \end{cases} \qquad p = \begin{cases} \hat{p} \text{ on } \Gamma_p^p, \\ \hat{p}^a \text{ on } \Gamma_a^p, \end{cases} \qquad V = \hat{V}^{pz} \text{ on } \Gamma_{pz}^V \qquad (80)$$

where $\hat{u}_i$, $\hat{u}_i^e$, $\hat{u}_i^{pz}$, $\hat{p}$, $\hat{p}^a$ and $\hat{V}^{pz}$ are given values. The relevant natural boundary conditions are expressed by Equations (57) (the second equation), (30), (42), (51), and (45).

## 8.7.2 Weak Form of the Coupled System

The weak form of coupled multiphysical system combines the weak forms for the corresponding problems presented in Section 8.4. The discussion of coupling interface conditions in Section 8.6 has presented very important results, namely that the coupling of two poroelastic domains, or a poroelastic domain to an elastic one, is naturally handled; i.e. the interface coupling integrals are zero, which results from the continuity of the fields of primary variables. Such a result is also straightforwardly obtained for elastic and piezoelectric domains. This is not the case when coupling to an acoustical domain. Therefore, define (for convenience) the following interface: $\Gamma_{\text{a-p,e,pz}} = \Gamma_{\text{a-p}} \cup \Gamma_{\text{a-e}} \cup \Gamma_{\text{a-pz}}$, which is a simple sum of all surfaces, edges and points where the acoustic domain is coupled to the poroelastic, elastic and piezoelectric domains. Now, the weak form of the coupled system reads

$$
\int_{\Omega_{pz}} \left( -\sigma_{ij}^{pz} \delta u_{i|j} + \omega^2 \varrho_{pz} u_i \delta u_i + D_i^{pz} \delta V_{|i} \right) + \int_{\Gamma_{pz}^t} \hat{t}_i^{pz} \delta u_i + \int_{\Gamma_{pz}^Q} \hat{Q}^{pz} \delta V
$$
$$
+ \int_{\Omega_e} \left( -\sigma_{ij}^e \delta u_{i|j} + \omega^2 \varrho_e u_i \delta u_i \right) + \int_{\Gamma_e^t} \hat{t}_i^e \delta u_i + \int_{\Omega_p} (\mathcal{P}) - \int_{\Gamma_p^p} \hat{p} \, n_i \, \delta u_i
$$
$$
+ \int_{\Omega_a} \left( -\frac{1}{\omega^2 \varrho_a} p_{|i} \delta p_{|i} + \frac{1}{K_a} p \, \delta p \right) + \int_{\Gamma_a^u} \hat{u}_i^a n_i^a \delta p + \int_{\Gamma_{\text{a-p,e,pz}}} n_i^a \left( \delta p \, u_i + p \, \delta u_i \right) = 0 \quad (81)
$$

where $\mathcal{P}$ stands for the integrand for a poroelastic domain and equals

$$
\mathcal{P} = -\sigma_{ij}^{ss} \delta u_{i|j} + \omega^2 \tilde{\varrho} \, u_i \, \delta u_i - \frac{\phi^2}{\omega^2 \tilde{\varrho}_{ff}} p_{|i} \, \delta p_{|i} + \frac{\phi^2}{\tilde{\lambda}_f} p \, \delta p
$$
$$
+ \phi \left( 1 + \frac{\tilde{\varrho}_{sf}}{\tilde{\varrho}_{ff}} \right) \left( \delta p_{|i} \, u_i + p_{|i} \, \delta u_i \right) + \phi \left( 1 + \frac{\tilde{\lambda}_{sf}}{\tilde{\lambda}_f} \right) \left( \delta p \, u_{i|i} + p \, \delta u_{i|i} \right) \quad (82)
$$

In the above equations $u_i$ are the displacements of a piezoelectric or elastic solid or of the solid phase of a poroelastic material, $V$ is the electric potential in the piezoelectric domain and $p$ is the pressure in the acoustic medium or in the pores of poroelastic medium. The variational equation (81) must be satisfied for *all* admissible variations (i.e. virtual or test functions) of primary variables: $\delta u_i$, $\delta V$ and $\delta p$. Furthermore,

$$
\sigma_{ij}^{pz} = C_{ijkl}^{pz} \frac{u_{k|l} + u_{l|k}}{2} + e_{kij}^{pz} V_{|k} , \qquad D_i^{pz} = e_{ikl}^{pz} \frac{u_{k|l} + u_{l|k}}{2} - \epsilon_{ik}^{pz} V_{|k} \quad (83)
$$

$$
\sigma_{ij}^{ss} = \mu_s \left( u_{i|j} + u_{j|i} \right) + \tilde{\lambda}_{ss} \, u_{k|k} \, \delta_{ij} \quad (84)
$$

and

$$
\sigma_{ij}^e = \mu_e \left( u_{i|j} + u_{j|i} \right) + \lambda_e \, u_{k|k} \, \delta_{ij} \qquad \text{or} \qquad \sigma_{ij}^e = C_{ijkl}^e \frac{u_{k|l} + u_{l|k}}{2} \quad (85)
$$

Notice that the terms in the first line of the functional (81) of the coupled multiphysical system refer to the piezoelectric subdomains and another two terms to the elastic subdomains; then there are two integrals pertaining to the poroelastic media and the last line brings terms relevant for the acoustical medium, where the last integral describes the interface coupling to other media. Obviously, all material parameters involved in the functional are functions of position defined on the relevant subdomains. For elastic solids, the first formula in (85) refers to *isotropic* elastic materials while the second one refers to elastic materials in general.

### 8.7.3 Galerkin Finite Element Approximation

The discrete equations for a finite element model will be obtained from the functional (81) by using finite element interpolants for the trial and test functions, as stated by the Galerkin method. Remember that $i, j, k, l \in \{1, 2, 3\}$ are indices referring to the coordinates of the system of reference. Now, new subscripts are introduced for the degrees of freedom of a discrete model, namely $m, n \in \{1, \ldots N_{DOF}\}$, where $N_{DOF}$ is the total number of degrees of freedom. For simplicity and to avoid any inconsistency, the summation convention is in use also for these subscripts.

Let $\mathcal{N}_{im}^u, \mathcal{N}_m^p, \mathcal{N}_m^V$ be the interpolants, i.e. the so-called *global shape functions* defined in the whole domain $\Omega$; they will be used to approximate the fields of displacements, pressure and electric potential, respectively. The shape functions, like all interpolants, satisfy the conditions

$$\mathcal{N}_{im}^u(\mathbf{x}_n) = \delta_{mn}, \qquad \mathcal{N}_m^p(\mathbf{x}_n) = \delta_{mn}, \qquad \mathcal{N}_m^V(\mathbf{x}_n) = \delta_{mn} \qquad (86)$$

where $\mathbf{x}_m$ is the point in $\Omega$ where the $m$th degree of freedom is localized. Although it refers to a nodal point of an FE mesh, it is not exactly the same since one node of the FE mesh may be attributed to several degrees of freedom; for example, in one nodal point of a piezoelectric medium three mechanical displacements and one electrical potential are set. It should be remembered that the subscripts $m$ and $n$ number the degrees of freedom and not the geometrical nodes of the FE mesh.

The trial functions (solutions) are approximated as linear combinations of the relevant shape functions, i.e.

$$u_i(\mathbf{x}) \approx \mathcal{N}_{im}^u(\mathbf{x}) \, q_m, \qquad p(\mathbf{x}) \approx \mathcal{N}_m^p(\mathbf{x}) \, q_m, \qquad V(\mathbf{x}) \approx \mathcal{N}_m^V(\mathbf{x}) \, q_m \qquad (87)$$

where $q_m$ are the degrees of freedom of the discrete model. They form the *global vector of degrees of freedom*, $\mathbf{q}$, and can be divided into five groups of components as follows:

$$\begin{aligned}
q_m \in \mathbf{q}^{u_i} \quad &\text{if } q_m = u_i(\mathbf{x}_m) \quad (i = 1, 2, 3) \\
q_m \in \mathbf{q}^p \quad &\text{if } q_m = p(\mathbf{x}_m) \\
q_m \in \mathbf{q}^V \quad &\text{if } q_m = V(\mathbf{x}_m)
\end{aligned} \qquad (88)$$

Here, $\mathbf{q}^{u_i}$ $(i = 1, 2, 3)$, $\mathbf{q}^p$ and $\mathbf{q}^V$ are subvectors of the vector $\mathbf{q}$, corresponding to the three mechanical displacements, pressure and electric potential, respectively; notice that the properties (86) have been used when defining the subvectors.

It is an obvious observation that the shape functions are zero on nonrelevant subdomains (e.g. $\mathcal{N}_m^V = 0$ on $\Omega / \Omega_{pz}$). Moreover, the definitions (87) show that some of them are zero in the whole domain $\Omega$, namely

$$\mathcal{N}^u_{im}(\boldsymbol{x}) \equiv 0 \text{ if } q_m \notin \mathbf{q}^{u_i}, \qquad \mathcal{N}^p_m(\boldsymbol{x}) \equiv 0 \text{ if } q_m \notin \mathbf{q}^p, \qquad \mathcal{N}^V_m(\boldsymbol{x}) \equiv 0 \text{ if } q_m \notin \mathbf{q}^V \quad (89)$$

Furthermore, the remaining shape functions are, in fact, nonzero only on several neighboring finite elements that share the same degree of freedom $q_m$ (and they are zero outside these elements). This feature is effectively used in finite element programs where the integrations are performed element by element and the results are assembled into a global matrix and vector.

The Galerkin method requires that the same shape functions are also used to approximate the corresponding test functions:

$$\delta u_i(\boldsymbol{x}) \approx \mathcal{N}^u_{im}(\boldsymbol{x})\,\delta q_m, \qquad \delta p(\boldsymbol{x}) \approx \mathcal{N}^p_m(\boldsymbol{x})\,\delta q_m, \qquad \delta V(\boldsymbol{x}) \approx \mathcal{N}^V_m(\boldsymbol{x})\,\delta q_m \qquad (90)$$

Remember that these fields of variations (or test functions) must satisfy the homogeneous boundary conditions on the essential boundaries, i.e.

$$\delta q_m = 0 \text{ on } \Gamma^u_{\text{p}} \cup \Gamma^p_{\text{a}} \cup \Gamma^u_{\text{e}} \cup \Gamma^u_{\text{pz}} \qquad (91)$$

and formally these conditions should also be satisfied by the approximations (90). However, to derive the governing equations, first the model is treated as if there were no boundaries where the primary dependent fields are prescribed. As a matter of fact, this is a standard approach in finite element methods where the essential boundary conditions are applied not until the system of discrete equations is derived.

Using the above approximations (87) and (90) for the functional (81) leads to the following equation:

$$\left(\tilde{A}_{mn}\,q_n - F_m\right)\delta q_m = 0 \qquad (92)$$

which must be satisfied for an arbitrary (yet admissible) set of components $\delta q_m$. This yields the following system of algebraic equations:

$$\tilde{A}_{mn}\,q_n = F_m \qquad (93)$$

where the *governing matrix* and the *right-hand-side vector* can be presented as assembled contributions of piezoelectric, elastic, poroelastic and acoustic subdomains, i.e.

$$\tilde{A}_{mn} = \tilde{A}^{\text{pz}}_{mn} + \tilde{A}^{\text{e}}_{mn} + \tilde{A}^{\text{p}}_{mn} + \tilde{A}^{\text{a}}_{mn} + A^{\text{a-p,e,pz}}_{mn} \qquad (94)$$

$$F_m = F^{\text{pz}}_m + F^{\text{e}}_m + F^{\text{p}}_m + F^{\text{a}}_m \qquad (95)$$

The obtained matrices and vectors contributing to the global system of discrete equations are the result of integrating – over the relevant subdomains, boundaries and interfaces – the terms approximated by the known (i.e. assumed) shape functions. The relevant integrals defining the component matrices and vectors are presented below.

Notice that in the formula for the system governing matrix (94) there is also a contribution, $A^{\text{a-p,e,pz}}_{mn}$, resulting from the coupling on the interface between the acoustic subdomain and the poroelastic and elastic (or piezoelectric) subdomains (the naturally handled coupling between the poroelastic and elastic subdomains provides no contribution). It will be apparent further below that the system matrix and its first four component matrices are frequency-dependent, while the interface–coupling matrix and the right-hand-side vector are not. Therefore, when carrying out a frequency analysis, these latter quantities (i.e. the boundary or interface terms resulting from the Neumann-type excitations or intersubdomain coupling) are to be computed only once, and should be used then for any computational frequency.

The piezoelectric, elastic, poroelastic and acoustic contribution matrices of Equation (94) are themselves composed of the following components:

$$\tilde{A}^{\text{pz}}_{mn} = K^{\text{pz}}_{mn} - \omega^2 M^{\text{pz}}_{mn} + L^{\text{pz}}_{mn} + B^{\text{pz}}_{mn} \tag{96}$$

$$\tilde{A}^{\text{e}}_{mn} = K^{\text{e}}_{mn} - \omega^2 M^{\text{e}}_{mn} \tag{97}$$

$$\tilde{A}^{\text{p}}_{mn} = \tilde{K}^{\text{p}}_{mn} - \omega^2 \tilde{M}^{\text{p}}_{mn} + \frac{1}{\omega^2} \tilde{P}^{\text{p}}_{mn} - \tilde{Q}^{\text{p}}_{mn} - \tilde{R}^{\text{p}}_{mn} - \tilde{S}^{\text{p}}_{mn} \tag{98}$$

$$\tilde{A}^{\text{a}}_{mn} = \frac{1}{\omega^2} P^{\text{a}}_{mn} - Q^{\text{a}}_{mn} \tag{99}$$

The frequency dependence is explicitly shown in the above formulas. However, in the case of the poroelastic subdomain matrix (98), the frequency dependence is also implicit because the component matrices depend on some frequency-dependent parameters of porelastic material. Moreover, the component matrices of the acoustic subdomain matrix (99) could also be frequency-dependent, i.e. $P^{\text{a}}_{mn} = \tilde{P}^{\text{a}}_{mn}(\omega)$ and $Q^{\text{a}}_{mn} = \tilde{Q}^{\text{a}}_{mn}(\omega)$. This happens when the acoustic medium is a fluid-equivalent model of a porus material with a rigid frame rather than a simple fluid (like air). Therefore, in the case of porous materials (both with rigid and elastic frames), the component matrices for corresponding subdomains must be recalculated for every computational frequency, whereas in the case of piezoelectric or elastic media, or perfectly elastic fluids, the component matrices need to be calculated only once and the corresponding subdomain matrices are then simply assembled for every computational frequency using Equations (96), (97) or (99), respectively.

The formulas and nomenclature (based on some physical interpretations) for all the component matrices and vectors will be given below. Moreover, these submatrices and subvectors of the discrete system will be visualized in a diagram (see Section 8.7.4 and Figure 8.3, page 299).

There are four subcomponents in the matrix (96) obtained for the piezoelectric subdomain. They are the *stiffness matrix*,

$$K^{\text{pz}}_{mn} = \tfrac{1}{2} \int_{\Omega_{\text{pz}}} C^{\text{pz}}_{ijkl} \left( \mathcal{N}^u_{kn|l} + \mathcal{N}^u_{ln|k} \right) \mathcal{N}^u_{im|j} \tag{100}$$

the *mass matrix*,

$$M^{\text{pz}}_{mn} = \int_{\Omega_{\text{pz}}} \varrho_{\text{pz}} \mathcal{N}^u_{im} \mathcal{N}^u_{in} \tag{101}$$

the *electric permittivity matrix*,

$$L^{\text{pz}}_{mn} = \int_{\Omega_{\text{pz}}} \epsilon^{\text{pz}}_{ik} \mathcal{N}^V_{m|i} \mathcal{N}^V_{n|k} , \tag{102}$$

and, finally, the *piezoelectric coupling matrix*,

$$
\begin{aligned}
B^{\mathrm{pz}}_{mn} &= \int_{\Omega_{\mathrm{pz}}} e^{\mathrm{pz}}_{kij}\,\mathcal{N}^{V}_{n|k}\,\mathcal{N}^{u}_{im|j} - \tfrac{1}{2}\int_{\Omega_{\mathrm{pz}}} e^{\mathrm{pz}}_{ikl}\left(\mathcal{N}^{u}_{kn|l}+\mathcal{N}^{u}_{ln|k}\right)\mathcal{N}^{V}_{m|i} \\
&= \int_{\Omega_{\mathrm{pz}}} e^{\mathrm{pz}}_{ikl}\left[\mathcal{N}^{V}_{n|i}\,\mathcal{N}^{u}_{km|l} - \tfrac{1}{2}\left(\mathcal{N}^{u}_{kn|l}+\mathcal{N}^{u}_{ln|k}\right)\mathcal{N}^{V}_{m|i}\right]
\end{aligned}
\tag{103}
$$

As shown in Equation (98), six component matrices are distinguished for the poroelastic subdomain, namely the *stiffness matrix* of the skeleton *in vacuo*,

$$
\tilde{K}^{\mathrm{p}}_{mn} = \int_{\Omega_{\mathrm{p}}}\left[\mu_{\mathrm{s}}\left(\mathcal{N}^{u}_{in|j}+\mathcal{N}^{u}_{jn|i}\right)\mathcal{N}^{u}_{im|j} + \tilde{\lambda}_{\mathrm{ss}}\,\mathcal{N}^{u}_{im|i}\,\mathcal{N}^{u}_{jn|j}\right]
\tag{104}
$$

the *mass matrix*,

$$
\tilde{M}^{\mathrm{p}}_{mn} = \int_{\Omega_{\mathrm{p}}} \tilde{\varrho}\,\mathcal{N}^{u}_{im}\,\mathcal{N}^{u}_{in}
\tag{105}
$$

the *kinetic* and *compressional energy matrices* of the fluid phase,

$$
\tilde{P}^{\mathrm{p}}_{mn} = \int_{\Omega_{\mathrm{p}}} \frac{\phi^{2}}{\tilde{\varrho}_{\mathrm{ff}}}\,\mathcal{N}^{p}_{m|i}\,\mathcal{N}^{p}_{n|i}\,, \qquad Q^{\mathrm{p}}_{mn} = \int_{\Omega_{\mathrm{p}}} \frac{\phi^{2}}{\tilde{\lambda}_{\mathrm{f}}}\,\mathcal{N}^{p}_{m}\,\mathcal{N}^{p}_{n}
\tag{106}
$$

and, finally, the *matrix of visco–inertial (or kinetic) coupling*,

$$
\tilde{R}^{\mathrm{p}}_{mn} = \int_{\Omega_{\mathrm{p}}} \phi\left(1+\frac{\tilde{\varrho}_{\mathrm{sf}}}{\tilde{\varrho}_{\mathrm{ff}}}\right)\left(\mathcal{N}^{p}_{m|i}\,\mathcal{N}^{u}_{in}+\mathcal{N}^{p}_{n|i}\,\mathcal{N}^{u}_{im}\right)
\tag{107}
$$

and the *matrix of elastic (or potential) coupling*,

$$
\tilde{S}^{\mathrm{p}}_{mn} = \int_{\Omega_{\mathrm{p}}} \phi\left(1+\frac{\tilde{\lambda}_{\mathrm{sf}}}{\tilde{\lambda}_{\mathrm{f}}}\right)\left(\mathcal{N}^{p}_{m}\,\mathcal{N}^{u}_{in|i}+\mathcal{N}^{p}_{n}\,\mathcal{N}^{u}_{im|i}\right)
\tag{108}
$$

These last two coupling matrices can be treated together since they share the same degrees of freedom (as a matter of fact they couple the displacement degrees of freedom with the pressure ones).

The elastic subdomain matrix (97) has two component matrices resulting from the stiffness and inertia of the elastic medium. These *stiffness* and *mass matrices* read as follows:

$$
K^{\mathrm{e}}_{mn} = \int_{\Omega_{\mathrm{e}}}\left[\mu_{\mathrm{e}}\left(\mathcal{N}^{u}_{in|j}+\mathcal{N}^{u}_{jn|i}\right)\mathcal{N}^{u}_{im|j} + \lambda_{\mathrm{e}}\,\mathcal{N}^{u}_{im|i}\,\mathcal{N}^{u}_{jn|j}\right]
\tag{109}
$$

$$
M^{\mathrm{e}}_{mn} = \int_{\Omega_{\mathrm{e}}} \varrho_{\mathrm{e}}\,\mathcal{N}^{u}_{im}\,\mathcal{N}^{u}_{in}
\tag{110}
$$

There are also two component matrices in the case of the acoustic subdomain matrix (99), namely the *kinetic* and *compressional energy matrices*, respectively:

$$P^{\mathrm{a}}_{mn} = \int_{\Omega_{\mathrm{a}}} \frac{1}{\varrho_{\mathrm{a}}} \mathcal{N}^p_{m|i} \mathcal{N}^p_{n|i}, \qquad Q^{\mathrm{a}}_{mn} = \int_{\Omega_{\mathrm{a}}} \frac{1}{K_{\mathrm{a}}} \mathcal{N}^p_m \mathcal{N}^p_n \tag{111}$$

As has already been mentioned, the acoustic medium contribution to the governing matrix (94) of the system (93) arises also from the interface coupling to the poroelastic and elastic (or piezoelectric) media. The relevant *interface coupling matrix* is computed as

$$A^{\mathrm{a\text{-}p,e,pz}}_{mn} = A^{\mathrm{a\text{-}p}}_{mn} + A^{\mathrm{a\text{-}e}}_{mn} + A^{\mathrm{a\text{-}pz}}_{mn} = \int_{\Gamma_{\mathrm{a\text{-}p,e,pz}}} n^{\mathrm{a}}_i \left( \mathcal{N}^p_m \mathcal{N}^u_{in} + \mathcal{N}^p_n \mathcal{N}^u_{im} \right) \tag{112}$$

Finally, the formulas for the component vectors of the right-hand-side vector (95) of the system of equations (93) must be provided. They are

$$F^{\mathrm{pz}}_m = F^{\mathrm{pz}t}_m + F^{\mathrm{pz}Q}_m = \int_{\Gamma^t_{\mathrm{pz}}} \hat{t}^{\mathrm{pz}}_i \mathcal{N}^u_{im} + \int_{\Gamma^Q_{\mathrm{pz}}} \hat{Q}^{\mathrm{pz}} \mathcal{N}^V_m$$

$$F^{\mathrm{e}}_m = \int_{\Gamma^t_{\mathrm{e}}} \hat{t}^{\mathrm{e}}_i \mathcal{N}^u_{im}, \qquad F^{\mathrm{p}}_m = - \int_{\Gamma^p_{\mathrm{p}}} \hat{p}\, n_i \mathcal{N}^u_{im}, \qquad F^{\mathrm{a}}_m = \int_{\Gamma^u_{\mathrm{a}}} \hat{u}^{\mathrm{a}}_i n^{\mathrm{a}}_i \mathcal{N}^p_m \tag{113}$$

These vectors arise from the Neumann boundary conditions of the piezoelectric, elastic, poroelastic and acoustic subdomains, respectively. Notice that in the case of an elastic medium, an imposed pressure or traction results in the Neumann condition, whereas it is a prescribed displacement in the case of an acoustic subdomain. As for the *biphasic* theory of poroelasticity, the mixed displacement–pressure formulation renders the imposed-pressure condition as a hybrid one, i.e. essential for the fluid phase and natural for the solid one. The imposed-displacement condition is essential for the solid phase and naturally handled by the fluid phase (thanks to the mentioned enhancement of the mixed formulation).

### 8.7.4 Submatrices and Couplings in the Algebraic System

The linear algebraic system of equations (93) constitutes a discrete model of a problem involving poroelastic, acoustic, elastic and piezoelectric media. Figure 8.2 presents a schematic diagram of such a problem. The coupling interfaces as well as the boundaries for essential and natural conditions are presented. In the case of the piezoelectric subdomain the boundary division relevant to electrical conditions is skipped. The interface couplings of poroelastic or acoustic media to a piezoelectric material are similar to the couplings of these media to an elastic material but, for clearness, they are not presented.

Figure 8.3 shows a diagram of the system of algebraic equations (93) where particular submatrices and subvectors are visualized. The system describes a discrete model of the considered multiphysical problem with an optimized ordering of degrees of freedom. Different interface couplings are manifested by the intersections of submatrices.

The two conceptual diagrams presented here (Figures 8.2 and 8.3) complete the theoretical study of this chapter. The discrete model (93) derived using the Galerkin method is ready for

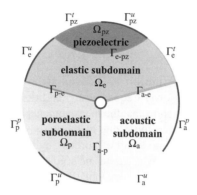

**Figure 8.2** Abstract configuration of a system made up of poroelastic, acoustic, elastic and piezoelectric media

finite element implementation. This allowed the development of numerical tools for accurate multiphysical modeling of active vibroacoustic problems involving poroelastic, acoustic, elastic and piezoelectric media. A virtual purpose for the development of such tools is modeling of hybrid noise attenuators (absorbers and insulators) in the form of active–passive liners, panels or composites of complex geometry. The next sections will bring results of such modeling of some developmental ideas and prototype designs.

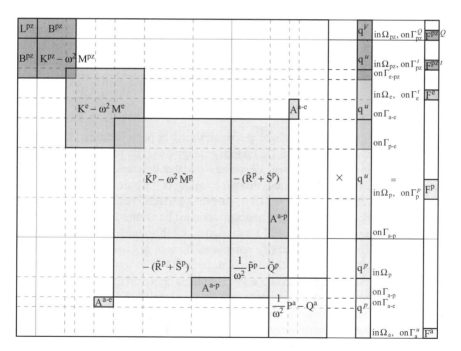

**Figure 8.3** Submatrices and subvectors of the global system of algebraic equations for a discrete model of a piezo–elasto–poroelastic structure coupled to an acoustic medium

## 8.8 Modeling of Poroelastic Layers with Mass Implants Improving Acoustic Absorption

### 8.8.1 Motivation

Recent experimental investigations report a significant improvement in the transmission loss of standard acoustic blankets at lower frequencies by addition of randomly placed masses to the poroelastic layers [38]. They show that the improvements by distributed masses (implants, inclusions) tend to be greater than those due to the mass effect alone. Therefore, there is a growing demand for advanced modeling of porous media with distributed masses, which should be at the same time sufficiently accurate and efficient to allow a reliable optimization of such poroelastic composites. Moreover, such modeling would be the first step for design of a new active composite where the vibrations of an elastic porous skeleton with distributed masses are controlled by active implants in order to get good acoustic performance at low frequencies.

This section presents the finite element modeling and frequency analysis of passive poroelastic layers with heavy solid implants, where an improvement of acoustical absorption is observed. Biot's theory of poroelasticity presented at the beginning of this chapter is used to model porous material and two techniques for considering mass implants are proposed and compared. The first technique fully uses the elasto–poroelastic subpart of the discrete model derived in Section 8.7 and the second one proposes and justifies a minor yet effective modification in the weak formulation of the problem. Finally, important results of analysis are presented and discussed to confirm the usefulness of further research concerning modeling and development of poroelastic composite noise attenuators.

### 8.8.2 Two Approaches in Modeling Small Solid Implants

As mentioned above, to model a domain of porous layer the mixed formulation of poroelasticity is used, while the solid implants are modeled in two ways. First, they can be very accurately modeled as small elastic subdomains in the poroelastic domain. Then, the Galerkin finite element model uses the following weak integral:

$$\mathcal{WF}_p + \mathcal{CI}_{p\text{-}e} + \mathcal{WF}_e = 0 \tag{114}$$

where $\mathcal{WF}_p$ and $\mathcal{WF}_e$ are the integrals of the weak forms of poroelasticity (25) and elasticity (26), respectively, whereas $\mathcal{CI}_{p\text{-}e}$ is the coupling integral (62) on the interface between the poroelastic and elastic domains. It should be remembered that the weak form $\mathcal{WF}_p$ of the mixed formulation of poroelasticity (25) ensures that the coupling of the poroelastic and elastic media is naturally handled [26] (i.e. $\mathcal{CI}_{p\text{-}e} = 0$). Thus, only the continuity between the solid phase displacements and the elastic subdomain displacements must be ensured, that is $u_i = u_i^e$ on $\Gamma_{p\text{-}e}$. However, since the implants are small, the finite element mesh around them becomes dense and the FE model is significantly enlarged (the poroelastic domain has four nodal degrees of freedom in three-dimensional models or three nodal degrees of freedom in two-dimensional models). However, the predominant effect of the solid implants (attached to the elastic skeleton of a porous medium) is caused by their mass since they are very small though heavy (and practically rigid) as compared to the poroelastic medium. Therefore, another approach may be proposed: it consists in adding some adequate inertial terms directly to the weak (variational) formulation of the poroelastic problem; thus instead of Equation (114) the following integral is used

$$\mathcal{WF}_p + \mathcal{MI} = 0 \quad \text{where } \mathcal{MI} = \int_{\Omega_p} \omega^2 \, m \, u_i \, \delta u_i \tag{115}$$

is the added (concentrated) mass term. Here $m = m(x)$ is a (local) distribution of additional mass added to the solid phase. In general, this approach (mathematically equivalent to the presence of concentrated masses in the given localizations) is effective if the mass is concentrated in points, and particularly in the nodes of the FE mesh. Thus, for a concentrated point-mass $M_0$ (added in the point $x_0$), it may be formally written $m(x) = M_0 \delta(x - x_0)$, where $\delta(.)$ is the Dirac delta function.

### 8.8.3 Acoustic Absorption of the Poroelastic Layer

The main purpose of the present analysis of poroelastic layers with solid implants is to assess how the heavy implants influence the acoustic absorption of layers. The acoustic absorption of a poroelastic layer fixed to a rigid wall and subject to a plane acoustic wave propagating in the air on to the layer surface at normal incidence will be computed as follows [12]. First, the acoustic impedance at normal incidence is determined at the interface between the poroelastic layer and the air:

$$Z = \frac{p}{v} \quad \text{where } v = j\omega u_1^t = j\omega \left[ (1 - \phi) u_1 + \phi U_1 \right] \tag{116}$$

Here, $v$ is the (amplitude of) velocity of the propagating wave at the layer–air interface (continuous across this boundary), whereas $p$ is the (amplitude of) wave pressure. Moreover, the expression (10) for the total displacement vector of a poroelastic medium is used where the porosity-dependent contributions of the displacements of both phases are involved. Now, the reflection coefficient in this point is computed as follows:

$$R = \frac{Z - Z_a}{Z + Z_a} \tag{117}$$

where $Z_a = \varrho_a c_a$ is the characteristic impedance of the air ($\varrho_a$ is the air density and $c_a$ the speed of sound). Finally, knowing the reflection coefficient, the acoustic absorption coefficient can be determined as

$$A = 1 - |R|^2 \tag{118}$$

This final property is real-valued (unlike the reflection coefficient $R$ and the impedance $Z$, which are complex).

### 8.8.4 Results of Analyses

Several finite element analyses of poroelastic layers with solid implants were carried out for the configuration presented in Figure 8.4 (left). Poroelastic material data for two different high-porosity polyurethane foams (termed A and B) were used for this configuration. The data are given in Table 8.1. It can be seen that the upper part of this table provides all parameters necessary for a much simpler model of porous material with a *rigid* frame. The additional parameters from the lower part of the table are used only by the advanced poroelastic model. Whatever the case, the data need to be completed by the parameters of the fluid in the pores. Such data for air are given in Table 8.2.

The thickness of the layer is 24 mm. At $x_1 = 0$ mm the layer is fixed to a rigid wall, whereas at $x_1 = 24$ mm the plane harmonic acoustic wave propagates on to the interface between the poroelastic layer and the air. At the depth of 4 mm from the incident surface and spaced by

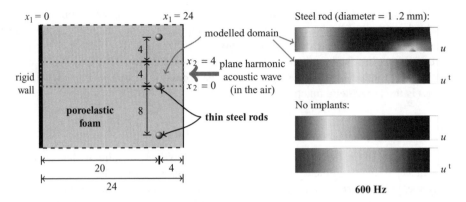

**Figure 8.4** (left) A 24 mm thick layer of poroelastic foam (fixed to a rigid wall) with small steel implants (regularly spaced steel balls or rods), with the modeled subdomain shown. (right) Solid phase ($u$) and total displacements ($u^t$) in the modeled subdomain of the poroelastic layer, with and without mass implants, at $f = 600\,\text{Hz}$

$\Delta x_2 = 8$ mm, thin steel rods are planted along the $x_3$ axis. Two cases of rod diameters were considered: 1.2 and 2 mm. The problem is modeled as a two-dimensional one in the $x_1 x_2$ plane. Moreover, the symmetry (regularity along the $x_2$ axis) makes it possible to model only a rectangular slice (of width $\Delta x_2 = 4$ mm) of the layer comprising only a half of one implant (see Figure 8.4).

There were two purposes of the analysis. First, the conformity between the accurately modeled solid implants (which involves a locally denser FE mesh) and the implants considered as additional weak mass terms was investigated. It was found that for the present configuration, the discrepancy between both models is small enough to allow the usage of the second, simpler model. An obvious conclusion is that the smaller the implants, the better is the conformity. Figure 8.4 (right) presents the solid phase and total displacements of the modeled slice (for two cases: the layer with the implants of diameter 1.2 mm and the layer without implants). The plots are independently scaled and so are only qualitative: the imposed boundary conditions, the presence (or absence) of mass implants, are visualized.

The second analysis consisted in determining the acoustic absorption of the poroelastic composite. To this end, the results of the FE analysis ($u^t$ at the layer's surface) were used

**Table 8.1** Poroelastic properties of two polyurethane foams

| Properties | | Foam A | Foam B |
| --- | --- | --- | --- |
| Porosity | $\phi$ | 0.97 | 0.99 |
| Tortuosity | $\alpha_\infty$ | 2.52 | 1.98 |
| Flow resistivity | $\sigma$ | $87\,\text{kN s/m}^4$ | $65\,\text{kN s/m}^4$ |
| Characteristic dimension of pores: | | | |
|     For viscous forces | $\Lambda$ | $37 \times 10^{-6}\text{m}$ | $37 \times 10^{-6}\text{m}$ |
|     For thermal forces | $\Lambda'$ | $119 \times 10^{-6}\text{m}$ | $121 \times 10^{-6}\text{m}$ |
| Solid phase mass density | $\varrho_\text{sph}$ | $31\,\text{kg/m}^3$ | $16\,\text{kg/m}^3$ |
| Shear modulus | $\mu_\text{s}$ | $55(1+\vartheta_j)\,\text{kPa}$ | $18(1+\vartheta_j)\,\text{kPa}$ |
| Loss factor | $\vartheta$ | 0.055 | 0.1 |
| Bulk Poisson's ratio | $\nu_\text{b}$ | 0.3 | 0.3 |

**Table 8.2**  Properties of the air as an interstitial fluid for porous material and as a simple acoustic medium (only upper part of the table)

| | | |
|---|---|---|
| Mass density | $\varrho_a$ | $1.225\,\text{kg/m}^3$ |
| Speed of sound | $c_a$ | $340\,\text{m/s}$ |
| Bulk modulus | $K_a$ | $1.42 \times 10^5\,\text{Pa} = \varrho_a\,c_a^2$ |
| Viscosity | $\eta$ | $1.84 \times 10^{-5}\,\text{Pa s}$ |
| Ratio of specific heats | $\gamma$ | $1.4$ |
| Ambient mean pressure | $P_a$ | $101\,325\,\text{Pa}$ |
| Prandtl's number | $B^2$ | $0.71$ |

by the analytical formulas for the impedance, the reflection and absorption coefficients (see Section 8.8.3). These formulas result from a one-dimensional analysis of the plane wave propagation, which is slightly violated if the solid implants are present. Therefore, the absorption coefficient was computed at two points of the layer surface: at $x_2 = 0\,\text{mm}$ and $x_2 = 4\,\text{mm}$ (see Figure 8.4, left), providing two limiting values. These values are plotted (for the range of frequency 25–800 Hz) in Figure 8.5 as curves (a) and (b), respectively. Moreover, curves (c) show the acoustic absorption for the homogeneous layer (i.e. no implants). The results were

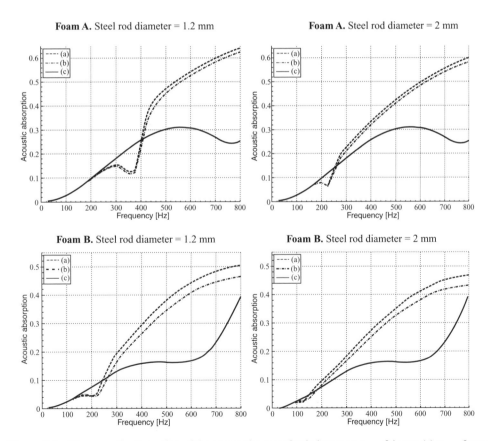

**Foam A.** Steel rod diameter = 1.2 mm

**Foam A.** Steel rod diameter = 2 mm

**Foam B.** Steel rod diameter = 1.2 mm

**Foam B.** Steel rod diameter = 2 mm

**Figure 8.5**  The acoustic absorption of the composite poroelastic layers at $x_1 = 24\,\text{mm}$: (a) $x_2 = 0\,\text{mm}$ (facing a rod) or (b) $x_2 = 4\,\text{mm}$ (between two rods). Also, (c) the absorption of the (homogeneous) layers with no implants. The frequency range $f = 25\text{–}800\,\text{Hz}$.

obtained for both versions of the poroelastic data: foam A (upper graphs) and foam B (lower graphs); and for both versions of the steel rod diameter: 1.2 mm (left graphs) and 2 mm (right graphs). Notice that in this latter case, the implant mass is 2.78 times greater than in the case of thinner rods.

Although the choice of configuration and materials was quite arbitrary, the following observations should prove to be general:

- The presence of mass implants significantly increases the acoustic absorption of porous layers (especially in the medium frequency).
- There is a lower frequency range, however, where the presence of implants deteriorate the absorption.
- This effect is significantly reduced if the implant mass is bigger; moreover, the range is then narrowed and shifted to even lower frequencies.
- Below this range (i.e. for very low frequencies) the mass implants have no noticeable effect (there is the same very poor performance of acoustic absorption).

### 8.8.5 Concluding Remarks

Layers of porous material with heavy solid implants may be modeled as poroelastic media with adequate point masses if the implants are very small (and sufficiently heavy). Such an approach should be very efficient for the optimization of composite configuration where the influence of the distribution of masses for the acoustic absorption of layers is analysed.

The presence of mass implants may significantly increase the acoustic absorption of porous layers, especially at medium frequency. It seems that the improvement by distributed masses (implants) may be greater than the one due to the mass effect alone (i.e. by a thin, heavy layer). Therefore, more numerical tests where this influence is analysed should be carried out.

In the lower frequency range the passive vibroacoustic attenuation by mass implants ceases to work and an active approach to the problem proves to be necessary. However, it seems that the most promising concept should combine active implants, distributed masses and possibly other solid implants; i.e. it should create an active poroelastic composite able to dissipate significantly the energy of acoustic waves also at low frequencies (where it should rely on an active control), whereas in the high- and medium-frequency range an excellent passive acoustic absorption would be guaranteed due to the designed absorbing properties of the poroelastic composite.

## 8.9 Designs of Active Elastoporoelastic Panels

### 8.9.1 Introduction

Accurate numerical tools developed on the basis of the model derived in Section 8.7 have been recently applied for modeling the multilayered panels that are widely used for limiting the transmission of acoustic waves. The panels are usually sandwiched structures made up of two elastic faceplates and a core of porous material. Passive panels are efficient enough at medium and high frequencies but exhibit a lack of performance at low frequency. To cope with this problem, the hybrid active–passive approach is applied: piezoelectric patches are added to the panel and behave as a secondary vibrational source, interfering with the low-frequency disturbance propagating in the panel [7,9]. Exact (though preliminary) modeling and analysis of such active sandwich panels were presented in References [7] to [9]. The panels were subjected to harmonic excitations modeled as a directly applied uniform pressure. The design of panels and the final qualitative results of the analysis will be briefly described in this section.

Another design of an active panel made up of a single elastic plate and a layer of poroelastic material will also be presented (see Section 8.9.3). This model will be thoroughly examined in Section 8.10 (an abridged discussion of preliminary investigations can also be found in Reference [11]). An important interaction of the panel with an acoustic medium (the air) will also be taken into account. The panel is active because piezoelectric patches are fixed to the elastic faceplate and form a piezo-actuator, which can be used to affect the bending vibrations of the faceplate; in this way, the vibroacoustic transmission through the panel can be controlled. In Section 8.10 the final accurate modeling of the entire hybrid panel and the results of relevant frequency analyses concerning the panel design are given. Finally, numerical testing of active and passive behaviour of the panel is performed.

## 8.9.2 Active Sandwich Panel

In Reference [7] a model of a prototype sandwich panel made up of two elastic faceplates and a poroelastic core was investigated. The panel is active, so an antisymmetric piezo-actuator composed of two co-located piezoelectric wafer elements is mounted on one of the elastic faceplates, which is termed the *active faceplate*. Both elastic faceplates are simply-supported. As a matter of fact, the whole prototype is a small cell of panel but larger panel surfaces can be obtained by juxtaposing the cells. Figure 8.6 presents the assembly and finite element mesh of one quarter of the panel cell, where advantage was taken of the symmetry by applying appropriate boundary conditions on the lateral faces.

The panel is subjected to a uniform pressure excitation (acting on the passive faceplate) and the bending vibrations of the active faceplate can also be excited by voltage applied to

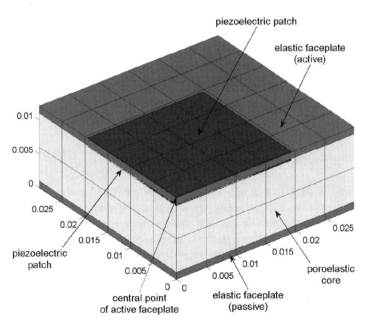

**Figure 8.6**   FE model of one quarter of an active sandwich panel

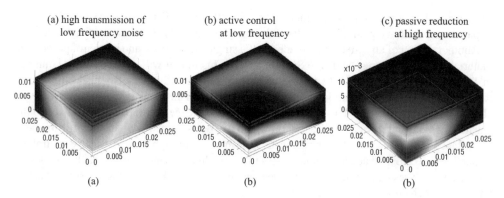

**Figure 8.7**   Results of (a) uncontrolled, (b) controlled and (c) passively reduced vibroacoustic transmission by the sandwich panel

the electrodes of the piezoelectric actuator to control low-frequency vibrations. Now, the principle of active reduction of vibroacoustic transmission by the active sandwich panel can be explained as follows. An incident acoustic wave propagates on to the passive faceplate. The wave is reflected and partially transmitted through the panel. The low-frequency (component of) acoustic wave is almost completely transmitted by the panel (see Figure 8.7(a)), whereas the transmission of high-frequency (component of) noise is passively reduced by the layer of poroelastic material (see Figure 8.7(c)). To reduce the vibroacoustic transmission of low-frequency (components of) noise, the piezoelectric actuator is used, counterexciting the panel vibrations, so that the normal displacements of the active faceplate are minimized (see Figure 8.7(b)), which means a reduction in the energy of the transmitted wave. A piezoelectric PVDF sensor, or even the piezoelectric actuator itself, can be used to measure the deflection of the active plate, enabling the active behavior of the panel to be controlled.

### 8.9.3  Active Single-Plate Panel

Another configuration of an active elastoporoelastic panel will be discussed at length in this section. It can be termed an active single-plate (or semi-sandwich) panel since it consists of:

- a single elastic plate;
- a poroelastic layer glued to the side of the plate exposed to noise;
- a piezoelectric actuator fixed to the elastic plate.

The investigations of such an assembly of a panel were presented in References [11] and [39]. As in the case of a sandwich panel, the proposed prototype is a small cell of an active single-plate panel so that larger panel surfaces are obtained by juxtaposing the panel cells. Besides being an active barrier limiting the transmission of acoustic waves, such a panel should also absorb the noise to some extent. Another important merit of the single-plate panel is that it will be nearly two times lighter than the similar sandwich panel.

A thorough examination of individual components of the panel and numerical tests of its behavior will be presented in the next section, so a complete description of its geometry and the material data of components are given below. The modeling of the panel will assume that

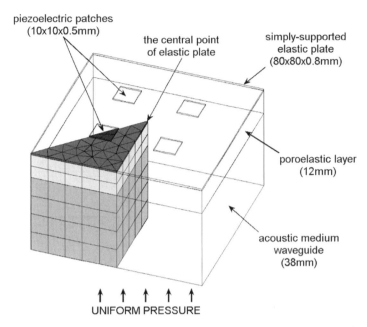

piezoelectric patches
(10x10x0.5mm)

the central point
of elastic plate

simply-supported
elastic plate
(80x80x0.8mm)

the central point
of elastic plate

poroelastic layer
(12mm)

acoustic medium
waveguide
(38mm)

↑ ↑ ↑ ↑ ↑
UNIFORM PRESSURE

**Figure 8.8** Dimensions of the system composed of a cell of the active elasto–poroelastic panel and acoustic medium waveguide, and the finite element mesh of the 1/8-slice model (a different brick element mesh was used for the full model of a system composed of the elastic plate and the 50 mm long acoustic waveguide; see, for example, Figure 8.9(d) below)

the poroelastic layer is coupled to a fragment of acoustic medium (air waveguide), which is subjected to an acoustic pressure excitation and transfers this excitation on to the panel. The assembly of the panel, as well as the finite element mesh of a modeled fragment of the system composed of the panel and the waveguide, are presented in Figure 8.8. Only a slice of the whole assembly is modeled after giving proper consideration to all possible symmetries.

The in-plane dimensions of the cell of the panel are 80 mm × 80 mm and the total thickness is 12.8 mm, where 0.8 mm is the thickness of the plate and 12 mm stands for the thickness of the poroelastic layer (see Figure 8.8). The plate of the panel cell is simply-supported. The elastic material of the plate is aluminum ($E_e = 70 \times 10^9$ N/m$^2$, $\nu_e = 0.33$ and $\varrho_e = 2700$ kg/m$^3$) and the poroelastic layer is made of a polyurethane foam with material properties given in Table 8.1 (foam A). The pores are filled with air. Table 8.2 gives all the necessary data for the air as the interstitial fluid and thus completes the material data for the elastic foam with air in the pores. The acoustic medium is a 80 mm × 80 mm × 38 mm waveguide of the air (the relevant material properties are in the upper part of Table 8.2). In the cases examined below, where only the aluminum plate or layer is analysed (i.e. with no poroelastic layer), the length of the acoustic waveguide is augmented by the thickness of the poroelastic layer and amounts to 50 mm.

In the complete assembly of the panel, piezoelectric transducers (0.5 mm thick) are glued to the upper surface of the elastic plate. They are through-thickness polarized so they stretch significantly in their plane when a voltage is applied to the electrodes, in this way inducing a bending deformation of the plate. The material of piezoelectric wafers is the transversely isotropic PZT4 ceramic, with data presented in Table 8.3. The material constants are given according to the contracted notation used in Section 8.4.3.

**Table 8.3** Material properties of (transversally isotropic) PZT4 ceramic (mass density $\rho_{pz} = 7500$ kg/m$^3$, $\epsilon_0 = 8.854 \times 10^{-12}$ F/M)

| Elastic constants | GPa | | Piezoelectric | C/m$^2$ |
|---|---|---|---|---|
| $\hat{C}_{11}^{pz} = \hat{C}_{22}^{pz}$ | 107.60 | | $\hat{e}_{15}^{pz} = \hat{e}_{24}^{pz}$ | 12.0 |
| $\hat{C}_{12}^{pz}$ | 63.12 | | $\hat{e}_{31}^{pz} = \hat{e}_{32}^{pz}$ | −9.6 |
| $\hat{C}_{13}^{pz} = \hat{C}_{23}^{pz}$ | 63.85 | | $\hat{e}_{33}^{pz}$ | 15.1 |
| $\hat{C}_{33}^{pz}$ | 100.40 | | | |
| $\hat{C}_{44}^{pz} = \hat{C}_{55}^{pz}$ | 19.62 | | Dielectric | $\epsilon_0$ |
| $\hat{C}_{66}^{pz} = \frac{1}{2}(\hat{C}_{11}^{pz} - \hat{C}_{12}^{pz})$ | 22.24 | | $\epsilon_{11}^{pz} = \epsilon_{22}^{pz}$ | 1936 |
| | | | $\epsilon_{33}^{pz}$ | 2109 |

Several configurations of piezoelectric transducers were checked, for example:

- one square patch 30 mm × 30 mm in the center of the upper side of the plate;
- four square patches of 10 mm × 10 mm fixed symmetrically around the center of the plate (as shown in Figure 8.8).

In the latter configuration it is assumed that the transducers are to be excited simultaneously by the same voltage, so they act in the same way on the plate and form one multipatch piezo-actuator. This approach gave a better performance (a bigger deflection of the center of the plate and a better deformation shape) using less piezoelectric material so it was used in the later analysis. Another effective configuration can be triangularly shaped piezoelectric actuators for vibroacoustic control systems, proposed in Reference [40].

## 8.10 Modeling and Analysis of an Active Single-Plate Panel

### 8.10.1 Kinds and Purposes of Numerical Tests

Several numerical tests concerning the modeling and design of particular components or the whole assembly of a panel cell were carried out. In some of them the panel or (only) the elastic plate is coupled to the acoustic medium (i.e. the air waveguide). Some of the finite element tests were matched with the corresponding analytical solutions. Three kinds of tests can be distinguished:

- *Plate tests* (eigenvalue and frequency-response analyses) are used to identify elastic plate resonances and to discover a specific resonance of the coupled system of the elastic plate and acoustic medium waveguide.
- *Multilayer tests* (analytical one-dimensional frequency-response analyses of multilayered media) are used to confirm the coupled resonance and to select the thickness and material of the poroelastic layer.
- *Panel tests* (frequency-response and single-frequency FE analyses) are used to test active and passive behavior (performance) of the panel.

The frequency responses were analyzed for a wide spectrum: up to 4500 Hz in the case of the panel components and up to 3500 Hz for the whole assembly of the panel.

## 8.10.2 Plate Tests

Figure 8.9 presents the results of modeling the elastic plate and its coupling with the acoustic medium. The frequency-response analyses for the following problems were performed:

- The plate was subjected to a concentrated force applied at the point (80 mm/4, 80 mm/4). The purpose of this analytical frequency-response solution is to reveal all resonances for all natural frequencies present in the considered spectrum (Figure 8.9(a)).
- The plate was subjected directly to a uniform pressure excitation (analytical and numerical solutions) (Figure 8.9(b) and (c)).
- The FE model of the plate coupled with the acoustic medium (the air) is subjected to a uniform acoustic pressure excitation (Figure 8.9(d)).

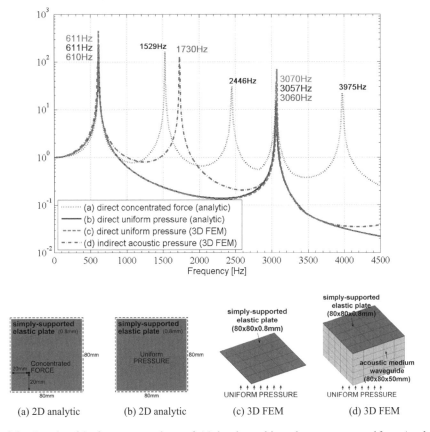

**Figure 8.9**   Results of the frequency analyses of: (a) the plate subjected to a concentrated force (analytical solution), (b,c) the plate subjected (directly) to the uniform pressure – analytical and FE solutions (the curves overlap), (d) the three-dimensional FE model of the plate coupled with the acoustic medium subjected to the excitation of uniform pressure, inducing a plane acoustic wave in the waveguide in the direction perpendicular to the plate

**Table 8.4**   Natural frequencies of the elastic plate: analytical and FE solutions (forms of deformation for the resonances (a) to (e) are shown in Figure 8.10)

| Number | Resonance | Mode | Analytic | Numeric | Error (%) |
|---|---|---|---|---|---|
| | | | Frequency (Hz) | | |
| 1 | (a) | (1,1) | 611 | 611 | 0.00 |
| 2, 3 | (b) | (1,2)(2,1) | 1529 | 1528 | 0.07 |
| 4 | (c) | (2,2) | 2446 | 2442 | 0.16 |
| 5, 6 | (d) | (1,3)(3,1) | 3057 | 3070 | 0.43 |
| 7, 8 | (e) | (2,3)(3,2) | 3975 | 3979 | 0.10 |
| 9, 10 | | (1,4)(4,1) | 5197 | 5292 | 1.83 |
| 11 | | (3,3) | 5503 | 5505 | 0.04 |
| 12, 13 | | (2,4)(4,2) | 6115 | 6187 | 1.18 |

In all the analyses the computed response was the maximum absolute value of the normal displacements (deflection) of the plate divided by the result obtained for the static case; for the problems with direct excitation (curves (a), (b) and (c) in Figure 8.9), this static solution was in fact identical. The presented results show that it is important to take into consideration the acoustic–elastic coupling (rather than applying the acoustic pressure directly), since a new 'coupled' resonance appears at 1735 Hz (Figure 8.9, curve (d)).

The numerical model of the plate is three-dimensional since the complete assembly of the hybrid panel also consists of the piezo-actuator, which produces bending deformation of the plate; in this case, only three-dimensional modeling of piezoelectric effects and the plate is suitable. The three-dimensional FE model of the plate was chosen to be a regular $5 \times 5 \times 1$ brick element mesh with cubic shape functions to approximate the displacement fields. An excellent conformity of at least the first eight natural frequencies is achieved for this FE model (with the error from 0 to less than 0.5 %). To check this, a relevant eigenvalue problem was solved: see Table 8.4 for analytical and numerical results and Figure 8.10 for the deformation

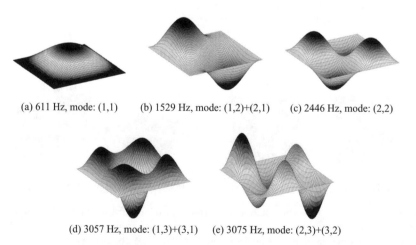

(a) 611 Hz, mode: (1,1)        (b) 1529 Hz, mode: (1,2)+(2,1)        (c) 2446 Hz, mode: (2,2)

(d) 3057 Hz, mode: (1,3)+(3,1)     (e) 3075 Hz, mode: (2,3)+(3,2)

**Figure 8.10**   Five resonance deformations of the elastic plate. Since the plate is rectangular, some of the deformations are combinations of two relevant eigenmodes. Notice that (b), (c) and (e) are antisymmetric deformations

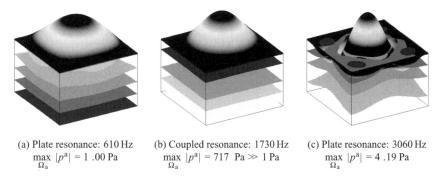

(a) Plate resonance: 610 Hz
$\max_{\Omega_a} |p^a| = 1.00$ Pa

(b) Coupled resonance: 1730 Hz
$\max_{\Omega_a} |p^a| = 717$ Pa $\gg 1$ Pa

(c) Plate resonance: 3060 Hz
$\max_{\Omega_a} |p^a| = 4.19$ Pa

**Figure 8.11** Three resonances of the coupled system of the elastic plate and acoustic waveguide (subjected to a time-harmonic pressure excitation of the amplitude of 1 Pa): (a,c) two plate resonances (the plate deflections tend to increase infinitely), (b) the coupled resonance of the whole system (not only the plate deflections but also the pressure $p^a$ in the acoustic medium $\Omega_a$ increase greatly).

shapes of the first five resonances. This is also confirmed by the frequency analysis (see curves (b) and (c) in Figure 8.9). It was checked that, for much denser meshes but using quadratic shape functions, this conformity is inferior. The comparison of the frequency analysis with the eigenproblem solution illustrates the fact that in the case of uniformly distributed loading, the resonances which cause antisymmetric modes of deformation are blocked (only *symmetric-shape* resonances, at 611 and 3057 Hz, appear in the relevant curve (b) of Figure 8.9). This is an important observation since the case of *uniform* loading applies to the examined problem of the small cell of the acoustic panel.

The acoustic medium waveguide (the air-gap) was modeled as a regular 5×5×5 brick element mesh with quadratic shape functions used to approximate the acoustic pressure field. The length of the waveguide is 50 mm and for the highest frequency considered (i.e. 4500 Hz) the acoustic wavelength equals 76 mm. This is several times more than 10 mm, which is the size of the waveguide finite element. Figure 8.11 shows the three resonances of the coupled system of the elastic plate and acoustic waveguide: the two (pure) plate resonances are at approximately 610 Hz (a) and 3060 Hz (c), and the coupled resonance of the elastic plate and acoustic cavity at approximately 1730 Hz (b). It should be noted that, in the latter case, not only the amplitude of maximal deflections of plate but also the amplitude of pressure in the acoustic waveguide tend to increase significantly at this resonance frequency of the coupled system.

### 8.10.3 Multilayer Analysis

Figure 8.12 shows the results obtained analytically for one-dimensional problems of wave propagation in multilayered media. Two configurations were examined:

- The two-layered medium has an acoustic layer of thickness of 50 mm coupled with a 0.8 mm thick layer of aluminum (Figure 8.12(a)).
- The three-layered medium has 38 mm of acoustic layer +12 mm of poroelastic layer +0.8 mm of aluminum (Figure 8.12(b)).

It should be noted that the total thickness for both configurations is the same. The excitation was a harmonic unit acoustic pressure applied on the acoustic layer and the results are the

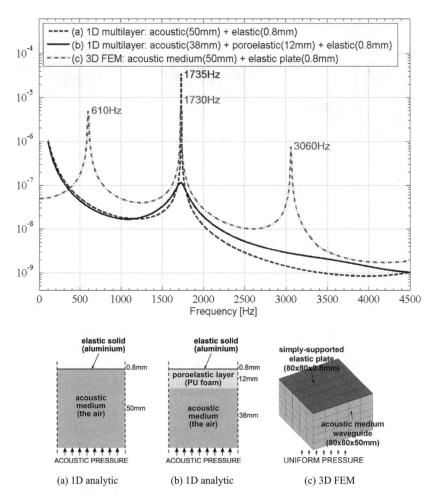

**Figure 8.12** Frequency analyses: (a) the results of a one-dimensional (analytical) analysis of a two-layered medium composed of air (50 mm) and aluminum (0.8 mm); (b) the results of a one-dimensional (analytical) analysis of a three-layered medium composed of air (38 mm), poroelastic material (12 mm) and aluminum (0.8 mm); (c) the same results as presented in Figure 8.9(d) but here 1 m is used as the reference for the logarithmic scale (the pressure amplitude was equal to 1 Pa)

amplitudes of displacements of the thin aluminum layer (1 m was used as a reference for the logarithmic scale). For comparison, the results obtained from the FE analysis of the aluminum plate coupled with the acoustic medium are recalled as curve (c) in Figure 8.12.

It can be seen that, in the case of one-dimensional multilayered solutions there is only one (coupled) resonance of the whole system. It was displayed for 1735 Hz, which almost exactly agrees with 1730 Hz of the coupled resonance of the three-dimensional FE model of the plate and acoustic medium; the small discrepancy may be, due to the fact that the analytical frequency-response analyses of multilayered media were carried out with a step of 5 Hz, whereas in the case of FE models, the computational frequencies were taken every 10 Hz. Thus, the coupled nature of this resonance is excellently confirmed. However, an even more important result is obtained when analysing the three-layered medium: the resonance is significantly attenuated by adding a well-chosen layer of poroelastic material (apart from the chosen one, two other foams and different thicknesses were tried). Though desirable damping properties of the poroelastic layer are revealed owing to this simple one-dimensional analysis, it cannot be stated how they affect the resonances relevant for the plate. To check this, a complete three-dimensional FE model must be analysed.

## 8.10.4 Analysis of Passive Behavior of the Panel

Figure 8.13 presents the frequency analysis for the plate linked with the poroelastic layer coupled to the acoustic medium. This time the piezoelectric patches are present but they are passive (the only effect is some locally added mass and stiffness). As a matter of fact, an analysis of the panel without patches was also carried out, but those results are not given here because they are similar and the conclusions are identical to the ones presented below.

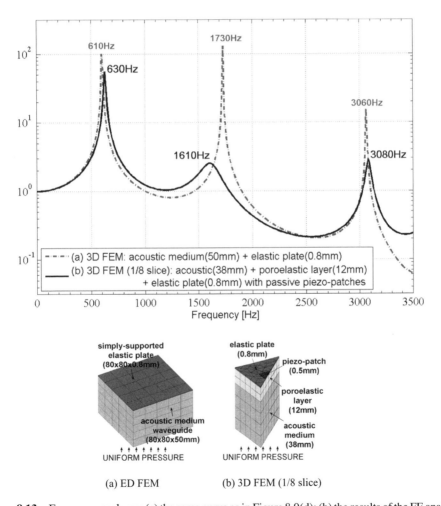

(a) ED FEM                    (b) 3D FEM (1/8 slice)

**Figure 8.13**   Frequency analyses: (a) the same curve as in Figure 8.9(d); (b) the results of the FE analysis of the complete assembly of the panel, i.e. the simply-supported plate (with passive piezo-patches) merged with the poroelastic layer connected to the acoustic medium (the limited 1/8-slice model was used)

On account of a considerable numerical cost, only one-eight of the active elasto–poroelastic panel was modeled (see Figure 8.8) by taking full advantage of the symmetries (suitable boundary conditions were imposed on the relevant lateral faces of the slice). This is a standard approach in statics but it might not be the completely right approach to use in dynamics (since antisymmetric modes cannot be represented). However, it is perfectly acceptable in this particular case of excitation by a *uniformly distributed* acoustical pressure where the contribution of antisymmetric modes tends to be insignificant, which can be seen from the frequency analysis of the simply-supported elastic plate subject to uniform pressure (see Figure 8.9(b)), collated with the solution of the eigenproblem for this plate (see Figure 8.10 to identify symmetric and antisymmetric modes of deformation). Figure 8.8 presents the FE mesh of this limited model: it is composed of parallelepiped elements with triangular bases. This time, for the poroelastic layer and acoustic medium, as well as for the elastic plate and the piezoelectric patch, the quadratic shape functions were used for all the dependent variables. The plate and the piezo-patch are very thin and were discretized with one layer of elements each, whereas for the 38 mm layer of the acoustic medium four layers of mesh elements were used and for the 12 mm layer of the poroelastic material two layers. This seems to ensure a sufficient convergence since this time the highest computational frequency was only 3500 Hz and for this frequency the wavelength in the air is 98 mm, while the shortest wavelength in the poroelastic medium is the shear wave with length equal to 12 mm.

Figure 8.13 compares the obtained results of the frequency analysis of passive behavior of the panel, i.e. the maximal amplitudes of the elastic plate deflections, with the results (presented already in Figures 8.9 and 8.12) of the FE analysis of the plate interacting directly with the acoustic medium. The static solution for this latter problem is the reference value for the logarithmic scale. The total thickness is the same for both configurations (providing that the thickness of piezoelectric patches is not considered). Notice that now the resonances for the active elasto–poroelastic panel are slightly shifted because of the locally added stiffness and mass of the piezoelectric patches. The most important observation is that the coupled resonance (now approximately 1600 Hz) and the higher resonance of the simply-supported plate (now approximately 3080 Hz) are attenuated when the poroelectric layer is present. On the contrary, the low-frequency resonance of the plate (the first eigenfrequency, now approximately 630 Hz) is fully manifested. For this lower frequency, an active control of vibrations is necessary.

Passive reduction of vibroacoustic transmission for the ('coupled resonance') frequency of 1610 Hz is presented in Figure 8.14. The deformations and amplitudes of vertical displacements are shown for the three following cases:

(a) the displacements of the elastic faceplate and the total displacements of the poroelastic layer
(b) the displacements of the elastic faceplate and of the fluid phase of the poroelastic layer;
(c) the displacements of the fluid phase of the poroelastic layer and of the acoustic medium.

In all three plots the same deformation scaling factor was used, but since the plate and solid phase deformations are hardly visible, an additional plot is given where the displacements of the plate and solid phase are rescaled 10 times. The amplitude of acoustic pressure of the exciting wave was 1 Pa (the system is linear, so for another value the results would be proportionally scaled). The maximal amplitude of normal (vertical) displacements of the fluid phase equals approximately $1.2 \times 10^{-6}$ m and the maximal amplitude of vertical displacements

Deformations and amplitude of vertical displacements:

| (a) plate and total displacements | (b) plate and solid phase | (c) fluid phase and acoustic medium |

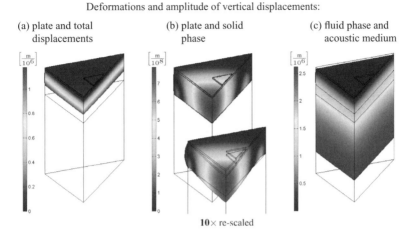

10× re-scaled

**Figure 8.14** Passive reduction of vibroacoustic transmission ($f = 1610\,\text{Hz}$)

in the acoustic waveguide is almost $2.6 \times 10^{-6}$ m, whereas the maximal deflection of the elastic plate is much less and reaches $8 \times 10^{-8}$ m. Notice also that the vibrations of the solid phase of the poroelastic layer are in antiphase to the wave propagating in the fluid phase (or in the air in the pores). The incident acoustic wave of this frequency is very poorly transmitted by the panel: its energy is dissipated by the poroelastic layer.

### 8.10.5 Test of Active Behavior of the Panel

The purpose of the test of active behavior of the panel is to determine the voltage excitation necessary for active reduction of vibroacoustic transmission and to check how it works by performing the vibroacoustic analysis with control. Harmonic steady-state analysis will be used for the test. However, since the problem is perfectly linear, the results obtained for different frequencies can be superposed. In practice, transient or multifrequency vibrations can be controlled by applying an appropriate signal to the piezoelectric actuator. Nevertheless, single frequency signals may also be used to control some predominant spectral components of vibrations.

Remember that the complete system considered here is composed of the active elasto–poroelastic panel and the waveguide of the acoustic medium. A plane (harmonic) acoustic wave propagates in the waveguide on to the poroelastic layer of the single-plate panel. The wave may be attenuated by the dissipative poroelastic layer, and partially reflected and transmitted through the panel. It has been shown in the previous section that for higher frequencies, the passive reduction of vibroacoustic transmission by poroelastic material is sufficient: the coupled resonance and the plate resonance around the higher eigenfrequency are well damped (see Figure 8.13). Nevertheless, the resonance vibrations at the first eigenfrequency of the plate are not attenuated and an active treatment must be applied with the use of the piezoelectric actuator.

### 8.10.5.1 Test Procedure

To examine the active reduction of vibroacoustic transmission, first, two single-frequency FE analyses must be carried out for different types of excitation:

- $p$ analysis: a (unit) acoustic pressure excites the panel through the acoustic medium (the piezoelectric actuator is passive).
- $V$ analysis: a (unit) voltage is applied on to the piezoelectric transducers (the zero pressure or air-impedance boundary condition is applied on to the acoustic medium).

In both cases, the deflection of the elastic plate, $w$, is computed to provide results necessary for further calculation. For this purpose the following quantities are defined:

$$w_p \overset{\text{def}}{=} \frac{w_{\text{plate centre}}^{(\hat{p}\neq 0,\ \hat{V}=0)}}{\hat{p}} \ (\text{m/Pa}), \qquad w_V \overset{\text{def}}{=} \frac{w_{\text{plate centre}}^{(\hat{p}=0,\ \hat{V}\neq 0)}}{\hat{V}} \ (\text{m/V}) \tag{119}$$

describing how the deflection in the center of the elastic plate depends on the harmonic excitation by acoustic pressure $\hat{p}$ or by voltage $\hat{V}$, respectively. The deflections are from steady-state responses for the *same* harmonic frequency $f$; obviously, the values $\hat{p}$ and $\hat{V}$ are the amplitudes of harmonic excitations.

Now, since the analysis is linear, the superposition principle can be used to calculate the deflection in the center of the plate for any given harmonic pressure, $\hat{p}$, and voltage, $\hat{V}$, simultaneously exiting the panel:

$$w_{\text{plate centre}} = w_p\,\hat{p} + w_V\,\hat{V} \tag{120}$$

For any low-frequency acoustic pressure excitation the largest deflection of the faceplate of the panel is always in the center. Therefore, a low vibroacoustic transmission is achieved if the deflection in the centre is small. This deflection can be used as a control parameter (measured by a sensor).

The piezoelectric actuator is designed to induce a similar shape of plate deformation as the one caused by the acoustic pressure excitation. Therefore, the 'transmission-controlling' voltage that should be applied to the piezo-actuator can be computed by requiring only that the deflection in the plate centre should equal zero, i.e.

$$w_{\text{plate centre}} = 0 \quad \longrightarrow \quad \hat{V} = -\frac{w_p}{w_V}\,\hat{p} = \alpha\,\hat{p} \tag{121}$$

Here, a *control coefficient* is defined as

$$\alpha \overset{\text{def}}{=} -\frac{w_p}{w_V} \ (\text{V/Pa}) \tag{122}$$

Now, the final harmonic analysis can be performed to complete the test.

- $\alpha$ analysis: the panel is simultaneously excited by the acoustic pressure $\hat{p}$ and by the 'controlling' voltage $\hat{V} = \alpha\,\hat{p}$.

The correct result of this analysis is that for any amplitude of acoustic pressure $\hat{p}$

$$|w_{p,\alpha}| \overset{\text{def}}{=} \left| \frac{w_{\text{plate centre}}^{(\hat{p}\neq 0,\ \hat{V}=\alpha\,\hat{p})}}{\hat{p}} \right| \text{(m/Pa)} \approx 0 \ll |w_p| \tag{123}$$

and in general:

$$\max \left| w_{\text{whole plate}}^{(\hat{p}\neq 0,\ \hat{V}=\alpha\,\hat{p})} \right| \ll |w_p\,\hat{p}| \tag{124}$$

which means that the vibrations of the elastic faceplate are reduced and so the vibroacoustic transmission is minimized. Obviously, the validity of the last inequality depends also on the form of vibrations excited by the piezoelectric actuator.

### 8.10.5.2 Results of the Test

Figures 8.15 and 8.16 illustrate the first two steps of the vibroacoustic control test (the $p$ analysis and $V$ analysis) performed for the (first resonance) frequency, $f = 630\,\text{Hz}$. The deformations and amplitudes of vertical displacements are shown for the same three cases as in Figure 8.14. The same primary deformation scaling factor as in Figure 8.14 was also used for all the plots in Figures 8.15 and 8.16.

Figure 8.15 showing the uncontrolled vibrations of the system confirms that for this resonance frequency there is a considerable vibroacoustic transmission through the panel (possibly that even augmented by the resonance vibrations of the plate). Notice that the maximal amplitudes of the plate and solid phase displacements are almost identical with the amplitudes of the fluid phase.

The plots in Figure 8.16 are very similar to the ones in Figure 8.15. They are, however, slightly different. The similarity results from the intended design of the piezoelectric actuator: the static or low-frequency bending deformations of the faceplate excited by the actuator should

Deformations and amplitude of vertical displacements ($p$-analysis):

(a) plate and total          (b) plate and solid          (c) fluid phase and
    displacements                   phase                    acoustic medium

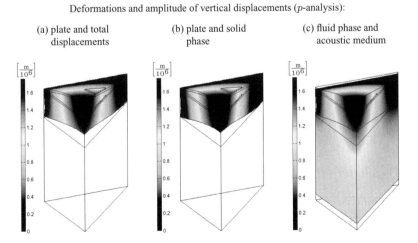

**Figure 8.15**   Uncontrolled vibrations for the unit acoustic pressure excitation ($f = 630\,\text{Hz}$)

Deformations and amplitude of vertical displacements ($V$-analysis):

(a) plate and total          (b) plate and solid          (c) fluid phase and
displacements                      phase                    acoustic medium

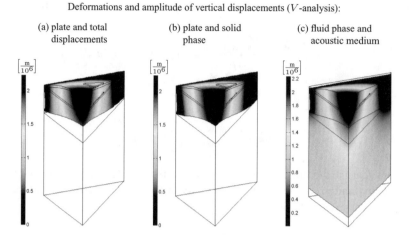

**Figure 8.16** Vibrations for the unit voltage excitation ($f = 630\,\text{Hz}$). (These plots are different though very similar to the ones presented in Figure 8.15)

be similar in shape to the deformation caused by the low-frequency acoustic pressure wave. This purpose is certainly achieved by the four symmetrically situated piezo-transducers (one piezo-patch set in the centre of the plate gave a worse bending shape).

Figure 8.17 presents the result of the $\alpha$ analysis, i.e. the same kind of deformations and amplitudes as in Figures 8.14 to 8.16, but for the system under the simultaneous excitation by the unit acoustic pressure and the adequate 'controlling' voltage (computed in the two previous analyses). This time, the deformations and vertical displacement amplitudes for the plate and the solid phase of the poroelastic layer are significantly smaller. Even the fluid phase and (consequently) the total displacements of the panel are smaller than those in the case of uncontrolled vibrations. Therefore, the plots in Figure 8.17 are presented in two versions regarding the magnitude of the deformation scale:

- (upper row) the same deformation scale as in Figures 8.14 to 8.16;
- (lower row) the deformation scaling factor is 10 or even 50 times greater than the one used for the plots in the upper row.

The 50 times re-scaled (middle) plot (b) in the lower row shows that the requirement used for the 'controlling' voltage computation is satisfied: the normal displacement in the central point of the elastic faceplate looks like being fixed, though it is only set in this zero position by the simultaneous action of the actuator and acoustic pressure wave. The remaining plots show that this entails a big reduction in deflection at other points of the plate. Thus, a very good reduction of vibroacoustic transmission achieved with the use of a piezo-actuator is confirmed.

## 8.10.6  Concluding Remarks

The results obtained from the frequency analyses of the semi-sandwich panel and its components allow the following conclusions to be drawn:

- The modelling of the interaction between the panel and the air is important because some acoustic–elastic coupling resonances may occur.

Deformations and amplitude of vertical displacements ($\alpha$-analysis):

(a) plate and total
displacements

(b) plate and solid
phase

(c) fluid phase and
acoustic medium

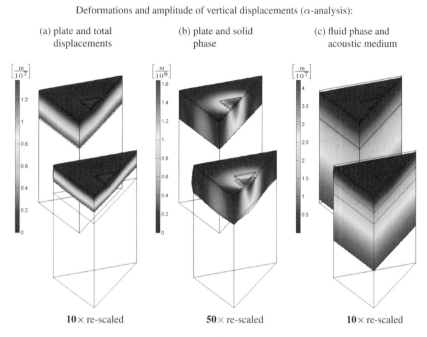

10× re-scaled                    50× re-scaled                    10× re-scaled

**Figure 8.17**   Vibroacoustic transmission actively reduced ($f = 630\,\text{Hz}$): the controlled vibrations for the simultaneous excitation by the unit pressure and adequate 'controlling' voltage. The corresponding plots shown in the upper and lower rows differ only in the scaling of deformations

- The poroelastic layer should be designed to attenuate high- and medium-frequency resonances and vibroacoustic transmission.
- The low-frequency transmission and resonance(s), especially around the first eigenfrequency of the elastic faceplate of the panel, require an active attenuation with the help of a piezo-actuator.

Moreover, these results, together with the analysis of active vibracoustic control, permit the conclusion to be made that the proposed model of the semi-sandwich panel should work properly for a wide frequency range as an efficient active–passive reducer of vibroacoustic transmission and also as an absorber of high-frequency noise. Nevertheless, a further and more advanced design consisting in some kind of parametric survey and configuration optimization must be performed with special attention paid to:

- a proper choice of poroelastic material and its thickness (for passive attenuation);
- an optimal design of a piezoelectric actuator and its localization (for active control).

Additional requirements are obvious: the panel should be light and thin, the piezoelectric actuator small.

## References

1. P. A. Nelson and S. J. Elliott, *Active Control of Sound*, Academic Press, 1992.
2. C. R. Fuller, S. J. Elliott and P. A. Nelson, *Active Control of Vibration*, Academic Press, 1996.

3. C. Guigou and C. R. Fuller, Control of aircraft interior broadband noise with foam-PVDF smart skin, *Journal of Sound and Vibration*, **220**(3), 1999, 541–557.
4. B. D. Johnson and C. R. Fuller, Broadband control of plate radiation using a piezoelectric, double-amplifier active-skin and structural acoustic sensing, *Journal of Acoustical Society of America*, **107**(2), February 2000, 876–884.
5. O. Lacour, M.-A. Galland and D. Thenail, Preliminary experiments on noise reduction in cavities using active impedance changes, *Journal of Sound and Vibration*, **230**(1), 2000, 69–99.
6. M.-A. Galland, B. Mazeaud and N. Sellen, Hybrid passive/active absorbers for flow ducts, *Applied Acoustics*, **66**, 2005, 691–708.
7. T. G. Zielinski, M.-A. Galland and M. N. Ichchou, Active reduction of vibroacoustic transmission using elasto-poroelastic sandwich panels and piezoelectric materials, in *Proceedings of the Symposium on the Acoustics of Poro-Elastic Materials SAPEM'05*, Lyon, 2005.
8. C. Batifol, T. G. Zielinski, M.-A. Galland and M. N. Ichchou, Hybrid piezo-poroelastic sound package concept: numerical/experimental validations, in *Conference Proceedings of ACTIVE 2006*, 2006.
9. C. Batifol, T. G. Zielinski, M. N. Ichchou and M.-A. Galland. A finite-element study of a piezoelectric/poroelastic sound package concept, *Smart Materials and Structures*, **16**, 2007, 168–177.
10. N. Sellen, M. Cuesta and M.-A. Galland, Noise reduction in a flow duct: implementation of a hybrid passive/active solution, *Journal of Sound and Vibration*, **297**, 2006, 492–511.
11. T. G. Zielinski, M.-A. Galland and M. N. Ichchou, Further modeling and new results of active noise reduction using elasto-poroelastic panels, in *Conference Proceedings of ISMA2006*, 2006.
12. J. F. Allard, *Propagation of Sound in Porous Media. Modelling Sound Absorbing Materials*. Elsevier, 1993.
13. T. G. Zielinski, Theoretical fundamentals of multiphysics approach for modelling of active–passive noise attenuators, *Journal of Sound and Vibration*, 2008 (submitted).
14. T. G. Zielinski, Modelling of poroelastic layers with mass implants improving acoustic absorption, in *Conference Proceedings of 19th International Congress on Acoustics ICA2007*, Madrid, 2007.
15. M. A. Biot, The theory of propagation of elastic waves in a fluid-saturated porous solid, *Journal of the Acoustics Society of America*, **28**, 1956, 168–191.
16. M. A. Biot, Mechanics of deformation and acoustics propagation in porous media, *Journal of Applied Physics*, **33**(4), April 1962, 1482–1498.
17. R. de Boer, *Theory of Porous Media: Highlights in Historical Development and Current State*, Springer, 1999.
18. K. Wilmanski, A few remarks on Biot's model and linear acoustics of poroelastic saturated materials, *Soil Dynamics of Earthquake Engineering*, **26**, 2006, 509–536.
19. M. Shanz, *Wave Propagation in Viscoelastic and Poroelastic Continua: A Boundary Element Approach*, Vol. 2, Lecture Notes in Applied and Computational Mechanics, Springer, 2001.
20. N. Atalla, R. Panneton and P. Debergue, A mixed displacement-pressure formulation for poroelastic materials, *Journal of the Acoustics Society of America*, **104**(3), September 1998, 1444–1452.
21. P. Debergue, R. Panneton and N. Atalla, Boundary conditions for the weak formulation of the mixed (u,p) poroelasticity problem, *Journal of the Acoustics Society of America*, **106**(5), November 1999, 2383–2390.
22. P. Göransson, A 3-D, symmetric, finite element formulation of the Biot equations with application to acoustic wave propagation through an elastic porous medium, *International Journal of Numerical Methods in Engineering*, **41**, 1998, 167–192.
23. N.-E. Hörlin, M. Nordström and P. Göransson, A 3-D hierarchical FE formulation of Biot equations for elasto-acoustic modelling of porous media, *Journal of Sound and Vibration*, **245**(4), 2001, 633–652.
24. R. Panneton and N. Atalla, An efficient finite element scheme for solving the three-dimensional poroelasticity problem in acoustics, *Journal of the Acoustics Society of America*, **101**(6), June 1997, 3287–3298.

25. R. Panneton and N. Atalla, Numerical prediction of sound transmission through finite multilayer systems with poroelastic materials, *Journal of the Acoustics Society of America*, **100**(1), July 1996, 346–354.

26. N. Atalla, M. A. Hamdi and R. Panneton, Enhanced weak integral formulation for the mixed (u,p) poroelastic equations, *Journal of the Acoustics Society of America*, **109**(6), June 2001, 3065–3068.

27. S. Rigobert, F. C. Sgard and N. Atalla, A two-field hybrid formulation for multilayers involving poroelastic, acoustic, and elastic materials, *Journal of the Acoustics Society of America*, **115**(6), June 2004, 2786–2797.

28. S. Rigobert, N. Atalla and F. C. Sgard, Investigation of the convergence of the mixed displacement–pressure formulation for three-dimensional poroelastic materials using hierarchical elements, *Journal of the Acoustics Society of America*, **114**(5), November 2003, 2607–2617.

29. J. N. Reddy, *Energy Principles and Variational Methods in Applied Mechanics*, John Wiley & Sons, Ltd, Chichester, 2002.

30. T. Ikeda, *Fundamentals of Piezoelectricity*, Oxford University Press, Oxford, 1990.

31. G. A. Maugin, *Continuum Mechanics of Electromagnetic Solids*, Elsevier, 1988.

32. A. Preumont, *Vibration Control of Active Structures*, Springer, Berlin, 2002.

33. J. N. Reddy, On laminated composite plates with integrated sensors and actuators, *Engineering Structures*, **21**, 1999, 568–593.

34. A. Benjeddou, Advances in piezoelectric finite element modeling of adaptive structural elements: a survey, *Computing Structures*, **76**, 2000, 347–363.

35. D. T. Blackstock, *Fundamentals of Physical Acoustics*, John Wiley & Sons, Ltd, Chichester, 2000.

36. I. Harari, A survey of finite element methods for time-harmonic acoustics, *Computing Methods and Applications in Mechanical Engineering*, **195**, 2006, 1594–1607.

37. L. L. Thompson, A review of finite-element methods for time-harmonic acoustics, *Journal of the Acoustics Society of America*, **119**(3), March 2006, 1315–1330.

38. M. R. F. Kidner, C. R. Fuller and B. Gardner, Increase in transmission loss of single panels by addition of mass inclusions to a poro-elastic layer: experimental investigation, *Journal of Sound and Vibration*, **294**, 2006, 466–472.

39. T. G. Zielinski, M.-A. Galland and M. N. Ichchou, Designs and analysis of smart panels for active–passive noise attenuation, *Smart Materials and Structures*, 2008.

40. P. Gardonio and S. J. Elliott, Smart panels with velocity feedback control systems using triangularly shaped strain actuators, *Journal of the Acoustics Society of America*, **117**(4), April 2005, 2046–2064.

# Acknowledgements

- Parts of Section 1.2 and Section 3.1 have been published as the following article:

P. Kołakowski. Structural Health Monitoring – a review with the emphasis on low-frequency methods. *Engineering Transactions*, 55(3):239–275, 2007.

The material has been re-used in this book with kind permission of IPPT-PAN Publications.

- Parts of Chapter 2 have been accepted for publication (May 2007) as the following article:

P. Kołakowski, M. Wikło, J. Holnicki-Szulc. The Virtual Distortion Method – a versatile reanalysis tool for structures and systems. *Structural and Multidisciplinary Optimization*, in press.

The material has been re-used in this book with kind permission of Springer Science and Business Media.

- Parts of Section 3.2 have been recommended for publication (November 2007) as the following article:

A. Świercz, P. Kołakowski, J. Holnicki-Szulc. Damage identification in skeletal structures using the Virtual Distortion Method in frequency domain. *Mechanical Systems and Signal Processing*, in revision.

- Parts of Section 3.3 have been recommended for publication (January 2008) as the following article:

A. Orłowska, P. Kołakowski, J. Holnicki-Szulc. Modelling and identification of delamination in double-layer beams by the Virtual Distortion Method. *Computers & Structures*, in revision.

- Parts of Section 3.4 have been published as the following article:

J. Holnicki-Szulc, P. Kołakowski, N. Nasher. Leakage detection in water networks. *Journal of Intelligent Material Systems and Structures*, 16:207–219, 2005.

The material has been re-used in this book with kind permission of Sage Publications.

- Parts of Section 3.5 have been recommended for publication (November 2007) as the following article:

M. Kokot, J. Holnicki-Szulc. Defect identification in electrical circuits via Virtual Distortion Method. Part I: Steady-state case. *Journal of Intelligent Material Systems and Structures*, in revision.

---

- Parts of Section 4.1 have been submitted for publication (January 2008) as the following article:

  K. Sekuła, J. Holnicki-Szulc. On-line impact load identification. *International Journal of Impact Engineering*, in review.

- Parts of Section 4.3 have been accepted for publication (January 2008) as the following article:

  Ł. Jankowski. Off-line identification of dynamic loads. *Structural and Multidisciplinary Optimization*, in press.

  The material has been re-used in this book with kind permission of Springer Science and Business Media.

- Parts of Section 5.2 have been published as the following article:

  J. Holnicki-Szulc, P. Pawłowski, M. Wikło. High-performance impact absorbing materials – the concept, design tools and applications. *Smart Materials and Structures*, 12:461–467, 2003.

  The material has been re-used in this book with kind permission of Institute of Physics Publishing.

- Parts of Section 5.4 have been published as the following article:

  G. Mikułowski, J. Holnicki-Szulc. Adaptive landing gear concept – feedback control validation. *Smart Materials and Structures*, 16:2146–2158, 2007.

  The material has been re-used in this book with kind permission of Institute of Physics Publishing.

- Parts of Section 5.4 have been recommended for publication (February 2008) as the following article:

  G. Mikułowski, Ł. Jankowski. Adaptive landing gear: optimum control strategy and potential for improvement. *Shock and Vibration*, in revision.

  The material has been re-used in this book with kind permission of IOS Press.

- Parts of Section 5.5 have been recommended for publication (November 2007) as the following article:

  C. Graczykowski, J. Holnicki-Szulc. Protecting offshore wind turbines against ship impacts by means of adaptive inflatable structures. *Shock and Vibration*, in revision.

  The material has been re-used in this book with kind permission of IOS Press.

- Parts of Section 5.6 have been published as the following article:

  M. Ostrowski, P. Griskevicius, J. Holnicki. Adaptive crashworthiness of front-end structure of motor vehicles. *SAE 2007 World Congress*, April 16–19, 2007 – Cobo Center – Detroit, Michigan, USA.

  The material has been reprinted with permission from SAE Paper # 2007-01-1180 © 2007 SAE International.

- Parts of Chapter 6 have been accepted for publication (January 2008) as the following articles:

  M. Wikło, J. Holnicki-Szulc. Optimal design of adaptive structures. Part I: Remodelling for impact reception. *Structural and Multidisciplinary Optimization*, in press.

  M. Wikło, J. Holnicki-Szulc. Optimal design of adaptive structures. Part II: Adaptation to impact loads. *Structural and Multidisciplinary Optimization*, in press.

  The material has been re-used in this book with kind permission of Springer Science and Business Media.

- Parts of Chapter 7 have been submitted for publication (August 2007) as the following article:

  A. Mróz, A. Orłowska, J. Holnicki-Szulc. Semi-active damping of vibrations. Prestress Accumulation-Release strategy development. *Shock and Vibration*, in review.

- Parts of Chapter 8 have been submitted for publication (November 2007) as the following articles:

  T.G. Zieliński. Theoretical fundamentals of multiphysics approach for modelling of active-passive noise attenuators. *Journal of Sound and Vibration*, in review.

  T.G. Zieliński, M.-A. Galland, M.N. Ichchou. Designs and analysis of smart panels for active-passive noise attenuation. *Smart Materials and Structures*, in review.

The authors would like to acknowledge the financial support from the following projects:

- Advanced aviation technologies – development of a flying demonstrator, INTEGRA, R10 005 02, 2007–2010,

- Structural health monitoring of critical structures, SHMCS, R03 015 02, 2007–2010,

- Integrated dynamic system of risk assessment, diagnostics and control of structures and technological processes, DIADYN, PBZ-KBN-105/T10/2003, 2005–2008,

- Intelligent materials made of metals, ceramics and polymers: design, production, properties, applications, MAT-INT, PBZ-KBN-115/T08/2004, 2005–2008,

- New methods for design of safe structures with identification of hazards and active adaptation in critical states, 3T11F00930, 2006–2009,

- New, ecological and safe technologies for energy production and conversion, EKO-ENERGIA network, 2007–2008,

granted by the State Committee for Scientific Research in Poland.

The assistance of INTECH - Foundation for Promoting the Transfer of Smart Technologies (http://www.fundacja-intech.org.pl/), in language verification and technical edition of the book, is gratefully acknowledged.

The authors sincerely thank Dr *Dariusz Wiącek* for design of the photo composition on the front cover.

# Index